Advances in

BOTANICAL RESEARCH

incorporating *Advances in Plant Pathology*

VOLUME 34

BIOTECHNOLOGY OF CEREALS

Advances in

BOTANICAL RESEARCH

incorporating *Advances in Plant Pathology*

Editor-in-Chief

Advances in

BOTANICAL RESEARCH

incorporating *Advances in Plant Pathology*

BIOTECHNOLOGY OF CEREALS

Edited by

P. R. SHEWRY
IACR–Long Ashton Research Station,
University of Bristol, Bristol, UK

P. A. LAZZERI
DuPont Wheat Transformation Laboratory,
c/o Rothamsted Experimental Station,
Harpenden, Hertfordshire, UK

and

K. J. EDWARDS
IACR–Long Ashton Research Station,
University of Bristol, Bristol, UK

Series Editor

J. A. CALLOW

School of Biological Sciences,
University of Birmingham,
Birmingham, UK

VOLUME 34

2001

ACADEMIC PRESS

A Harcourt Science and Technology Company

San Diego   San Francisco   New York   Boston
London   Sydney   Tokyo

Academic Press
A Harcourt Science and Technology Company
Harcourt Place, 32 Jamestown Road, London NW1 7BY, UK
http://www.academicpress.com

Academic Press
A Harcourt Science and Technology Company
525 B Street, Suite 1900, San Diego, California 92101-4495, USA
http://www.academicpress.com

ISBN 0-12-005934-7

A catalogue record for this book is available from the British Library

Typeset by Helius, Brighton, East Sussex
Printed in Great Britain by MPG Books Limited, Bodmin, Cornwall

01 02 03 04 05 06 MP 9 8 7 6 5 4 3 2 1

CONTENTS

Cereal Genomics

K. J. EDWARDS and D. STEVENSON

Exploiting Cereal Genetic Resources

R. J. HENRY

Transformation and Gene Expression

P. BARCELO, S. RASCO-GAUNT, C. THORPE and P. A. LAZZERI

Opportunities for the Manipulation of Development of Temperate Cereals

J. R. LENTON

Manipulating Cereal Endosperm Structure, Development and Composition to Improve End-use Properties

P. R. SHEWRY and M. MORELL

Resistance to Abiotic Freezing Stress in Cereals

M. A. DUNN, G. O'BRIEN, A. P. C. BROWN, S. VURAL
and M. A. HUGHES

Genetics and Genomics of the Rice Blast Fungus *Magnaporthe grisea*: Developing an Experimental Model for Understanding Fungal Diseases of Cereals

N. J. TALBOT and A. J. FOSTER

Impact of Biotechnology on the Production of Improved Cereal Varieties

R. G. SOLOMON and R. APPELS

Overview and Prospects for Cereal Biotechnology

P. R. SHEWRY, P. A. LAZZERI and K. J. EDWARDS

The colour plate appears between pages 168 and 169

CONTRIBUTORS TO VOLUME 34

R. APPELS   *CSIRO Plant Industry, PO Box 1600, Canberra, ACT 2601, Australia*

P. BARCELO   *DuPont Wheat Transformation Laboratory, c/o Rothamsted Experimental Station, Harpenden, Hertfordshire AL5 2JQ, UK*

A. P. C. BROWN   *School of Biochemistry and Genetics, University of Newcastle upon Tyne, Newcastle upon Tyne NE2 4HH, UK*

M. A. DUNN   *School of Biochemistry and Genetics, University of Newcastle upon Tyne, Newcastle upon Tyne NE2 4HH, UK*

K. J. EDWARDS   *IACR–Long Ashton Research Station, Department of Agricultural Sciences, University of Bristol, Long Ashton, Bristol BS41 9AF, UK*

A. J. FOSTER   *School of Biological Sciences, University of Exeter, Washington Singer Laboratories, Perry Road, Exeter EX4 4QG, UK*

R. J. HENRY   *Cooperative Research Centre for Molecular Plant Breeding, Centre for Plant Conservation Genetics, Southern Cross University, PO Box 157, Lismore NSW 2480, Australia*

M. A. HUGHES   *School of Biochemistry and Genetics, University of Newcastle upon Tyne, Newcastle upon Tyne NE2 4HH, UK*

P. A. LAZZERI   *DuPont Wheat Transformation Laboratory, c/o Rothamsted Experimental Station, Harpenden, Hertfordshire AL5 2JQ, UK*

J. R. LENTON   *IACR–Long Ashton Research Station, Department of Agricultural Sciences, University of Bristol, Long Ashton, Bristol BS41 9AF, UK*

M. MORELL   *CSIRO Division of Plant Industry, Institute of Plant Production and Processing, PO Box 1600, Canberra, ACT 2601, Australia*

G. O'BRIEN   *School of Biochemistry and Genetics, University of Newcastle upon Tyne, Newcastle upon Tyne NE2 4HH, UK*

S. RASCO-GAUNT   *DuPont Wheat Transformation Laboratory, c/o Rothamsted Experimental Station, Harpenden, Hertfordshire AL5 2JQ, UK*

P. R. SHEWRY   *IACR–Long Ashton Research Station, Department of Agricultural Sciences, University of Bristol, Long Ashton Bristol BS41 9AF, UK*

R. G. SOLOMON   *CSIRO Plant Industry, PO Box 1600, Canberra, ACT, 2601, Australia*

D. STEVENSON   *IACR–Long Ashton Research Station, Department of Agricultural Sciences, University of Bristol, Long Ashton, Bristol BS41 9AF, UK*

N. J. TALBOT   *School of Biological Sciences, University of Exeter, Washington Singer Laboratories, Perry Road, Exeter EX4 4QG, UK*

C. THORPE   *DuPont Wheat Transformation Laboratory, c/o Rothamsted Experimental Station, Harpenden, Hertfordshire AL5 2JQ, UK*

S. VURAL   *School of Biochemistry and Genetics, University of Newcastle upon Tyne, Newcastle upon Tyne NE2 4HH, UK*

CONTENTS OF VOLUMES 24–33

Contents of Volume 25

THE PLANT VACUOLE

J. A. RAVEN

The Vacuole and Cell Senescence

P. MATILE

Protein Bodies: Storage Vacuoles in Seeds

G. GALILI and E. M. HERMAN

Compartmentation of Secondary Metabolites and Xenobiotics in Plant Vacuoles

M. WINK

Solute Composition of Vacuoles

R. A. LEIGH

The Vacuole and Carbohydrate Metabolism

C. J. POLLOCK and A. KINGSTON-SMITH

Vacuolar Ion Channels of Higher Plants

G. J. ALLEN and D. SANDERS

The Physiology, Biochemistry and Molecular Biology of the Plant Vacuolar ATPase

U. LÜTTGE and R. RATAJCZAK

The Molecular and Biochemical Basis of Pyrophosphate-Energized Proton Translocation at the Vacuolar Membrane

R.-G. ZHEN, E. J. KIM and P. A. REA

The Bioenergetics of Vacuolar H$^+$ Pumps

J. M. DAVIES

Transport of Organic Molecules Across the Tonoplast

E. MARTINOIA and R. RATAJCZAK

Contents of Volume 28

Contents of Volume 31

Contents of Volume 32

Plant Protein Kinases

Contents of Volume 33

PREFACE

Cereals are the major crops in the world in terms of yield area under production and geographic distribution, providing food for human consumption, feed for livestock and raw material for industry. Although they have been, and continue to be, the subject of major research programmes at the basic and strategic levels, the large genome size and low transformation efficiencies of all of the major cereals except rice have limited attempts to isolate key genes and to manipulate plant development and composition. This situation is now changing, with the development of efficient transformation systems for wheat and maize and the establishment of large-scale EST sequencing projects in commercial and public laboratories. It is, therefore, appropriate to review these developments, and the major targets for cereal manipulation. This took place at a symposium held at Long Ashton Research Station (Bristol, UK) in September 1999, bringing together workers from plant biotechnology laboratories in the public and private sectors with biochemists, physiologists and the end-user industries. This symposium has formed the basis for the present volume, which discusses the tools, targets and future prospects in cereal biotechnology.

Peter R. Shewry
Keith J. Edwards
Paul A. Lazzeri

Cereal Genomics

KEITH J. EDWARDS and DAVID STEVENSON

IACR–Long Ashton Research Station, Department of Agricultural Sciences, University of Bristol, Long Ashton, Bristol BS41 9AF, UK

Advances in Botanical Research Vol. 34
incorporating Advances in Plant Pathology
ISBN 0-12-005934-7

I. INTRODUCTION

It is remarkable that much of the world's food supply consists of products derived from just a handful of crops which all belong to the relatively small *Poaceae* family (Devos and Gale, 1997). Together, rice, barley, wheat and maize provide the human race with the majority of its calorific requirements. It is the ability of the cereals to synthesize large qualities of calorie-rich carbohydrates that makes them so important (Kent and Evers, 1994). During the latter half of the 20th century, the total yield from each of the four cereals discussed herein, has increased from 650 million tons in 1950 to 1800 million tons in 1995. This dramatic increase has been due to both conventional plant breeding and improved agricultural practices. While it is clear from these figures that we have been remarkably successful in improving cereal yields, when the simultaneous increase in world population is taken into consideration, the figures are less optimistic. Current information suggests that the amount of grain available per individual reached a peak in the mid-1980s and has since shown evidence of declining (data from World Agriculture Towards 2010: FAO (Alexandratos, 1995). This trend has often been referred to as the population time bomb (Brown, 1994). The propensity for developing countries to move away from consuming grain to consuming animal protein produced from feed grain, means that the amount of grain available per head of population will probably continue to show a significant deterioration in the near future (Briggs, 1998). Given this scenario, it is clear that, while there is still much to be gained from both conventional plant breeding and continued improvements in agricultural practices, further dramatic developments will have to take place if the world is not to run short of food in the early years of the new millennium. It has been suggested that the rapidly emerging area of genomics may be the ultimate provider of these developments (Phillips and Freeling, 1998).

Genomics is essentially the study of the genome (McKusick *et al.*, 1993). Clearly the subject area of genomics is vast and it has therefore become convenient, if not necessary, to divide the area into several subsections, including molecular markers, expressed sequence tags (ESTs), structural genomics and functional genomics. Molecular markers are important in germplasm evaluation, the map-based cloning of genes, and, of course, molecular breeding. ESTs (Adams *et al.*, 1991) and associated expression arrays, are of use in gene discovery and characterization. Structural genomics covers the physical organization of the genome and how this relates to function. Finally, functional genomics include both gene function and the tools used to examine gene function, such as transposon tagging and gene inactivation via antisense technology (Ecker and Davis, 1986).

During the 1990s, the major emphasis of plant genomics has been the *Arabidopsis* genome sequencing project (Bevan, 1997). However, with the completion of the project in 2000 (Kaul *et al.*, 2000; Theologis *et al.*, 2000;

Salanoubat *et al.*, 2000; Tabata *et al.*, 2000), it has been suggested that in the first years of the 21st century, cereal genomics will become the most highly resourced area of plant biotechnology. Early indications in both the private and public sectors suggest that a large amount of funding is finding its way into programmes involving cereal genomics. For instance, in 1998, of the $50 million made available by the National Science Foundation (NSF) to plant genomics, about $40 million was awarded to ten projects involving one or more of the four cereals discussed here. In the 1999 NSF funding round, a further eight cereal projects have been given a total of $30 million (http://www.nsf.gov/bio/pubs/awards/genome99.html). In addition, the French government has announced their own $140 million plant genomics programme, which contains a substantial cereal component. It is clear that whilst these developments are impressive, they are only the beginning of the drive to carry out public-based cereal genomics programmes. However, given this massive investment in cereal genomics, it is fair to examine both what has already been delivered and what might be delivered in the near future. The following is a brief survey of the current status of cereal genomics.

II. MOLECULAR MARKERS

A. INTRODUCTION

Before the development of molecular markers, breeders relied on phenotypic markers, such as kernel colour or plant height. Even now, such a low-technology approach should not be underestimated: breeding complex traits, such as yield, are to a significant degree still carried out using simple breeder observations. The first true molecular markers were isoenzymes (MacDonald and Brewbaker, 1972). However, isoenzymes are difficult to work with and are unevenly distributed on the genetic map (Nielsen and Scandalios, 1974). In the early 1980s, restriction fragment length polymorphisms (RFLPs) were developed to overcome these problems (Helentjaris *et al.*, 1986). RFLPs can be seen as the first universal molecular marker. RFLPs are superior to both isoenzymes and phenotypic markers in that they can be chosen to represent any part of the genome and they are both co-dominant and multiallelic (Brettschneider, 1998). Given these advantages, it is not surprising that in the years following their development, RFLP-based genotyping and molecular breeding were adopted by most of the major plant breeding companies and academic laboratories. However, 15 years on from these early developments, molecular breeders and those involved in genotyping now have a multitude of different molecular markers that can be called upon. Briefly, these markers fall into two distinct groups:

1. multiallelic, co-dominant, single locus markers, such as RFLPs and microsatellites (Weber and May, 1989); and

2. multilocus, dominant, single allele markers, such as amplified fragment length polymorphisms (AFLPs) (Vos *et al.*, 1995; Ridout and Donini, 1999) and the now almost discontinued randomly amplified polymorphic DNAs (RAPDs) (Welsh and McClelland, 1990).

For the purpose of this review, only RFLP, AFLP and microsatellite markers will be discussed further.

B. RICE

Collectively, there are nine genetic maps of rice consisting of 2275 RFLPs and 12 isoenzymes at 1157 loci. A total of 457 QTLs have been mapped on to one or more of these populations (http://ars-genome.cornell.edu/rice/whatsnew.html). Several groups have independently used AFLPs to both map the rice genome and to carry out diversity and genotyping studies (Zhu *et al.*, 1999). In addition, some attempts have been made to integrate AFLP markers into the RFLP maps (Cho *et al.*, 1998). Several hundred microsatellite markers have been developed for rice (McCouch *et al.*, 1997). By January 2001, 1059 microsatellite markers were publically available. The primer sequences, their map position, motif description, number of alleles and polymorphism information content of these microsatellites will be made available through the University of Cornell Web page (S. McCouch, personal communication). With the recent and continuing publication of large amounts of both EST sequence data and bacterial artificial chromosome (BAC) sequence data, it is logical to suppose that the number of microsatellite markers will increase significantly during the next 3 years.

C. BARLEY

Several groups, notably in North America, Europe and Australia, have been active in the generation of barley RFLP markers. For instance, the Langridge group at the Waite Institute has generated three comprehensive barley RFLP maps consisting of various barley quantitative trait loci (QTLs) including emergence, leaf number, plant height, kernel weight and seed dormancy (http://wheat.pw.usda.gov/ggpages/maps.html; Langridge *et al.*, 1995). However, as in rice, it is clear that the drive for new barley RFLP markers has slowed down considerably in the last few years, due in part to the development of AFLP and microsatellite-based technologies. Within the barley genomics community, a number of groups have independently used AFLPs to both map the genome and carry out diversity and genotyping studies (Ellis *et al.*, 1997; Yin *et al.*, 1999). AFLPs have been used to generate extremely high-density maps of the region surrounding the *Mla* locus, presumably as the first step to cloning the gene (Schwarz *et al.*, 1999). Thanks to the effort of various European Community and North American initiatives, several hundred microsatellite markers are now publically available (Struss and Plieske, 1998). The majority of these have been generated through the laboratory of Waugh *et al.*, via the EU FAIR programme (http://synteny.nott.ac.uk/barley.html; Russell *et al.*, 1997; Donini *et al.*,1998; Mano *et al.*, 1999). As yet, there is no major public effort to generate either large numbers of barley ESTs or genomic sequences; therefore it is probable that the worldwide effort by the barley community to generate barley microsatellites specifically will continue for the foreseeable future.

D. WHEAT

Given its hexaploid genome and very low level of sequence polymorphism, wheat is probably the most difficult of the cereals in which to develop and use molecular markers (Merezhko, 1998). This is largely due to the fact that hexaploid wheat is no more than 10 000 years old (Dvorak *et al.*, 1998). Studies have shown that the various genotypes of wheat vary by as little as 1 bp in 1000 bp (Bryan *et al.*, 1999), an observation that is confirmed by RFLP analysis (Mingeot and Jacquemin, 1999). Furthermore, RFLP markers usually detect homologous sequences present on all three sets of homologous chromosomes. Despite these difficulties, until recently RFLPs were widely used for molecular breeding (Parker *et al.*, 1998). Currently, approximately 2000 RFLP markers are generally available (http://wheat.pw.usda.gov/). Given the low level of polymorphism, it is not surprising that AFLPs have only been of limited use for both molecular breeding and varietal identification (Ma and Lapitan, 1998; Hartl *et al.*, 1999). The low level of polymorphism observed in wheat has led to a number of active microsatellite development programmes. The public database currently contains references to approximately 500 wheat microsatellites (http:// wheat.pw.usda.gov/; Roder *et al.*, 1998; Korzun *et al.*, 1999), with many more being held in the hands of private companies (http://www.agrogene.com/ microsatellites/Default.htm). Given the intrinsic property of microsatellites to detect higher levels of polymorphism than either RFLPs or AFLPs, it is clear that, at least for the foreseeable future, microsatellites will be the genotyping method of choice for wheat (Plaschke *et al.*, 1995).

E. MAIZE

Phenotypically, maize is the most variable of all the food cereals. This variability continues all the way down to the DNA level. Because of this, large numbers of RFLP, AFLP and microsatellite markers have been developed for maize (Senior *et al.*, 1996; Pejic *et al.*, 1998; Davis *et al.*, 1999). As of September 1999, the Maizedb held information on more than 2000 mapped RFLP probes and approximately 630 microsatellite primer sets representing

589 distinct loci (http://www.agron.missouri.edu/ssr.html). Several groups have also utilized AFLPs as a means of generating targeted markers for specific regions of the genome (Pe *et al.*, 1999). In 1998, the NSF provided the funds for a maize microsatellite programme centred on the University of Columbia, Missouri, USA. This programme is designed to both utilize microsatellite-enriched libraries and maize ESTs (generated by another NSF award to Stanford University) eventually to generate several thousand mapped microsatellite markers (http://www.cafnr.missouri.edu/mmp/).

F.   MOLECULAR MARKERS: THE NEXT STEP

As previously described, cereal marker-assisted breeding and genome mapping both rely upon the availability of polymorphic genetic markers. It is clear from recent work that microsatellite markers have become the marker of choice for cereal molecular breeders. Unfortunately, the current methods for detecting microsatellites all rely upon electrophoretic separation of DNA in agarose or polyacrylamide gels. Although developments in fluorescent DNA fragment analysis have made it possible to analyse many microsatellite loci simultaneously and capture the resulting data automatically (Heyen *et al.*, 1997), the advent of these semiautomated systems and refinements, such as capillary gel electrophoresis (Gonen, *et al.*, 1999), gel-based technology is still labour intensive and time consuming for the large-scale genotyping required in both experimental genome analysis and marker-assisted breeding programmes.

The requirements for a high throughput non-gel-based genotyping system include increased scope for automation and a simple binary scoring system that can be reliably read by machine with no human intervention. The differential hybridization between probe DNA and allele-specific oligonucleotides (Guo *et al.*, 1994), which underpins both DNA genotyping chips and matrix-assisted laser desorption/ionization–time of flight (MALDI-TOF)- based genotyping, could provide the basis for such a system (Guo *et al.*, 1994; Griffin *et al.*, 1999). Both DNA genotyping chips and MALDI-TOF-based genotyping first require the development of allele-specific oligonucleotides (ASOs; Guo *et al.*, 1994). ASO technology is based upon the principle that, when hybridized under appropriate conditions, synthetic DNA oligonucleotide probes (15–21 bases) will anneal to their complementary sequences only if they are perfectly matched. Under the correct conditions, a single base-pair mismatch can be sufficient to prevent the formation of a stable probe–target duplex. Hybridization and non-hybridization can then be monitored via a suitable detection system. This two-state hybridization–non-hybridization is binary in nature and therefore ideal for automated scoring.

Because ASOs have the potential to provide a quick, cheap, multiallelic and multilocus test, they should be in regular use within all cereal genotyping laboratories. Unfortunately, whilst they are in regular use for the detection of certain human genetic diseases (Saiki *et al.*, 1989), they are not in regular use for non-human genotyping. The reason for this becomes apparent when one considers the enormous cost of developing ASOs; for each locus, a mapped single copy probe has to be sequenced and suitable PCR primers designed. These primers must then be used to amplify the corresponding fragment from all the other possible genotypes. These fragments must then be sequenced and the sequences compared with one another to determine ASOs for each of the possible alleles. Thus, when one considers that, in an average plant genotyping laboratory, 100 different loci might routinely be screened, then the amount of work required to develop ASOs for each locus and each possible allele becomes considerable.

For cereals, the amount of work required to produce ASOs would be considerably reduced if existing molecular markers could be used. These markers would already have been mapped and therefore could be chosen based upon their known and useful map position. Current RFLP markers could therefore offer such a short cut. Unfortunately, in a recent study of wheat RFLPs, Bryan and co-workers found that the sequence variation present between different wheat RFLP "alleles" was insufficient to design ASOs (Bryan *et al.*, 1999). However, recent results from both our own laboratory (Mogg *et al.*, 1999) and the laboratory of Powell (http://www.scri.sari.ac.uk), suggests that the flanking regions of microsatellites could provide a rich source of ASOs. If these results are confirmed, then it is probable that within the next 10 years, ASO-based molecular breeding and genotyping in conjunction with either DNA chips or MALDI-TOF, will become the method of choice for cereals.

III. EXPRESSED SEQUENCE TAGS

A. INTRODUCTION

The random sequencing of cDNA clones to generate expressed sequence tags is a phenomenon of the 1990s. Unfortunately, a large number of the EST sequencing programmes appear to be little more than opportunistic in nature, with little regard being shown to either coordinate the genotypes used to make the libraries or accurately stage the tissue used to generate the original mRNA. Given this situation, it is possible that the information generated could be of little use in determining which genes are expressed at a specific time or at what level they are expressed. If this pessimistic view is correct, then it is probable that, before long, a large number of these programmes will have to be repeated.

In any review of cereal EST programmes, it is difficult to ignore the large-scale sequencing programmes being undertaken by the major agri-biotechnology and seed companies; unfortunately, it is difficult to determine the exact number of ESTs that have been sequenced in these programmes or find referenced information within the scientific journals. Therefore, for the purposes of this review, we will only discuss the information generated by publicly funded programmes, which have or intend to publish both the number of ESTs generated and their sequences.

B. RICE

As of September 1999, the DDBJ contains 26 264 public rice ESTs (Sasaki, 1998; Yamamoto and Sasaki, 1997). Given the number of large high-profile public sequencing efforts, particularly in Asia, this figure is probably an underestimation of the total number of clones sequenced.

C. BARLEY AND WHEAT

The public databases contain relatively few wheat and barley ESTs. Presumably, this is because there are, as yet, no large-scale public sequencing programmes for either wheat or barley. However, the International Triceae EST Cooperative (ITEC) is attempting to change this situation (http://wheat.pw.usda.gov/genome). Currently, the ITEC database consists of more than 11 000 ESTs derived from a number of different developmental stages of both wheat and barley. Whilst these sequences are not as yet available to the public, they are available to any researcher who deposits sequences in the database. By January 2001, 64 000 ESTs had been released to the public databases. In addition to this communal effort, the NSF has recently announced that it is to fund a wheat EST programme at the University of California (http://wheat.pw.usda.gov/wEST/insf/title.html). It is clear from both these proposed programmes and from programmes thought to be in the pipeline, that the situation with regard to both barley and wheat ESTs will change rapidly over the next few years.

D. MAIZE

Given its commercial importance, it is not surprising that several commercial maize EST programmes have been in place for several years and have between them generated several hundred thousand EST sequences. A small semipublic maize EST programme, which began as an informal collaboration between the Universities of Florida and Arizona, has resulted in a few hundred ESTs being submitted to the public databases (Shen *et al.*, 1994). However, in 1998, the NSF awarded $12 million to Stanford University to sequence maize ESTs. To date, this programme has been remarkably successful. For instance, between April and October 1999, the Stanford group submitted over 30 000 EST to the public databases (http://

www.zmdb.iastate.edu/zmdb/EST_project.html). Libraries used to date on this programme, include leaf primordia, apical meristem, stressed root, endosperm, ear, tassel and mixed anther and pollen. Given this considerable effort, it is clear that public-sector maize researchers will soon have as many ESTs at their disposal as those in the private sector.

E. EST ARRAYS

In themselves ESTs are of limited value and the vast majority submitted to the public databases have so far been of little use. It has been suggested that, when generated in very large numbers, EST sequence data can be used to describe the expression profile of specific tissues at various developmental stages. However, given the undetermined quality of most of the libraries used, the results from such studies should be treated with caution. With the advent of "DNA chips", the information gained from EST sequencing programmes becomes much more relevant (Schena, 1996). A considerable number of reviews have been written on expression chips, probably more than actual scientific reports (Granjeaud et al., 1999). EST-based expression chips have the advantages that, unlike Northern blots, they allow the simultaneous comparison of the expression levels of several thousand different genes. Given these possibilities, several different strategies have been developed for expression arrays, including nylon membranes, glass slides or high-density oligonucleotides (Zhao et al., 1995; Duggan et al., 1999; Lockhart et al., 1996).

To date, no reports or conference abstracts have been published that utilize cereal ESTs in expression arrays. However, given recent NSF funding, it is only a matter of time before such reports become commonplace (http://www.nsf.gov/bio/pubs/awards/genome99.html).

IV. STRUCTURAL GENOMICS

A. INTRODUCTION

Structural genomics is obviously concerned with the study of the physical structure of the genome and how this affects both the evolution of the genome and the expression of the gene complement. Given current technology, the main focus of those involved in cereal structural genomics has been the generation of an integrated genetic and physical map. Once generated, the next step is the complete sequencing of the genome. Given the descriptive nature of structural genomics, it is not surprising that it has not received the same level of general scientific interest as functional genomics or EST-based programmes. Nevertheless, given current developments, it is the field of structural genomics where the most significant advances will probably be made in the next few years.

B. SYNTENY

It is impossible to talk about cereal structural genomics without first mentioning synteny. It has been known for several years that, although the genome size of the different cereals varied enormously, the order of homologous genes on the different cereal chromosomes was very similar (Moore, 1995; Gale and Devos, 1998). This conservation of homologous gene order is presumably a consequence of the cereals having a common ancestor (Kellogg, 1998). This observation has allowed cereal geneticists to treat the cereals as one single species for mapping and cloning purposes (McCouch, 1998). This syntenic relationship has been shown to extend to the level of a few hundred kilobases (Chen et al., 1997). The observation that barley, wheat and maize have a similar gene order to rice, but that rice has a smaller genome, has led to rice being used as a model system for the mapped-based cloning of genes that are of interest in the other cereals (Foote et al., 1997). In turn, this development (not forgetting the agricultural importance of rice in its own right) has led to an enormous amount of resource being directed to the characterization of the rice genome. Therefore, it is logical that we start our cereal structural genomics survey with rice.

C. RICE

Rice has a relatively small genome size (4.3×10^8 bp; Martinez et al., 1994). Rice also has the most complete set of structural genomics related resources of any cereal. Several rice BAC and yeast artificial chromosome (YAC) libraries are available (Umehara et al., 1995; Zhang et al., 1996; see also http://genome1.bio.bnl.gov/pub/maize/riceproject.html). A number of programmes are utilizing these resources for both the generation of whole genome contigs (Kurata et al., 1997; Zhang and Wing, 1997) and the characterization of specific regions of the genome (Wang et al., 1995; Umehara et al., 1996). Given the advanced state of the physical mapping of the rice genome, it is not surprising that the complete sequence of the rice genome is set to be determined in approximately 2004 (http://www.staff.or.jp/seqcollab.html; see also comments in Nature 1999, **401**,102).

D. BARLEY

Barley has a genome size of approximately 5×10^9 (Kleine et al., 1993). A number of barley YAC and BAC libraries have been constructed (Kleine et al., 1993; Yu et al., 1999). In both cases, these libraries have been used for map-based procedures, both to clone genes directly from barley and via synteny using rice as a model species (Kleine et al., 1993). No attempt has yet been made to construct either a complete physical map of the barley genome or carry out large-scale genome sequencing.

E. WHEAT

Given the large size of the hexaploid wheat genome (16×10^{12} bp; Gill and Gill, 1994), it is not surprising that, as yet, no complete (three genome equivalents or more) YAC or BAC libraries have been reported. Partial hexaploid wheat YAC libraries have been constructed. However, recently, there have been reports of the construction of BAC libraries for *T. urartu*, *T. monococcum*, *A. speltoides* and *A. tauschii*, which are presumed to represent the A, B and D genomes, respectively, of hexaploid wheat (Liavetzky *et al.*, 1999; Dweikat and Ohm, 1999; Moullet *et al.*, 1999). To date, only the construction and partial characterization of these libraries have been reported and no reports have been received of these libraries being used to generate partial or complete physical maps. That said, given currently developments, it is probably only a matter of time before a complete BAC library is constructed for hexaploid wheat.

F. MAIZE

Maize has a genome size of approximately 2.5×10^9 (Bennett and Smith, 1976). Although maize is a diploid, its genetic characteristics suggest that it is almost certainly an ancient tetraploid. A three-genome equivalent YAC library for the inbred line LH82 has been reported (Edwards *et al.*, 1992), along with several BAC libraries for the inbred B73 (Teofilas *et al.*, 1999). In addition, the author's laboratory has just completed the construction of a three-genome BAC library for the inbred F2 (O'Sullivan *et al.*, 1999). As part of the 1998 NSF initiative, Coe and co-workers, have received funds to both generate a ten-genome equivalent BAC library for B73 and generate a complete physical map within 5 years (http://www.nsf.gov/bio/pubs/awards/genome98.html). Given its commercial importance to world agriculture and the above resources, it is only a matter of time, perhaps just 5 years, before the complete sequence of the maize genome is determined.

V. FUNCTIONAL GENOMICS

A. INTRODUCTION

Functional genomics in cereals may be regarded as the return of the prodigal son. Many of the molecular tools that we now take for granted were first discovered in maize in the 1930s and 1940s (McClintock, 1948, 1950). Maize transposable elements have now been used extensively as "gene machines" in dicotyledonous species, such as *Arabidopsis thaliana* (Long *et al.*, 1993; Bhatt *et al.*, 1996; Carol *et al.*, 1999) and tomato (Takken *et al.*, 1998), as well as in the monocot, maize (Das and Martienssen, 1995). More recently, activation/dissociation (Ac/Ds) has been used in rice (Takeshi *et al.*, 1997), barley (McElroy *et al.*, 1997) and wheat (Takumi, 1998). The benefits of transposon tagging systems in any species are obvious. Transposons provide the following benefits for genetic analysis of particular traits.

1. They permit rapid production of mutations for functional analysis (McClintock, 1950).
2. They serve as molecular tags permitting the rapid isolation of the mutated gene (Das and Martienssen, 1995).
3. They may provide a reversible phenotype aiding molecular analysis of the trait (Carol *et al.*, 1999).
4. Transposons may alter the expression of the mutated gene in such a way as to provide altered, yet non-lethal phenotypes in otherwise critically functioning genes (Malone *et al.*, 1993).

In recent years, transposons have been used to an increasing extent in gene function analysis in cereals. The following section aims to give an overview of the recent work that has taken place in this increasingly broad area of cereal functional genomics.

B. RICE

The small size of the rice genome is currently being exploited in a number of genome sequencing projects (Sasaki, 1998). However, until recently, no useful transposon systems had been identified within rice. Although some work had been done with antisense technology (Fujisawa *et al.*, 1999), the lack of transposon systems has delayed the broad introduction of functional genomics in this model species. Recently, a family of transposons has been discovered in rice which are homologous to the *Mutator* (*Mu*) transposon of maize (Yoshida *et al.*, 1998; Ishikawa *et al.*, 1999). Yoshida called this element OsMuDR. OsMuDR was found to be present as a single copy element in the Japanese cultivar of *Oryza sativa*.

Whilst isolating SINE1 retroposon family members, Motohashi *et al.* (1996) found evidence of a En/Spm transposon-like elements in rice. One SINE element was found to harbour a 1536-bp insertion flanked by 13-bp imperfect inverted repeats. The sequence of the repeats (beginning with 5′-CACTA-3′) was indicative of an insertion by an En/Spm class transposon.

Very little information is available on the mobility of the OSMuDR or En/Spm elements and there is considerable debate as to their transmission vertically through the germline versus horizontally by other means. The acquisition of sequence data by studies such as those discussed may help resolve this question.

Hirochika (1997) studied the number and activity of retrotransposons elements in rice. This work indicated that there were approximately 1000 retrotransposons in rice, which fell into 32 families. Amongst these, Tos17 (a *copia*-like element) was actively transcribed under tissue-culture conditions in both *japonica* and *indica* cultivars. The elements do not appear to be transcribed under non-stressful growth conditions. Analysis of the target sites of this transposable element indicates that it is a significant cause of somaclonal variation in rice. The Tos17 element is now being used as a gene-tagging tool in rice (Hirochika *et al.*, 1996).

Three groups have recently published work on Ac/Ds tagging in rice (Takeshi *et al.*, 1997; Enoki *et al.*, 1999). Both groups have used the maize Ac element to produce insertional mutations. Chin *et al.* (1999) used a modified Ds element to the same end. In their analysis of the sequences flanking transposed Ac elements, Enoki *et al.* found that 21 of the 99 elements examined had inserted into coding regions. Sequences flanking four insertions produced matches with rice cDNA sequences (Enoki *et al.*, 1999).

In addition to transposon tagging, gene knockouts for use in functional genomics can be obtained by post-transcriptional silencing. For example, Fujisawa *et al.* (1999) produced transgenic plants, which expressed an antisense copy of the α-subunit of the rice heterotrimeric G-protein. The affected plants had little or no detectable levels of target gene expression and were both physically dwarfed and set abnormally small seed. Clearly, antisense technology provides an opportunity to develop functional genomics tools in rice.

Functional genomics in rice is an exciting area, progress is rapid and a number of important papers have appeared this year. Surveying rice functional genomics could, in many ways, be regarded as similar to surveying *Arabidopsis* functional genomics in the mid-1990s.

C. BARLEY AND WHEAT

In the public sector, the problems associated with transforming barley and the hexaploid nature of the wheat genome have, to date, precluded the type of gene knockout system described elsewhere in this section. Most transposon-induced mutations are recessive and are therefore of limited use in a hexaploid species. However, transposon-based systems are applicable to gene knockouts in diploid wheat. Takumi (1998) used microprojectile bombardment to transform diploid and hexaploid wheat cell lines with Ac and Ds constructs. A stabilized Ac transposase source was used to drive the excision of a Ds element on a separate vector. The Ds element was located within a hygromycin B resistance gene (*hph*). Excision resulted in the acquisition of hygromycin resistance by the transformed cells. Similar work using the Ac/Ds system in barley (McElroy *et al.*, 1997) indicates that these experiments are at a rudimentary stage. However, once sufficient numbers of transformed plants carrying mobile genetic elements are created, the prospects for functional genomics in these cereals should improve significantly.

D. MAIZE

Maize may be regarded as the matriarch of all cereals when one discusses functional genomics. Not only has there been more research on transposons in maize than in any other cereal, but it is from maize that most of the transposons utilized in other systems originate (McClintock, 1948, 1950). Transposons from maize are now found in *Arabidopsis* (Carol *et al.*, 1999), tomato (Takken *et al.*, 1998) and yeast (Weil and Cains, 1999).

McClintock first identified Ds in the 1940s as a revertible mutation at the *Waxy* locus (McClintock, 1948, 1950). Further work has shown that Ac, and its internally deleted and non-autonomous derivative Ds, has a number of properties that make them suitable as molecular tools in functional genomics that are listed below.

1.   They preferentially transpose to linked sites.
2.   They have a preference for insertion in hypomethylated, genic sequences.
3.   Transposition is self-limiting in maize, as Ac copy number increases the transposition rate of Ds falls.
4.   Ac/Ds insertion mutations are reversible.

The preference of Ac/Ds to transpose to linked sites is only of benefit when the target gene of interest lies close to the location of the Ac/Ds element. Thus, when using Ac or Ds to tag your gene of interest, it is beneficial to identify elements that are in proximity to that gene. A number of genes have been tagged with Ac/Ds (reviewed in Nevers *et al.*, 1986; Peterson, 1993; Cains *et al.*, 1997). One way around Ac/Ds limited capacity to tag genes in close proximity is to introduce further copies of Ac or Ds elements throughout the genome (Brutnell and Langdale, 1999).

En/Spm elements have a history as long as that of Ac/Ds (McClintock, 1948; Nevers *et al.*, 1986; Masson *et al.*, 1991). Again, Spm tends to transpose to linked sites.

The *Mutator* transposon was originally identified by Robertson as a heritable high forward-mutation rate, exhibited by lines derived from a single maize stock (Robertson, 1978). *Mu* is currently the most commonly used transposon for the so-called "maize gene machine" (Martienssen *et al.*, 1999; Stevenson *et al.*, 1999; Walbot *et al.*, 1999). *Mu* has a number of useful similarities and differences from Ac/Ds and Spm (Bennetzen, 1993):

1. *Mu* replicates during transposition;
2. *Mu* has a high forward-transposition rate;
3. *Mu* preferentially transposes to hypomethylated sites; and
4. *Mu* transposes to unlinked sites.

Unlike Ac, *Mu* encodes two proteins, MURA and MURB. MURA encodes the transposase and MURB, a probable regulator of transposition activity (Donlin *et al.*, 1995). Like Ac/Ds and En/Spm, *Mu* exists as a family of related elements. *MuDR* is the autonomous element, while a large number of related, internally deleted, elements also exist. These deletion derivatives can be mobilized when MURA (transposase) is supplied *in trans*.

Currently both Stevenson and Martienssen are using *Mu* to tag large numbers of genes in maize (Martienssen *et al.*, 1999; Stevenson *et al.*, 1999). Raizada and co-workers have recently modified *Mu* to form "RescueMu" to produce a maize transposon system analogous to the binary transposon systems in *Arabidopsis* (http://www.zmdb.iastate.edu/). In this system, an ampicillin resistance marker and *E. coli* origin of replication have been added to a *Mu1* element. This *Mu* element is located in a marker gene (the *Lc* transcriptional activator gene involved in anthrocyanin production). Mobilization of the *Mu1* element, by transposase supplied *in trans*, results in purple spotting on the leaves of transgenic plants.

E. CONCLUSIONS

Cereal functional genomics is at the stage of developing the tools required to operate on a similar level to *Arabidopsis* functional genomics. For both rice and maize, the molecular tools are now in place and EST sequencing projects are providing the raw material for the various gene knockout programmes. Given this scenario and further publicly funded initiatives, it is clear that we should expect to see rapid advances in the area of functional genomics in the near future.

VI. FUTURE PROSPECTS

Cereal genomics has progressed rapidly in the last 5 years. From a time when the rice genome was considered to be far too complex to consider large-scale physical mapping and sequencing, we have moved on to the point where several groups are now racing to complete the full rice genome sequence, to generating contig maps for maize and to determining the sequence and function of the full complement of cereal ESTs. Given these developments and future possibilities, the next 10 years will almost certainly be the time when the study of the cereal genome really does come of age.

ACKNOWLEDGEMENTS

IACR–Long Ashton receives grant-aided support from the Biotechnology and Biological Sciences Research Council of the United Kingdom. DS was funded by a BBSRC-Agri-Foods Committee award to KJE.

REFERENCES

Adams, M. D., Kelley, J. M., Gocayne, J. D., Dubnick, M., Polymeropoulos, M. H., Xiao, H., Merril, C. R., Wu, A., Olde, B., Moreno, R. F., Kerlavage, A. R., Mccombie, W. R. and Venter, J. C. (1991). Complementary-DNA sequencing-expressed sequence tags and human genome project. *Science* **252**, 1651–1656.

Alexandratos, N. (1995). "World Agriculture: Towards 2010". Wiley, Chichester.

Bennett, M. D. and Smith, J. B. (1976). Nuclear DNA amounts in Angiosperms. *Transactions of the Royal Society* **274**, 227–274.

Bennetzen, J. L. (1993). The *Mutator* transposable element system of maize. *Current Topics in Microbiology and Immunology* **204**, 195–229.

Bevan, M.W. (1997). The *Arabidopsis* genome project. *FASEB Journal* **11**, 860.

Bhatt, A. M., Page, T., Lawson, E. J. R., Lister, C. and Dean, C. (1996). The use of Ac as an insertional mutagen in *Arabidopsis*. *Plant Journal* **9**, 935–945.

Brettschneider, R. (1998). RFLP analysis. *In* "Molecular Tools for Screening Biodiversity" (A. Karp, P. G. Isaac and D. S. Ingram, eds), pp. 83–95. Chapman and Hall, London.

Briggs, S.P. (1998). Plant genomics: More than food for thought. *Proceedings of the National Academy of Sciences (USA)* **95**, 1986–1988.

Brown, L. R. (1994). "Full House: Reassessing the Earth's Population Carrying Capacity" (Worldwide Environmental Alert Series). W.W. Norton and Company.

Brutnell, T. and Langdale, J. (1999). Shotgun mutagenesis: a mapped-based approach to gene tagging in maize. *41st Annual Maize Genetics Conference*, T28.

Bryan, G. J., Stephenson, P., Collins, A., Kirby, J., Smith J. B. and Gale M. D. (1999). Low levels of DNA sequence variation among adapted genotypes of hexaploid wheat. *Theoretical and Applied Genetics* **99**, 192–198.

Cains, R., Saedler, H. and Lonnig, W. E. (1997). Plant transposable elements. *Advances in Plant Pathology* **27**, 331–470.

Carol, P., Stevenson, D. S., Bisanz C., Breitenbach, J., Sandmann, G., Mache, R., Coupland, G. and Kuntz, M. (1999). Mutations in the *Arabidopsis* gene IMMUTANS cause a variegated phenotype by inactivating a chloroplast terminal oxidase associated with phytoene desaturation. *Plant Cell* **11**, 51–68.

Chen, M., SanMiguel, P., deOliveira, A. C., Woo, S. S., Zhang, H., Wing, R. A. and Bennetzen, J. L. (1997). Microcolinearity in sh2-homologous regions of the maize, rice and sorghum genomes. *Proceedings of the National Academy of Sciences (USA)* **94**, 3431–3435.

Chin, H.G., Choe, M. S., Lee, S. H., Park, S. H., Koo, J. C., Kim, N. Y., Lee, J. J., Oh, B. G., Yi, G. H., Kim, S. C., Choi, H. C., Cho, M. J. and Han, C. D. (1999). Molecular analysis of rice plants harboring an Ac/Ds transposable element-mediated gene trapping system. *Plant Journal* **19**, 615–623.

Cho, Y.G., McCouch, S. R., Kuiper, M., Kang, M. R., Pot, J., Groenen, J. T. M. and Eun, M. Y. (1998). Integrated map of AFLP, SSLP and RFLP markers using a recombinant inbred population of rice (*Oryza sativa* L.). *Theoretical and Applied Genetics* **97**, 370–380.

Das, L. and Martienssen, R. A. (1995). Site-selected mutagenesis at the Hcf106 locus in maize. *Plant Cell* **7**, 287–294.

Davis, G. L., McMullen, M. D., Baysdorfer, C., Musket, T., Grant, D., Staebell, M., Xu, G., Polacco, M., Koster, L., MeliaHancock, S., Houchins, K., Chao, S. and Coe, E. H. (1999). A maize map standard with sequenced core markers, grass genome reference points and 932 expressed sequence tagged sites (ESTs) in a 1736-locus map. *Genetics* **152**, 1137–1172.

Devos, K. M. and Gale, M. D. (1997). Comparative genetics in the grasses. *Plant Molecular Biology* **35**, 3–15.

Donini, P., Stephenson, P., Bryan, G. J. and Koebner, R. M. D. (1998). The potential of microsatellites for high throughput genetic diversity assessment in wheat and barley. *Genetic Resources and Crop Evolution* **45**, 415–421.

Donlin, M. J., Lisch, D. and Freeling, M. (1995). Tissue-specific accumulation of MURB, a protein encoded by MuDR, the autonomous regulator of the *Mutator* element. *Plant Cell* **7**, 1989–2000.

Duggan, D. J., Bittner, M., Chen, Y., Meltzer, P. and Trent, J. M. (1999). Expression profile using cDNA microarrays. *Nature Genetics* **21**, 10–14.

Dvorak, J., Luo, M., Yang, Z. and Zhang, Y. H. (1998). The structure of the *Aegilops task* genepool and the evolution of hexaploid wheat. *Theoretical and Applied Genetics* **97**, 657–670.

Dweikat, I. and Ohm, H. (1999). Construction of a bacterial artificial chromosome library in wheat *Triticum urartu*. *Plant and Animal Genome VII Conference*, P100.

Ecker, J. R. and Davis, R. W. (1986). Inhibition of gene-expression in plant cells by expression of antisense RNA. *Proceedings of the National Academy of (USA)* **83**, 5372–5376.

Edwards, K. J., Thompson, K. H., Edwards, D., deSaizieu, A., Sparks, C., Thompson, J. C., Greenland, A. J., Eyers, M. and Schuch, W. (1992). Construction and characterisation of a yeast artificial chromosome library containing 3 haploid maize genome equivalents. *Plant Molecular Biology* **19**, 299–300.

Ellis, R.P. J., McNicol, W., Baird, E., Booth, A., Lawrence, P., Thomas, B. and Powell, W. (1997). The use of AFLPs to examine genetic relatedness in barley. *Molecular Breeding* **3**, 359–369.

Enoki, H., Izawa, T., Kawahara, M., Komatsu, M., Koh, S., Kyozuka, J. and Shimamoto, K. (1999) Ac as a tool for the functional genomics of rice. *Plant Journal* **19**, 603–613.

Foote, T., Roberts, M., Kurata, N., Sasaki, T. and Moore, G. (1997). *Genetics* **147**, 801–807.

Fujisawa, Y., Kato, T., Ohki, S., Ishikawa, A., Kitano, H., Sasaki, T., Asahi, T. and Iwasaki, Y. (1999). Suppression of the heterotrimeric G protein causes abnormal morphology, including dwarfism, in rice. *Proceedings of the National Academy of Sciences (USA)* **96**, 7575–7580.

Gale, M. D. and Devos, K. M. (1998). Comparative genetics in the grasses. *Proceedings of the National Academy of Sciences (USA)* **95**, 1971–1974.

Gill, K. S. and Gill, B. S. (1994). Mapping in the realm of polypoidy – the wheat model. *Bioessays* **16**, 841–846.

Gonen, D., Veenstra VanderWeele, J., Yang, Z., Leventhal, B. L. and Cook, E. H. (1999). High throughput flourescent CE-SSCP SNP genotyping. *Molecular Psychiatry* **4**,339–343.

Granjeaud, S., Bertucci, F. and Jordan, B. R. (1999). Expression profiling: DNA arrays in many guises. *BioEssays* **21**,781–790.

Griffin, T. J., Hall, J. G., Prudent, J. R. and Smith, L. M. (1999). Direct genetic analysis by matrix-assisted laser desorption/ionization mass spectrometry, *Proceedings of the National Academy of Sciences (USA)* **96**, 6301–6306.

Guo, Z., Guilfoyle, R. A., Thiel, A. J., Wang, R. and Smith, L. M. (1994). Direct fluorescence analysis of genetic polymorphisms by hybridisation with oligonucleotide arrays on glass support. *Nucleic Acids Research* **22**, 5456–5465.

Hartl, L., Mohler, V., Zeller, F. L., Hsam, S. L. K. and Schweizer, G. (1999). Identification of AFLP markers closely linked to the powdery mildew resistance genes *Pm1c* and *Pm4a* in common wheat (*Triticum aestivum* L.). *Genome* **42**, 322–329.

Helentjaris, T., Slocum, M., Wright, S., Schaefer, A. and Nienhuis, J. (1986). Construction of genetic linkage maps in maize and tomato using restriction fragment length polymorphism. *Theoretical and Applied Genetics* **72**, 761–769.

Heyen, D. W., Beever, J. E., Da, Y., Evert, R. E., Green, C., Bates, S. R. E., Ziegle, J. S. and Lewin, H. A. (1997). Exclusion probabilities of 22 bovine microsatellite markers in fluorescent multiplexes for semiautomated parentage testing. *Animal Genetics* **28**, 21–27.

Hirochika, H. (1997). Retrotransposons of rice: their regulation and use for genome analysis. *Plant Molecular Biology* **35**, 231–240.

Hirochika, H., Sugimoto, K., Otsuki, Y., Tsugawa, H. and Kanda, M. (1996) Retrotransposons of rice involved in mutations induced by tissue culture. *National Academy of Sciences (USA)* **93**, 7783–7788.

Ishikawa, R., Senda, M., Akada, S., Harada, T. and Niizeki, M. (1999). Structural differences of *RiceMutator* and maize *Mutator* elements and the *RiceMutator*. *41st Annual Maize Genetics Conference*, P120.

Kaul, S. *et al.* (2000). Analysis of the genome sequence of the flowering plant Arabidopsis thaliana. *Nature* **408**, 796–815.

Kellogg, E. A. (1998). Relationship of cereal crops and other grasses. *Proceedings of the National Academy of Sciences (USA)* **95**, 2005–2010.

Kent, N. L. and Evers, A. D. (1994). "Kent's Technology of Cereals". Pergamon, Oxford.

Kleine, M., Michalek, W., Graner, A., Herrmann, R. G. and Jung C. (1993). Construction of a barley (*Hordeum vulgare*) YAC library and isolation of a HOR1-specific clone. *Molecular and General Genetics* **240**, 265–272.

Korzun, V., Roder, M. S., Wendehake, K., Pasqualone, A., Lotti, C., Ganal, M. W. and Blanco, A. (1999). Integration of dinucleotide microsatellites from hexaploid bread wheat into a genetic linkage map of durum wheat. *Theoretical and Applied Genetics* **98**, 1202–1207.

Kurata, N., Umehara, Y., Tanoue, H., and Sasaki, T. (1997). Physical mapping of the rice genome with YAC clones. *Plant Molecular Biology* **35**, 101–113.

Langridge, P., Karakousis, A., Collins, N., Kretschmer, J. and Mannings, S. (1995). A consensus linkage map of barley. *Molecular Breeding* **1**, 389–395.

Liavetzky, D., Muzzi, G., Wing, R. and Dubcovsky, J. (1999). *Plant and Animal Genome VII Conference*, P99.

Lockhart, D. J., Dong, H., Byrne, M. C., Follettie, M. T., Gallo, M. V., Chee, M. S., Mittmann, M., Wang, C., Kobayashi, M., Horton, H. and Brown, E. L. (1996). Expression monitoring by hybridisation to high-density oligonucleotide arrays. *Nature Biotechnology* **14**, 1675–1680.

Long, D., Swinbourne, J., Martin, M., Wilson, K., Sundberg, E., Lee, K. and Coupland, G. (1993). Analysis of the frequency of inheritance of transposed Ds elements in *Arabidopsis* after activation by a CaMV 35S promoter fusion to the Ac transposase gene. *Molecular and General Genetics* **241**, 627–636.

Ma, Z. Q. and Lapitan, N. L. V. (1998). A comparison of amplified and restriction fragment length polymorphism in wheat. *Cereal Research Communications* **26**, 7–13.

MacDonald, T. and Brewbaker, J. L. (1972). Isoenzyme polymorphism in flowering plants. *Journal of Heredity* **63**, 11–14.

Malone, M. E., Fassler, F. S. and Winston, F. (1993). Molecular and genetic characterisation of *Spt4*, a gene important for transcription initiation in *Saccharomyces cerevisiae*. *Molecular and General Genetics* **237**, 449–459.

Mano, Y, SayedTabatabaei, B. E., Graner, A., Blake, T., Takaiwa, F., Oka, S. and Komatsuda, T. (1999). Map construction of sequence-tagged sites (STSs) in barley (*Hordeum vulgare* L.) *Theoretical and Applied Genetics* **98**, 937–946.

Martinez, C. P., Arumuganathan, K., Kikuchi, H. and Earle, E. D. (1994). Nuclear-DNA content of 10 rice species as determined by flow-cytometry. *Japanese Journal of Genetics* **69**, 513–52.

Martienssen, R. A., Volbrecht, E., Rabinowicz, P., May, B., Senior, L., Stein, L., Freeling, M. and Alexander, D. (1999). Target-selected mutagenesis in maize using Robertson's *Mutator* transposons. *41st Maize Annual Genetics Conference*, T22.

Masson, P., Banks, J. A. and Federoff, N. (1991). Structure and function of the maize Spm transposable element. *Biochimie* **73**, 5–8.

McClintock, B. (1948). "Yearbook of the Carnegie Institute of Washington, No. 47". Carnegie Institute of Washington, Washington, DC.

McClintock, B. (1950). The origin and behaviour of mutable loci in maize. *Proceedings of the National Academy of Sciences (USA)* **36**, 344–355.

McCouch, S. M. (1998). Towards a plant genomics initiative: thoughts on the value of cross-species and cross-genera comparisons in the grasses. *Proceedings of the National Academy of Sciences (USA)* **95**, 1983–1985.

McCouch, S. R., Chen, X. L., Panuad, O., Temnykh, S., Xu, Y. B., Cho, Y. G., Huang, N., Ishii, T. and Blair, M. (1997). Microsatellite marker development, mapping and applications in rice genetics and breeding. *Plant Molecular Biology* **35**, 89–99.

McElroy, D., Louwerse, J. D., McElroy, S. M. and Lemaux, P. G. (1997). Development of a simple transient assay for Ac/Ds activity in cells of intact barley tissue. *Plant Journal* **11**, 157–165.

McKusick, V. A., Kucherlapati, R. S. and Ruddle, F. H. (1993). Genomics – stock taking after 5 years. *Genomics* **15**: 1–2.

Merezhko, A. E. (1998). Impact of plant genetics resources on wheat breeding. *Euphytica* **100**, 295–303.

Mingeot, D. and Jacquemin, J. M. (1999). Mapping of RFLP probes characterised for their polymorphism on wheat. *Theoretical and Applied Genetics* **98**, 1132–1137.

Mogg, R., Hanley, S. and Edwards, K. J. (1999). Generation of maize allele specific oligonucleotides from the flanking regions of microsatellite markers. *Plant and Animal Genome Conference*, P491.

Moore, G. (1995). Cereal genome evolution: pastoral pursuits with "Lego" genomes. *Current Opinion in Genetics and Development* **5**, 717–724.

Motohashi, R., Ohtsubo, E. and Ohtsubo, H. (1996). Identification of Tnr3, a suppressor-Mutator enhancer-like transposable element from rice. *Molecular and General Genetics* **250**, 148–152.

Moullet, O, Zhang, H. B. and Lagudah, E. S. (1999). Construction and characterisation of a large DNA insert library from the D genome of wheat. *Theoretical and Applied Genetics* **99**, 305- 313.

Nevers, P., Shepherd, N. S. and Saedler, H. (1986). Plant transposable elements. *Advances in Botanical Research* **12**, 103–203.

Nielsen, G. and Scandalios, J. G. (1974). Chromosomal location by use of trisomics and new alleles of an endopeptidase in *Zea mays. Genetics* **77**, 679–686.

O'Sullivan, D., Ripoll, P. J. and Edwards, K. J. (1999). A maize BAC library from the European flint line F2. *Plant and Animal Genome VII Conference*, P96.

Parker, G. D., Chalmers, K. J., Rathjen, A. J. and Langridge, P. (1998). Mapping loci associated with flour colour in wheat (*Triticum aestivum* L.). *Theoretical and Applied Genetics* **97**, 238- 245.

Pe, M. E., Fink, R., Gatti, E., Binelli, G. and Isaac, P. G. (1999). Isolation of *GaNS1*, a gene with post-meiotic expression in maize. *Plant and Animal Genome VII Conference*, P254.

Pejic, I., Ajmone-Marsan, P., Morgante, M., Kozumplick, V., Castiglioni, P., Taramino, G. and Motto, M. (1998). Comparative analysis of genetic similarity among maize lines detected by RFLPs, RAPDs, SSRs and AFLPs. *Theoretical and Applied Genetics* **97**, 1248–1255.

Peterson, P. A. (1993) Transposable elements in maize: their role in creating plant genetic variability. *Advances in Agronomy* **51**, 79–124.

Phillips, R. L. and Freeling, M. (1998). Plant genomics and our food supply: an introduction. *Proceedings of the National Academy of Sciences (USA)* **95**, 1969–1970.

Plaschke, J., Ganal, M. and Roder, M. (1995). Detection of genetic diversity in closely related bread wheat using microsatellite markers. *Theoretical and Applied Genetics* **91**, 1001–1007.

Ridout, C. J. and Donini, P. (1999). Technical focus: use of AFLP in cereals research. *Trends in Plant Science* **4**, 76–79.

Robertson, D. S. (1978). Characterisation of a mutator system in maize. *Mutational Research* **21**, 21–28.

Roder, M. S., Korzun, V., Wendehake, K., Plaschke, J., Tixier, M. H., Leroy, P. and Ganal, M. W. (1998). A microsatellite map of the wheat genome. *Genetics* **149**, 2007–2023.

Russell J., Fuller, J., Young, G., Thomas, B., Taramino, G., Macaulay, M., Waugh, R. and Powell, W. (1997). Discriminating between barley genotypes using microsatellite markers. *Genome* **40**, 442–450.

Saiki, R. K., Walsh, P. S., Levenson, C. H. and Erlich, H. A. (1989). Genetic analysis of amplified DNA with immobilized sequence-specific oligonucleotide probes. *Proceedings of the National Academy of Sciences (USA)* **86**, 6230–6234.

Salanoubat, M. *et al.* (2000). Sequence and analysis of chromosome 3 of the plant *Arabidopsis thaliana. Nature* **408**, 820–822.

Sasaki, T. (1998). The rice genome project in Japan. *Proceedings of the National Academy of Sciences (USA)* **95**, 2027–2028.

Schena, M. (1996). Genome analysis with gene expression microarrays. *BioEssays* **1**, 427–431.

Schwarz, G., Michalek, W., Mohler, V., Wenzel, G. and Jahoor, A. (1999). Chromosome landing at the Mla locus in barley (*Hordeum vulgare* L.) by means of high resolution mapping with AFLP markers. *Theoretical and Applied Genetics* **98**, 521–530.

Senior, M.L., Chin, E. C. L., Lee, M., Smith, J. S. C. and Stuber, C. W. (1996). Simple sequence repeat markers developed from maize sequences found in the GENBANK database: map construction. *Crop Science* **36**, 1676–1683.

Shen, B., Carneiro, N., Torresjerez, I., Stevenson, B., McCreery, T., Helentjaris, T., Baysdorfer, C., Almira, E., Ferl, R. J., Habben, J. E. and Larkins, B. (1994).

Partial sequencing and mapping of clones from 2 maize cDNA libraries. *Plant Molecular Biology* **26**, 1085–1101.

Stevenson, D. S., Forsyth, A., Holdsworth, M. and Edwards, K. J. (1999). A hybridisation based screen for identifying gene knockouts in maize. *41st Annual Maize Genetics Conference*, P93.

Struss, D. and Plieske, J. (1998). The use of microsatellite markers for detection of genetic diversity in barley populations. *Theoretical and Applied Genetics* **97**, 308–315.

Tabata, S. et al. (2000). Sequence and analysis of chromosome 5 of the plant *Arabidopsis thaliana*. *Nature* **408**, 823–826.

Takeshi, I., Oshini, T., Nakano, T., Ishida, N., Enoki, H., Hashimoto, H., Itoh, K., Terada, R., Wu, C. Y., Miyazaki, C., Endo, T., Iida, S. and Shimamoto, K. (1997). Transposon tagging in rice. *Plant Molecular Biology* **35**, 219–229.

Takken, F. L. W., Schipper, D., Nijkamp, H. J. J. and Hille, J. (1998). Identification and Ds-tagged isolation of a new gene at the Cf-4 locus of tomato involved in disease resistance to *Cladosporium fulvum* race 5. *Plant Journal* **14**, 401–411.

Takumi, S. (1998). Hygromycin-resistant calli generated by activation of maize Ac/Ds transposable elements in diploid and hexaploid wheat cultured cell lines. *Genome* **39**, 1169–1175.

Teofilas, S., He, L., Chang, Y. L. and Zhang, H. B. (1999). Generation and characterisation of a BAC library from the maize inbred line B73 for maize genomics research. *Plant and Animal Genome VII Conference*, P97.

Theologis, A. *et al.* (2000). Sequence and analysis of chromosome 1 of the plant *Arabidopsis thaliana*. *Nature* **408**, 816–820.

Umehara, Y., Inagaki, A., Tanque, H., Yasukochi, Y., Nagamura, Y., Saji, S., Otsuki, Y., Fujimura, T., Kurata, N. and Minobe, Y. (1995). Construction and characterisation of a rice YAC library for physical mapping *Molecular Breeding* **1**, 79–89.

Umehara, Y., Tanoue, H., Kurata, N., Ashikawa, I., Minobe, Y. and Sasaki, T. (1996). An ordered yeast artificial chromosome library covering over half of rice chromosome 6. *Genome Research* **6**, 935–942.

Vos, P., Hogers, R., Bleeker, M., Rijans, M., Van de Lee, T., Hornes, M., Frijters, A., Pots, J., Peleman, J., Kuiper, M. and Zabeau, M. (1995). AFLP: a new technique for DNA fingerprinting. *Nucleic Acids Research* **23**, 4407–4414.

Walbot, V., Chandler, V. L., Galbraith, D., Larkins, B., Freeling, M., Hake, S., Schmidt R., Smith, L., Brendel, V. and Sachs, M. M. (1999). Maize gene discovery, DNA sequencing and phenotypic analysis. *41st Annual Maize Genetics Conference*, P129.

Wang, G. L., Holsten, T. E., Song, W. Y., Wang, H. P. and Ronald, P. C. (1995). Construction of a rice bacterial artificial chromosome library and identification of clones linked to the XA-21 disease resistance locus. *Plant Journal* **7**, 525–533.

Weber, J. and May, P. E. (1989). Abundant class of human DNA polymorphisms which can be typed using the polymerase chain reaction. *American Journal of Human Genetics* **44**, 388–396.

Weil, C. and Cains, R. (1999). Ac/Ds transposition in yeast cell. *41st Annual Maize Genetics Conference*, P134.

Welsh, J. and McClelland, M. (1990). Fingerprinting genomes using PCR with arbitrary primers. *Nucleic Acids Research* **18**, 7213–7218.

Yamamoto, K. and Sasaki, T. (1997). Large scale EST sequencing in rice. *Plant Molecular Biology* **35**, 135–144.

Yin, X., Stam, P., Dourleijn, C. J. and Kropff, M. J. (1999). AFLP mapping of quantitative trait loci for yield-determining physiological characters in spring barley. *Theoretical and Applied Genetics* **99**, 244–253.

Yoshida, S., Kasai, Y., Tamaki, K., Watanabe, K., Fujino, M. and Nakamura, C. (1998). Stimulation of albino regeneration from rice tissue culture by proline under high osmosis: a possible relationship with an endogenous transposable element Os-MuDR. *Biotechnology and Biotechnological Equipment* **12**, 3–7.

Yu, Y., Tomkins, J., Frisch, D., Waugh, R., Brueggeman, R., Kudrna, D., Kleinhofs, A. and Wing, R. (1999). Construction and characterisation of a barley (*Hordeum vulgare*) cv. Morex BAC library. *Plant and Animal Genome VII Conference*, P102.

Zhang, H.B., Choi, S. D., Woo, S. S., Li, Z. K. and Wing, R. A. (1996). Construction and characterisation of two rice bacterial artificial chromosome libraries from the parents of a permanent recombinant inbred mapping population. *Molecular Breeding* **2**, 11–24.

Zhang, H.B. and Wing, R. A. (1997). Physical mapping of the rice genome with BACs. *Plant Molecular Biology* **35**, 115–127.

Zhao, N., Hashida, H., Takahashi, N., Misumi, Y. and Sasaki, Y. (1995). High-density cDNA filter analysis: a novel approach for large-scale quantitative analysis of gene expression. *Gene* **156**, 207–213.

Zhu, J. H., Stephenson, P., Laurie, D. A., Li, W., Tang, D. and Gale, M. D. (1999). Towards rice genome scanning by map-based AFLP fingerprinting. *Molecular and General Genetics* **261**, 184–195.

Exploiting Cereal Genetic Resources

ROBERT J. HENRY

*Cooperative Research Centre for Molecular Plant Breeding, Centre for
Plant Conservation Genetics, Southern Cross University, PO Box 157,
Lismore NSW 2480, Australia*

Advances in Botanical Research Vol. 34
incorporating Advances in Plant Pathology
ISBN 0-12-005934-7

I. INTRODUCTION

The cereals are a group of grasses with large seeds used as basic human and animal foods. The following species are usually defined as the major cereals: barley, maize, millets, oat, rice, rye, sorghum and wheat (Kent and Evers, 1994). The crops of these species cultivated today are the products of domestication over many thousands of years. Plant breeding has had an impact mainly in the last century (Law, 1995). Modern biotechnology, which is now starting to influence these species, has allowed an acceleration of the rate of plant breeding (Henry, 1995).

These cultivated plants and the related plants remaining in wild populations (Nevo, 1998) together with seeds in germplasm and breeding collections (Merezhko, 1998) represent the genetic resource available for exploitation in cereal production. Other plants, and indeed other organisms generally, have also become a genetic resource with the development of cereal transformation techniques. This review outlines the available genetic resources and the potential for their exploitation using modern molecular techniques.

A. IMPORTANCE OF CEREALS AND CEREAL GENETIC RESOURCES

Cereals are the source of most human food and are produced in large quantities internationally (Fig. 1). Successful exploitation of the available genetic resources in plant improvement, especially for the major cereal species (wheat, rice and maize), is essential to world food security. Genetic improvement of cereals is necessary to ensure food production grows to match continuing world population growth. Efficient cereal production may also contribute to reducing the impact of agriculture and food production on the global environment (Pimentel et al., 1995). Competition of agriculture with forests and other land uses is minimized by maximizing the efficiency of cereal production per unit area of land. These considerations indicate the essential role of effective exploitation of cereal genetic resources in the enhancement of cereal production and protection of the global environment.

A first step in more effective use of cereal genetic resources is to define and document the resources that are available. The cereal gene pool (Harlan and Wet, 1971) may be defined at several levels (Fig. 2). The primary gene pool of each cereal species includes the populations of the species itself with no barriers to gene transfer by normal sexual crossing. The secondary gene pool may be defined to include other species, usually within the same genus with which successful hybridization is possible by conventional means, even if recovery of fertile hybrid individuals is low. The tertiary gene pool extends to more distantly related species for which special techniques such as embryo

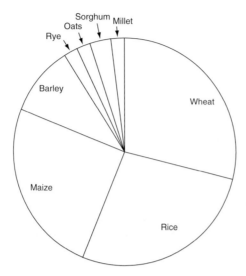

Fig. 1. Major cereals in agricultural production. Relative levels of annual production of major cereal species (Kent and Evers, 1994). Cultivated crops represent a major genetic resource.

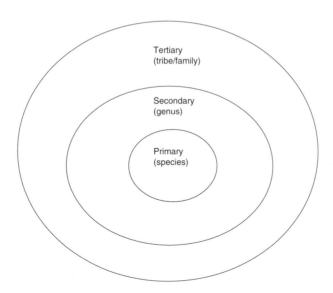

Fig. 2. A schematic representation of the concept of a gene pool.

rescue are necessary to achieve successful gene transfer. All of these resources may be of value in cereal improvement.

B. EXPLOITATION OF GENETIC RESOURCES

The process of exploiting cereal genetic resources may involve exploration, classification, collection, evaluation and utilization (Fig. 3). Conservation of cereal genetic resources can be *ex situ* (in germplasm collections) or *in situ* in the wild. Conservation by cultivation is another limited strategy. Exploitation of cereal genetic resources using biotechnology may be achieved by analysis and manipulation of cereal genomes using DNA analysis tools and by recombinant DNA techniques. Analysis of cereal genomes is a challenging task because of the comparatively large genome size of some of the species. The sizes of cereal genomes vary from 400 Mbp in rice to 16 500 Mbp in wheat (Fig. 4). Biotechnology has application in the selection of superior genotypes within populations, in recombining cereal genomes in desirable ways and in expanding the range of genes available in the breeding gene pool.

1.   Approaches to Selecting within the Cereal Gene Pool
Genetic improvement of cereals has been the result of human selection during the process of domestication. Conventional plant breeding has relied

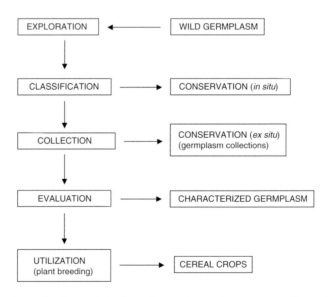

Fig. 3.   Steps in the exploitation of cereal genetic resources. Stages in which molecular methods may be employed are highlighted in bold (based upon Henry, 1997).

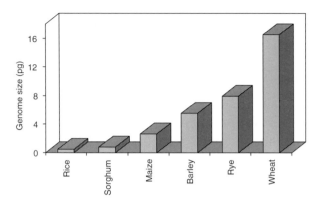

Fig. 4. Relative sizes of cereal genomes (Henry, 1997).

on selection of desirable phenotypes (Sorrells and Wilson, 1997). Modern breeding of species such as wheat involves the evaluation of large numbers of traits (Table I) that are required for commercial production. Selection for traits such as end-product quality often requires large samples to allow milling and product production for quality assessment. Small scale tests are essential for testing at early stages in breeding (Table I). The large number and complexity of these traits may severely restrict the range of germplasm that can be exploited in breeding. Strategies for selection in breeding may include mass selection or analysis of individuals. Mass selection for disease resistance by infecting plants with the pathogen allows very large numbers of plants to be quickly culled from breeding populations. In contrast, analysis of individuals for complex traits may require very laborious analysis of large amounts of plant material. Examples of this type of trait are the end-product tests such as loaf volume in wheat.

Molecular markers are increasingly important tools for selection in cereal breeding (Henry, 1997). Linkage of markers to important traits allows their use to select for the trait using DNA from seedlings at any stage in the breeding programme. The increasing amount of data available on sequences of cereal genes is allowing the analysis of the gene itself rather than relying on a linked marker. Direct selection for the desirable allele by DNA analysis may be considered to be the application of perfect markers.

Traditional DNA analysis has relied upon electrophoretic separation of DNA and other relatively complex protocols. Application of techniques developed initially for human diagnostics may allow DNA analysis in selection of cereals to be dramatically simplified and automated (Shepherd and Henry, 1988).

The process of domestication of cereals apparently applied selection to corresponding genetic loci in different species (Paterson et al., 1995). Large seed size is desirable in all species. Further exploitation of cereal genetic

TABLE I

Quality testing of wheat at early generations in the breeding process indicating the large number of complex traits required for commercial success. (Adapted from Wrigley and Morris, 1996)

Quality attribute	Test material	Method or equipment
Colour (red/white)	Whole grains	Visual: image analysis
		NaOH soak
		Colour meter
		Near-infrared transmittance (NIT)
		Image analysis
Hardness	Whole grains	Visual
		Perten
		SKCS 4100
		NIT
	Wholemeal	Near-infrared reflectance
Grain size	Grain	Particle size index
	Grain	Visual
		Thousand-kernel weight
		Perten SKCS 4100
Protein content	Grain	NIT, Kjeldahl
	Meal, or flour	NIT, Dumas
Milling	Grain	Quadrumat Junior
		US Department of Agriculture micro mill
Dough strength	Grain	Micro-mixograph
	Flour (wholemeal)	Size exclusion high-performance liquid chromatography (size distribution)
	Half grain, meal	
	Half grain, meal	Sodium dodecyl sulphate (SDS) gel electrophoresis (glutenin subunits)
	Half grain, meal	Antibody test kit
Baking quality	Flour	SDS sedimentation
		Zeleny sedimentation
		Pelshenke fermentation
Noodle, starch manufacture	Meal, flour, starch	Rapid visco-analyser (RVA) pasting
Baking water absorption	Flour	Alkaline water absorption
Weather damage potential	Meal, flour	RVA
		Falling number
Milling quality	Grain	Laboratory scale mill
Flour colour	Flour	Colour meter
Dough-mixing properties	Flour	Mixograph
		Farinograph
		Extensigraph
Pan-bread baking	Flour	Bake test (a range)
Cookie baking	Flour	Bake test
Noodles	Flour	Small-scale simulations
	Flour, meal, starch	RVA
		Amylograph
Flat bread	Flour, meal	Small-scale simulations

resources may be achieved by more intense selection at these loci or at new loci introduced by widening the gene pool.

2.   Recombining Cereal Genomes

Recombination of genes within the cereal species is the primary approach to the generation of new genetic variation for selection in plant improvement. Controlled pollination allows recombination of genes from different individuals. Breeders deliberately emasculate the flowers of cereals by removing the anthers prior to pollen maturity and conduct controlled pollination. Selfing in self-pollinating species such as wheat allows segregation and selection for desirable combinations of the genes from the two parents within the populations at F2 and subsequent generations. Molecular markers provide important tools to improve the efficiency of this process.

Larger genetic variation may be introduced by recombining genes from different species. Introgression of genes from other cereal species or wild relatives is an important approach to genetic improvement in cereals. Synthetic hexaploid wheats may be generated to capture a wider range of genes from wild diploid species (Appels and Lagudah, 1990). Molecular markers allow this process to be monitored and achieved with a minimum alteration of loci other than that being targeted. The use of rye in wheat improvement is an important example of this process. The IBL/IRS wheat rye translocation, IB(R) substitution and IAL/IRS translocation contribute to improved performance in yield and disease resistance in wheat (Rabinovich, 1998). In many cases this can be achieved by normal sexual crossing. For example, introgression of genes from wild barley (*Hordeum chilense*) into wheat may be accelerated by the development of polymerase chain reaction (PCR) markers specific for detection of wild barley chromosomes (Hernandez *et al.*, 1999). Embryo rescue may be used to achieve wider crosses (Brar and Khush, 1997). Cellular techniques are useful in some cases (Liu *et al.*, 1999). Cell culture of addition lines has been used to facilitate the transfer of genes from barley to wheat (Henry *et al.*, 1993). Cell fusion provides opportunities to recombine the genomes of different species. Wild relatives of rice have been used as sources of genes for rice improvement by fusion of protoplasts of *Orzya* species outside the range that can be crossed conventionally (Jelodar *et al.*, 1999; Liu *et al.*, 1999). Transformation with genes from the same or other species or alteration of the pattern of expression of genes already present are options for more radical genetic changes.

3.   Expanding the Cereal Gene Pool

Generation of mutants is a method for expanding the cereal gene pool by introducing new variation. Although much of this variation is deleterious, some useful characters may be introduced (Molina-Cano *et al.*, 1999).

Transformation techniques may be used to introduce new genes into the cereal gene pool. The cereals themselves are an important but not exclusive source of genes for cereal transformation. Examples of genes introduced into the maize gene pool by transformation are given in Table II.

Novel genes not found in nature can be generated by modification of cereal or other genes.

The manipulation of expression of a cereal gene by co-suppression on antisense approaches may be considered an expansion of the cereal gene pool as the genotype and phenotype resulting may be novel.

II.   CEREAL GENETIC RESOURCES

A.   INTRODUCTION

The cereals are members of the Poaceae (Grass family) and can be grouped within the following tribes: Triticeae, Aveneae, Poeae, Oryzeae, Chorideae, Paniceae, Maydeae and Andropogonea (Devos and Gale, 1997). The evolution of the Poaceae is depicted in Fig. 5, and the basic structure of cereal genomes is indicated in Table III.

Cereal germplasm collections are maintained as seed stores in many locations (Table IV). Collections of cereal mutants are important genetic resources because they allow analysis of gene function in addition to being a source of genetic variation for cereal improvement. Wild populations and in some cases cultivated crops contain important genetic resources not always represented in germplasm collections. DNA banks have not yet become major stores of cereal germplasm but represent an important option for the future (Adams, 1997). Large insert libraries such as yeast artificial chromosome (YAC) and beast artificial chromosome (BAC) libraries and other genomic DNA libraries and cDNA libraries may be considered important resources especially when the clones have been sequenced and/or mapped. Expressed sequence tags (EST) and sequence databases must also be considered as important genetic resources.

TABLE II

Examples of genes introduced into maize genomes by transformation (Malik, 1999; Zhong et al., 1999)

Gene	Purpose	Source
Bacillus thuringiensis toxin	Insect resistance	Bacteria
Glufosinate resistance	Weed control	Bacteria
Sulfonylurea resistance	Weed control	Tobacco
Glyphosate resistance	Weed control	Bacteria
Male sterility	Plant breeding	Bacteria
Aprotinin	Therapeutic agent (pharmaceutical)	Bovine

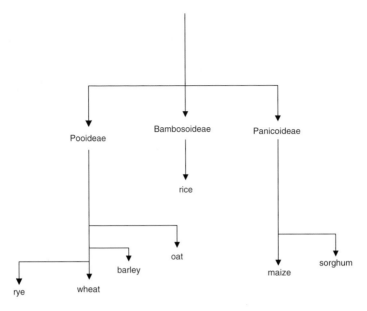

Fig. 5.   Evolution of the Poaceae.

TABLE III
Cereal genome organization

Species	Chromosome number (n)	Ploidy
Wheat	21	6×
Rice	12	2×
Barley	7	2×
Oats	21	6×
Sorghum	10	2×
Maize	10	4×

Wheat, rice and maize dominate world cereal production (Fig. 1). A slightly longer list of cereal species are regionally important.

B.   WHEAT

Wheat is the cereal produced in the greatest quantities internationally. Cultivated common wheat (*Triticum aestivum*) is a hexaploid (AABBDD) whereas durum wheat (*T. durum*) is a tetraploid. Polyploidy is a common feature associated with plant domestication (Hilu, 1993). Wild diploid progenitor species (Abdel-Aal *et al.*, 1998) are important sources of genes for wheat improvement. Some of these species are cultivated to a limited

TABLE IV
Some important cereal germplasm collections; for more information, see Chin (1994)

Organization	Location
CIMMYT (Centro International de Mejoramiento de Maizy Trigo)	Mexico
IRRI (International Rice Research Institute)	Philippines
ICARDA (International Centre for Agricultural Research in Dry Areas)	Syria
ICRISAT (International Crop Research Institute for the Semi Arid Tropics)	India
ICGR (Institute of Crop Germplasm Resources)	China
N.I. Vavilov Institute	Russia
Nordic Genebank	Sweden
NIAR (National Institute of Agrobiological Resources)	Japan

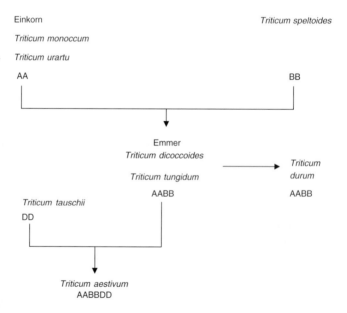

Fig. 6.   Relationships between some cultivated and wild wheats.

extent (e.g. *T. speltoides*). The relationships between some species in this group are depicted in Fig. 6 (Dvorak *et al.*, 1988, 1993). Hexaploid wheat is not genetically diverse, apparently due to the genetic bottleneck created by the events lending to polyploidy.

Genetic resources for wheat include the variation that exists within the polyploids (*T. aestivum* and *T. durum*), diploids (Castagna *et al.*, 1997; Dvorak *et al.*, 1998) especially other *Triticum* spp. and *Aegilops* spp.) and more distant genera. Variation available within *T. aestivum* probably

TABLE V
The genus Oryza. *Adapted from Vaughan (1994)*

Species	Other names(s)	Chromosome number	Genome group
O. sativa complex			
O. sativa L.		24	AA
O. nivara Sharma et Shastry	O. rufipogon (annual form)	24	AA
O. rufipogon Griff.	O. perennis O. rufipogon (perennial form)	24	AA
O. glaberrima Steud.		24	AA
O. barthii A. Chev.	O. breviligulata	24	AA
O. longistaminata Chev. et Roehr.	O. barthii	24	AA
O. meridionalis Ng		24	AA
O. officinalis complex	Also called O. latifolia complex or group C		
O. officinalis Wall ex Watt	O. minuta	24	CC
O. minuta Presl. Et Presl.	O. officinalis	48	BBCC
O. rhizomatis Vaughan		24	CC
O. eichingeri Peter		24	CC
O. punctata Kotschy ex Steud	O. schweinfurthiana for the tetraploid form	24, 48	BB, BBCC
O. latifolia Desv.		48	CCDD
O. alta Swallen		48	CCDD
O. grandiglumis (Doell) Prod.		48	CCDD
O. australiensis Domin		24	EE
O. brachyantha Chev. Et Roehr.		24	FF
O. schlechteri Pilger		48	Unknown
O. ridleyi complex			
O. ridleyi Hook. f.		48	HHJJ
O. longiglumis Jansen		48	Unknown
O. meyeriana complex			
O. meyeriana (Zoll. Et Mor. ex Steud.) Baill.		24	Unknown

remains poorly exploited by breeders (Merezhko, 1998). The development of synthetic hexaploids is allowing the exploitation of the wider genetic diversity in the wild diploid relatives.

The hexaploid nature of common wheat provides a buffer to the adverse effects of the introduction of genes from wild relatives that are often

experienced in other species. The wild relatives of wheat are closer
genetically to wheat than the wild relatives of barley are to barley (excluding
Hordeum spontaneum).

C. RICE

Rice is largely consumed as a whole-grain food. The constraint of acceptable
grain appearance and flavour limit the range of genotypes exploited. *Oryza
sativa* and *Oryza glaberrima* are among 22 species (Table V) in the genus
(Vaughan, 1994). Interspecific hybrids between *Oryza* species are possible,
although not always fertile. Embryo rescue has been used to achieve
introgression of genes from *Oryza* species. The Oryzeae tribe contains 12
genera representing the tertiary gene pool of rice (Table VI). *Zizania
palustris* (North American wild rice) is also cultivated as a food crop. The
Australian *Potamophila parviflora* has also been evaluated as a source of
genes for rice improvement (Abedinia *et al.*, 1998). Weedy rice found in
many rice producing areas may represent an important genetic resource
(Cho *et al.*, 1995; Suh *et al.*, 1997). Genes contributing to success as a weed
may be advantageous in crop varieties. Genes for adaptation to the local
environment are likely to be found in rice growing as a weed.

D. BARLEY

Cultivated barley (*Hordeum vulgare*) is derived from wild barley (*Hordeum
vulgare* ssp. *spontaneum*) and the wild barley gene pool is readily accessible

TABLE VI
The tribe Oryzeae

Genus	Species (number)	Distribution	Chromosome number ($2n$)
Oryza	22	Pantropical	24, 48
Leersia	17	Worldwide	24, 48, 60, 96
Chikusichloa	3	China, Japan	24
Hygroryza	1	Asia	24
Porteresia	1	South Asia	48
Zizania	3	Europe, Asia North America	30, 34
Luziola	11	North and South America	24
Zizaniopsis	5	North and South America	24
Rhynchoryza	1	South America	24
Maltebrunia	5	Tropical and Southern Africa	Unknown
Prosphytochloa	1	Southern Africa	Unknown
Potamophila	1	Australia	24

TABLE VII
The genus Hordeum *(Bothmer* et al., *1995)*

Species	Gene Pool	Description
Section *Hordeum*		Annual Mediterranean region
H. vulgare	Primary	
H. bulbosum	Secondary	
H. murinum	Tertiary	
Section *Anisolepis*		Short annual perennial South America
H. pusillum	Tertiary	
H. intercedens	Tertiary	
H. euclaston	Tertiary	
H. flexuosum	Tertiary	
H. muticum	Tertiary	
H. chilense	Tertiary	
H. cordobense	Tertiary	
H. stenostachys	Tertiary	
Section *Critesion*		Long annual perennial America
H. pubiflorum	Tertiary	
H. halophilum	Tertiary	
H. comosum	Tertiary	
H. jubatum	Tertiary	
H. arizonicum	Tertiary	
H. procerum	Tertiary	
H. lechleri	Tertiary	
Section *Stenostachys*		European, Asiatic and North America
H. marinum	Tertiary	
H. secalinum	Tertiary	
H. capense	Tertiary	
H. bogdanni	Tertiary	
H. roshevitzii	Tertiary	
H. brevisubulatum	Tertiary	
H. brachyantherum	Tertiary	
H. depressum	Tertiary	
H. guatemalense	Tertiary	
H. erectifolium	Tertiary	
H. tetraploidum	Tertiary	
H. fuegianum	Tertiary	
H. parodii	Tertiary	
H. patagonicum	Tertiary	

for barley improvement. Restriction fragment length polymorphism (RFLP) evidence indicates that barley may have been domesticated in the Fertile Crescent and in Morocco (Molina-Cano *et al.*, 1999). Members of the *Hordeum* genus (Bothmer *et al.*, 1995) are listed in Table VII. *Hordeum bulbosum* is easily crossed with barley, although crosses with this species usually result in haploid plants owing to the elimination of the *H. bulbosum* chromosomes. This approach is used to produce homozygous double haploid lines for breeding purposes. Other species of *Hordeum* require embryo rescue to recover hybrids that are likely to be sterile, preventing the ready exploitation of genes from these species.

Beer and whisky are produced from malted barley. Malting performance is acceptable in a relatively narrow range of barley genotypes (Edney, 1996). *Hordeum spontaneum* represents a genetically diverse source of diverse phenotypes for barley improvement (Henry and Brown, 1987; Weining and Henry, 1995).

E. MAIZE

Archaeological records of maize (*Zea mays*) date back 7000 years. The origins of maize and domestication of the species in central America has recently been clarified by molecular analysis. Maize originated from a wild plant, teosinte, with a very different appearance Selection of maize with short branches tipped with ears rather than long branches with tassels at the tip involved a gene *teosinte brancheli* (*tbi*) that determines this genetic difference. Analysis of this gene in cultivated maize has shown that domestication was associated with selection of regulatory elements in this gene and that the variation in the coding region of the gene is similar in maize and teosinte (Wang *et al.*, 1999) This suggests that domestication captured the variation in regions not subject to human selection probably by back crossing of cultivated selections with wild teosinte populations over a period of hundreds of years.

A large amount of genetic diversity exists in land races of maize (Louette *et al.*, 1997) and efforts have been made to conserve this diversity in traditional maize growing communities. Elite commenced varieties have a much narrower genetic base, which could benefit from the introgression of more diverse germplasm (Tallury and Goodman, 1999). Domestication has resulted in the loss of some characters and the retention of others (Table VIII).

TABLE VIII
Domestication of maize (Wang et al., *1999)*

Primitive maize	Domesticated maize	Common traits
Slender cobs	Loss of dormancy	Unisexual inflorescences
Short ears	Increased grain size	Tassle and ear
Hard grains	Loss of hard casing	C4, photosynthesis
Tassels on long branches	Ears on short branches	
Brownish colour		

TABLE IX
The genus Sorghum *(S. Dillon, personal communication)*

Species	Other names	Country of origin/ distribution
S. amplum		Australia
S. angustum		Australia
S. arundinaceum	*S. bicolor* subsp. *arundinaceum*, *S. lanceolatum*, *S. macrochaeta*, *S. pugionifolium*, *S. stapfii*, *S. usambarense*, *S. verticilliflorum*, *S. virgatum*, *S. vogelianum*	Africa: Angola, Benin, Botswana, Burkina Faso, Cameroon, Central African Republic, Chad, Cote D'Ivoire, Egypt, Equatorial Guinea, Ethiopia, Gabon, Gambia, Ghana, Guinea, Kenya, Liberia, Malawi, Mali, Mauritania, Mozambique, Namibia, Niger, Nigeria, Senegal, Sierra Leone, Somalia, South Africa, Sudan, Swaziland, Tanzania, Uganda, Zaire, Zambia, Zimbabwe Naturalized in the Americas, Australia, and India
S. bicolor	*S. basutorum*, *S. caffrorum*, *S. caudatum*, *S. cernuum*, *S. conspicuum*, *S. coriaceum*, *S. dochna*, *S. durra*, *S. elegans*, *S. gambicum*, *S. guineense*, *S. japonicum*, *S. margaritiferum*, *S. malaleucum*, *S. mellitum*, *S. membranaceum*, *S. miliiforme*, *S. nervosum*, *S. nigricans*, *S. notabile*, *S. roxburghii*, *S. saccharatum*, *S. simulans*, *S. splendidum*, *S. subglabrescens*, *S. technicum*, *S. vulgare*	Cultivated throughout tropic, subtropic and warm temperate regions
S. brachypodum		Australia

TABLE IX *Continued*

Species	Other names	Country of origin/ distribution
S. bulbosum		Australia
S. ecarinatum		Australia
S. exstans		Australia
S. grande		Africa, Asia, Australia, Mexico
S. halepense	S. controversum, S. miliaceum	Africa: Egypt, Libya Asia – Temperate: Afghanistan, Armenia, Azerbaijan, Iran, Iraq, Israel, Jordan, Kazakhstan [south], Kyrgyzstan, Lebanon, Syria, Turkey, Turkmenistan, Uzbekistan Asia – Tropical: India, Pakistan Naturalized in warm temperate regions, exact native range obscure
S. interjectum		Australia
S. intrans		Australia
S. laxiflorum		Australia, Papua New Guinea, Philippines (?)
S. leiocladum		Africa, Asia, Australia, Mexico
S. macrospermum		Australia
S. matarankense		Africa, Asia, Australia, Mexico
S. nitidum	S. fulvum	Africa, Asia, Australia, Mexico
S. plumosum		Australia
S. propinquum		Asia – Tropical: India (South), Indochina, Malaysia (Malaya), Myanmar, Philippines, Sri Lanka, Thailand
S. purpureosericeum	S. dimidiatum	Africa: Nigeria to Ethiopia, Uganda, Kenya, Tanzania Asia – Tropical: India

TABLE IX   *Continued*

Species	Other names	Country of origin/ distribution
S. stipoideum		Australia
S. timorense	*S. australiense,* *S. brevicallosum*	Africa, Asia, Australia, Mexico
S. versicolor		Africa: Kenya, Mozambique, Namibia, South Africa – Natal and Transvaal, Tanzania probably also Zambia, Zimbabwe, Malawi, Botswana
S. × *almum*		Southern America: Argentina, Paraguay, Uruguay Natural hybrid arising from cultivated and weedy *Sorghum* in Argentina, naturalized elsewhere
S. × *drummondii*	*S. bicolor* subsp. *drummondii, S. hewisonii, S. niloticum, S. sudanense, S. vulgare* var. *drummondii, S. vulgare* var. *sudanense*	May occur as a weed wherever sorghum is cultivated. One form cultivated for forage (Sudangrass)

F.   SORGHUM

Sorghum (*Sorghum bicolor*) is a more tropical species than the other cereals. The *Sorghum* genus (Acheampong *et al.*, 1984; Sun *et al.*, 1994) contains about 25 species (Table IX). Sorghum ranks fourth in world cereal production and is a stable food in Africa and India. Sorghum was probably domesticated in Africa about 5000 years ago. The *Sorghum bicolor* and *Sorghum propinquum* ($2n = 20$) represent the primary gene pool of sorghum. The tetraploid *Sorghum halepense* (Johnson grass) is a fodder crop that may be an autotetraploid of *Sorghum propinquum* (Acheampong *et al.*, 1984). Sugarcane, with a much higher ploidy, and maize (also within the Panicoideae) may provide a distant source of genes for sorghum improvement. Recent molecular studies suggest that the genus may need significant revision (Dillon, personal communication).

G. OAT

Oat (*Avena sativa*) is a hexaploid species with a basic chromosome number of 21. Smaller amounts of *Avena byzantina* and *Avena nuda* are cultivated. These species are able to be crossed readily and may be considered one species. The tetraploid *Avena strigosa* is also cultivated.

H. RYE

Cultivated rye (*Secale cereale*) is outcrossing while many wild rye species are selfing. Chromosome translocations may restrict gene introduction from some wild species. *Secale vavilovii* may be an important source of genetic variation for rye improvement and has been used successfully as a source of genes for self-fertility. Triticale is a product of wheat rye hybridization. Triticale has both wheat and rye as genetic resources and ideally combines the hardiness of rye with the desirable grain characteristics of wheat. Wheat rye translocations have been used extensively in wheat improvement (Rabinovich, 1998).

I. OTHERS

Millets are cereals of minor importance. Pearl (*Pennisetum glaucum*) and finger (*Eleusine corocana*) millets are produced mainly in Africa and Asia (Table X).

III.   APPLICATION OF GENOME ANALYSIS TO EXPLOITATION OF CEREAL GENETIC RESOURCES

A. INTRODUCTION

Analysis of cereal genomes is a primary requirement for their successful exploitation. Variations within and between cereal species can be exploited by selection and recombination guided by analysis of genes and genetic markers. The purpose of DNA analysis in cereal genetic resource exploitation may include identification of genes or genotypes in the collection and maintenance of genetic stocks, identification of breeding lines and varieties, selection of parents and genes for use in breeding, screening of breeding lines for the presence of desirable genes or traits, evaluation of seed purity, and control of ownership of genetic materials. Many of these objectives can be achieved using molecular markers, others require characterization of genes and genome arrangement.

TABLE X
Major types of millets. (Adapted from Rooney, 1996)

Species	Common names	Cytogenetic origin	Location grown
Pennisetum glaucum, *P. americanum,* *P. typhoides*	Pearl bajra cattail, bulrush, candlestick, sanyo, munga, seno, souna	West Africa	West Africa
Eleusine corocana	Finger, raji, Arica, bird's foot, rapoke, Hansa	Originated in Africa and domesticated in India Eastern Asia	East and Central Africa, India, China
Setaria italica	Foxtail, Italian, kangni, navane, German, Siberian, Hungarian		Asia (China, India, Japan), North Africa, Southeast Europe, Near East
Panicum milliaceum	Proso, common, Hershey, panivarigu, broomcorn, hog, samai, Russia	China	East Asia, India, Egypt
Echinochloa frumentacea, *E. crus-galli,* *E. utilis*	Japanese, barnyard, sanwa, Kweichou, kudiraivali, Sawan, Korean	Java/Malaysia	East Asia, India, Egypt
Paspalum scrobiculatum, *P. commersoni*	Kodo, varagu, bastard, ditch, naraka	Africa or India	India
Eragrostis tef	Teff	Ethiopia	East Africia (Ethiopia)
Digitaria exilis, *D. iburua*	Fonio, fundi, hungry rice, acna, crabgrass, rashan	Domesticated in Nigeria	West Africia (Savanna)

B. MOLECULAR MARKERS

Molecular marker methods have application in cereal genetic resource exploitation as tools for the selection of desirable genotypes in marker-assisted selection. Fingerprinting of cereal germplasm using molecular markers allows identification of individual accessions and may aid the elimination of duplicates in collections. Analysis of genetic relationships using markers can be used in the selection of parents. Accelerated back crossing and introgression of genes from novel sources are also important specific applications of markers.

Many different types of molecular marker have been used for cereal analysis and these differ in some important ways.

1. RFLP

Restriction fragment length polymorphism analysis was the first DNA-based marker method applied widely in cereal genetic analysis (LeCorre and Bernard, 1995). One of the great advantages of this technique is that hybridization of DNA, unlike many PCR-based methods, does not require perfect homology. This allows the same probe to be used across many or all of the cereals in comparing homologous loci. This allows the possibility of mapping homologous loci across cereal species providing an important tool for comparative mapping.

RFLP analysis has been used to study genetic diversity in cereal varieties (Maroof *et al.*, 1994). RFLP have been produced for species such as wheat with low polymorphism (Mingeot and Jacquemin, 1999). These maps are often generated in wide crosses and the polymorphism is not present in narrower crosses used in plant breeding. This had led to a search for more polymorphic markers for use in breeding.

2. Simple Sequence Repeat

Microsatellite or simple sequence repeat (SSR) analysis has been more recently developed for cereals (Varshney *et al.*, 1998). This technique retains the advantages of co-dominant analysis as found with RFLP analysis but allows PCR-based analysis with the advantages of greater sensitivity and potential for automation. SSR markers are ideal for assessment of diversity within cereal population (Struss and Plieske, 1998). Inter-simple sequence repeat (ISSR) amplification has been used successfully to fingerprint rice (Blair *et al.*, 1999). The coverage of the genome with molecular markers is often uneven and maps with large numbers of loci often have large regions without markers. A targeted approach to the isolation of SSR markers has been developed to provide more uniform genome coverage. BAC clones from the regions with few or no markers may be used to create small insert libraries to search for SSR loci (Cregan *et al.*, 1999). This approach should allow useful SSR maps of cereal genomes to be generated. EST databases may be a source of SSR markers with wide utility (Scott *et al.*, 2000).

3. Amplification Fraction Length Polymorphism

Amplification fragment length polymorphism (AFLP) analysis allows the screening of very large numbers of markers. AFLP markers are dominant but do not require prior knowledge of sequences at specific loci as in SSR analysis. AFLP analysis may be used to determine relationships between varieties (Ellis *et al.*, 1997) and for identification. AFLP markers have been used to produce genetic maps, often in combination with SSR or RFLP markers (Boivin *et al.*, 1999).

4. Single Nucleotide Polymorphisms

Single nucleotide polymorphisms (SNPs) may provide a valuable genetic marker system for use in cereals. This polymorphisms may be found at high frequency in plant genomes. Analysis of EST sequences especially the untranslated regions allows identification of SNP sites for use as genetic markers (Gu *et al.*, 1998). Detailed SNP analysis allows the selection of the exact allele required. These markers also provide a detailed record of evolution at specific loci.

5. Other Markers

Random amplified polymorphic DNA (RAPD) markers have been widely used but may be replaced in most cases with the more reproducible AFLP markers. DNA amplification fingerprinting (DAF) (Caetano-Anolles *et al.*, 1992) is a variation on the RAPD method that has had limited application in cereals. Semi-random approaches include the use of random amplified microsatellite polymorphism (Davila *et al.*, 1999) or intron–exon splice junctions (Weining and Langridge, 1991).

Molecular markers may be converted to simpler formats for routine applications. RFLP markers may be assayed more easily by conversion to PCR assays (Lee and Penner, 1997). Conversion of RFLP markers may be difficult. Mano *et al.* (1999) analysed 43 sequence tagged site (STS) primer pairs derived from RFLP markers in barley and found two length polymorphisms with 15 more polymorphisms being revealed by the use of an extensive range of restriction enzymes. AFLP markers from barley have also proved difficult to convert to PCR markers (Shan *et al.*, 1999). PCR analysis may detect polymorphism in gene families such as the storage protein genes of cereals (O'Ovidio *et al.*, 1990). Ribosomal gene markers may be useful in establishing evolution relationships between cereals (Taketa *et al.*, 1999).

C. IDENTIFICATION AND DISTINCTION OF GENOTYPES

The DNA-based identification of cereal genotypes is an important tool in the exploitation of cereal genetic resources. Fingerprinting of germplasm provides a method for better management of genetic stocks. Duplicate lines in germplasm collections can be eliminated, allowing room for storage of a greater number of unique genotypes. Accurate pedigree data can be obtained by use of DNA fingerprinting (Ko and Henry, 1994; Ko *et al.*, 1996; Henry *et al.*, 1997; Garland *et al.*, 1999) to confirm the identity of parents. Errors in the labelling of breeding lines and germplasm collections can be

TABLE XI

Some reports of molecular analysis of heterosis in cereals

Species	Method	Reference
Wheat	RFLP	Barbosaneto *et al.* (1996)
	STS	Martin *et al.* (1995)
	RAPD	Liu *et al.* (1999)
	Differential display	Sun *et al.* (1999)
	AFLP	Barrett and Kidwell (1998)
	STS, AFLP	Burkhamer *et al.* (1998)
	RFLP, RAPD	Perenzin *et al.* (1998)
Rice	RFLP	Zhao *et al.* (1999)
	RFLP, SSR	Maroof *et al.* (1997)
	SSR	Liu and Wu (1998)
	Differential display RFLP, SSR	Xiong *et al.* (1998)
	RFLP	Wang *et al.* (1998)
	RFLP	Zhang *et al.* (1996)
	RFLP	Xiao *et al.* (1995)
	RFLP, SSR	Zhang *et al.* (1997)
	RFLP, SSR	Zhang *et al.* (1995)
	SSR, RAPD	Xiao *et al.* (1996)
	RFLP, SSR	Zhang *et al.* (1996)
Maize	RFLP	Devienne *et al.* (1994)
	RFLP	Ragot *et al.* (1995)
	RFLP	Nair *et al.* (1995)
	RFLP	Burstin *et al.* (1995)
	RAPD	Lanza *et al.* (1997)
Sorghum	RFLP	Jordan *et al.* (1998)

AFLP, amplification fragment length polymorphism; RAPD, random amplified polymorphic DNA; RFLP, restriction fragment length polymorphism; SSR, simple sequence repeat; STS, sequence tagged site.

corrected using DNA analysis (Poulsen *et al.*, 1996). New techniques may improve the efficiency of genotyping (Shepherd and Henry, 1988). Microsatellites currently provide a useful molecular reference method for cereal genotyping (Akagi *et al.*, 1997; Taramino *et al.*, 1997).

D.   ANALYSIS OF GENETIC VARIATION

Molecular analysis of genetic relationships may be useful in the identification of parents for use in breeding. Choice of parents that are less closely related may increase the chance of obtaining a favourable heterosis or hybrid vigour (Table XI). Success in predicting hybrid performance from molecular analysis of genetic distance between parents has been mixed (Tsaftaris, 1995; Zhang *et al.*, 1995). The success of this analysis may depend upon the extent of genetic variation in the germplasm being evaluated

(Yu *et al.*, 1997). The molecular basis of heterosis may vary in different cases. Yu *et al.* (1998) concluded that epistasis was a key determinant of heterosis in rice.

Molecular methods are often better indications of genetic relationships than morphological traits. Analysis of molecular markers has shown visual inspection to be a more reliable measure of genetic relationships than classical phenotype analysis in maize lines (Dillmann and Guerin, 1998).

Analysis of genetic variation in germplasm collections can indicate the extent to which available genetic resources have been adequately collected and suggest gaps in collections. For example, if accessions from a particular region show little variation, an analysis of the effectiveness of collection in that region may be required. More intense collection in areas of high variation may also be indicated.

Evaluation of genetic diversity in current cultivars provides an estimate of the genetic vulnerability of the crop to changes in the biotic or abiotic environment. Parsons *et al.* (1997) found different markers suggested different relationships in rice but this may be due to the distribution of the markers chosen in the genome. Generally, different markers give similar estimates.

E. MARKER-ASSISTED SELECTION

Selection of superior phenotypes by analysis of genotype requires either a specific knowledge of the desirable alleles at the loci influencing the trait or of markers linked to the trait. The growth in data on plant gene sequences allows more genes to be assayed directly. Linkage analysis, especially for quantitative traits, is a laborious process requiring the availability of suitable populations segregating for the trait of interest and data on the variation in the trait within the population across environments.

These populations need to be recombinant inbred lines (RILs) or preferably doubled haploid (DH) (Howes *et al.*, 1998) to allow replication of the genotype across environments. Accurate analysis of phenotype is essential for successful linkage analysis. The bulked segregant analysis (BSA) approach (Michelmore *et al.*, 1991) may allow more rapid identification of a linked marker than conventional mapping approaches.

Microsatellite and other molecular maps of cereal genomes (Chen *et al.*, 1997) provide a framework for use in marker-assisted selection. Maps with large numbers of markers allow a choice of alternative markers in the region of interest to increase the chances of finding useful polymorphisms in any particular cross.

Marker-assisted selection has been applied extensively to genes for disease resistance in cereals (Ordon *et al.*, 1997). The development of molecular markers for single disease resistance genes has been a relatively simple process (Poulsen *et al.*, 1995). The advantages of using markers to select for disease resistance genes include the ability to select for the gene in the absence of the pathogen, to allow for breeding in advance of the arrival of a disease in a particular region and to combine several genes to provide more durable resistance.

Application of marker-assisted selection to quantitative trail loci (QTL) (Hayes *et al.*, 1996; Han *et al.*, 1997) may require more complex strategies. Zhu *et al.* (1999) suggest that traits such as yield in barley may require analysis of the best combinations of QTL alleles rather than simple combination of the best alleles from different backgrounds. Detailed molecular analysis offers the possibility of dissecting and understanding complex genetic interactions (Li, 1998).

F.   EXPLOITING SYNTENY

The evolutionary relationships between cereal genomes can be used in the exploitation of cereal germplasm (Devos and Gale, 1997; McCouch, 1998). Conservation of gene order on the chromosomes can be used to locate genes

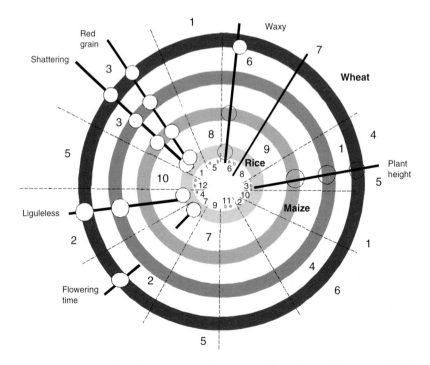

Fig. 7.   Relationships between cereal genomes (based upon Devos and Gale, 1997) showing loci for important traits.

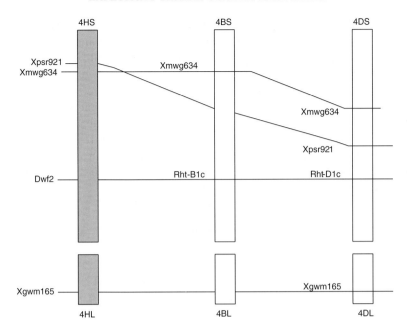

Fig. 8.   Comparative map of barley and wheat dwarfing genes (Ivandic *et al.*, 1999).

in species in which they have not been mapped. Understanding of the major chromosomal rearrangements that have taken place during the divergence of the major species is central to the successful application of this approach. The relationships between the genomes of the major cereals is depicted schematically in Fig. 7. A specific example of dwarfing genes is given in Fig. 8.

The rice genome may be a useful model for cereal genome analysis (Gale and Devos, 1998). Complete sequencing of the rice genome will allow rapid analysis of other cereal genomes. Gene-rich areas may be interposed with gene-poor areas, making the distances between some genes in large genomes no larger than in small genomes. Success with using synteny to map genes across cereal species may be case specific. Conservation may be very high in some parts of the genome and low in others.

Species-specific genes may prove to be exceptions to the background of synteny along the chromosome. Evolutionary forces during speciation may result in some very important genes being differentiated.

The dwarfing genes have had an enormous impact on cereal improvement. Modern semi-dwarf cultivars developed in the green revolution have enhanced yields associated with improved harvest index. The *Dwf* 2 6A insensitive dwarfing gene from barley has been mapped on the short arm of chromosome 4H, a position that corresponds to the *Rht*-B1 and *Rht*-D1 loci in wheat (Ivandic *et al.*, 1999) (Fig. 8). Comparative analysis of the barley and wheat genomes may in this case provide the tools necessary to clone these loci in the much larger wheat genome.

G. EXPRESSED GENES

The large size of many cereal genomes has resulted in attention being focused on the expressed sequences. The sequencing of large numbers of cDNA clones or ESTs has been undertaken in many species. The main focus of research in this area has been the need to attribute function to the genes detected. Expression analysis using microarrays (Risi *et al.*, 1997; Lemieux *et al.*, 1998) may help to assign function. Functional analysis of cereal genes may also be achieved by complementation of mutants using transformation. Other approaches include the use of transposable elements as found in maize. Transformation is being used to transfer these genetic elements into other cereal species (e.g. rice, barley) for genetic analysis (Pereira, 1998).

H. PHYSICAL MAPPING

Conventional karyotyping provides relatively low-resolution maps of the cereal chromosomes. The development of large insert libraries (e.g. YAC, BAC and PAC) has accelerated the development of physical maps. These clones are important tools in map-based cloning of genes in cereals and this process can be extended across species by comparative mapping. Contiguous collections of clones covering entire chromosomes can be ordered by restriction analysis, end sequencing or cDNA hybridization. Techniques for *in situ* hybridization (FISH or GISH) using a variety of probes allows the position of specific genes to be visualized on the chromosome.

I. GENOMIC SEQUENCING

Sequencing of cereal genomes has focused initially on the relatively small rice genome. The rice genome provides considerable cereal-specific information beyond that found in the sequence of the model *Arabidopsis* genome. The sequencing of the rice genome is proceeding by analysis of the sequence of BAC clones covering the chromosomes in tiles. These tiles are selected from highly redundant libraries providing on average ten-fold or more genome coverage. A complete genome sequence will be an extremely valuable tool in cereal genome analysis and exploitation of genetic resources.

IV. APPLICATION OF RECOMBINANT DNA TECHNIQUES TO THE EXPLOITATION OF CEREAL GENETIC RESOURCES

A. INTRODUCTION

Cereal transformation techniques are now widely available (Christou, 1996; Birch, 1997), allowing the introduction of specific desirable genes. Genetic engineering of cereals offers approaches to transfer or modify genes from the same species, another cereal or an unrelated species. New "synthetic" genes, not found in nature, may also be added. The most novel applications of transformation are usually associated with the introduction of new genes into the cereal gene pool.

The cereal species have many unique properties that make them suitable for growing in specific areas or for the production of particular products. The ideal cereal would include characteristics combined from several different current cereals. The development of cereals combining elements from the entire cereal gene pool should be possible in the future.

B. USEFUL GENES FROM CEREALS

Genes from cereal species provide specific opportunities to modify the expression of cereal genes. Over-expression or uncontrolled expression of genes may produce useful traits. For example, rice thamatin-like proteins expressed constitutively confer resistance to *Rhizoctonia sulani*, the cause of sheath blight disease (Datta *et al.*, 1999). Reduction in the level of expression or elimination of cereals genes can also be approached in this way. Movement of genes between species may allow useful characters to be introduced from the cereals. Transposon tagging systems may be transferred from maize to other cereal species by transformation (Izawa *et al.*, 1997; Takumi *et al.*, 1999). Cereals are very important sources of regulatory sequences for control of genes in transgenic cereals (McKinnon and Henry, 1995).

C. NOVEL GENES FOR CEREAL TRANSFORMATION

Genes from non-cereal sources may allow more radical expansion of the cereal gene pool. Other plants and other organisms may be important sources of novel genes. Plant proteins from diverse sources (Shewry and Lucas, 1997) may be used to engineer resistance to pests and diseases. Genes for herbicide and insect resistance have been transferred from bacteria. Viruses provide a source of genes for the production of virus-resistant cereals. Manipulation of cereal quality may be possible with a wide range of novel genes (Henry, 1999). The objective of attempts to expand the cereal gene pool in this way may be to improve nutritional or processing quality. The introduction of genes encoding proteins with altered amino-acid composition may reduce the amino-acid imbalance in diets reliant on cereals. Novel thermostable proteins (enzymes) may improve processing quality of cereals (Vickers *et al.*, 1996). For example, more heat-stable hydrolytic enzymes could improve the efficiency of conversion of barley to beer at high temperatures. Cereals may become delivery vehicles for vaccines or contraceptives with the expression of appropriate antigenes or animal proteins. Pharmacetical or other high-value compounds might also be produced in cereals by expression of novel genes (Malik, 1999).

D.   INTROGRESSION OF OTHER GENOMES

Expansion of the cereal genome may be achieved by introgression of significant parts of the genomes of non-cereal species. Introduction of more than a single gene or of genes of unknown identity may be achieved by techniques of somatic hybridization or direct DNA introduction in processes that are intermediate between those involved in conventional crossing and traditional genetic engineering with one or a few known genes. Options include the introduction of large clones (e.g. BACs) or genomic DNA. Transfer of DNA from wild rice (*Zizania palustris*) to rice by this method has been attempted (M. Abedinia, unpublished). Somatic hybrids between rice and *Zizania latifolia* have also been generated with evidence for the introduction of genes from *Zizania* into rice (Liu *et al.*, 1999). These techniques may be extended to the introduction of genes from a wide range of sources expanding greatly the effective cereal gene pool.

ACKNOWLEDGEMENTS

The author thanks the Australian Grains, Sugar and Rural Industries Research and Development Corporations (GRDC, SRDC and RIRDC) for support.

REFERENCES

Abdel-Aal, E. S. M., Sosulski, F. W. and Hucl, P. (1998). Origins, characteristics and potentials of ancient wheats. *Cereal Foods World* **43**, 708–715.

Abedinia, M., Henry, R. J. and Clark, S. C. (1998). *Potamophila parviflora* -R.Br a wild rice from Eastern Australia: distribution and phylogeny. *Genetic Resources and Crop Evolution* **45**, 399–406.

Acheampong, E., Anishetty, N. M. and Williams, J. T. (1984). "A World Survey of Sorghum and Millets Germplasm". IBPGR Secretariat, Rome, 1–41.

Adams, R. P. (1997). Conservation of DNA: DNA banking. *In* "Biotechnology and Plant Genetic Resources" (J. A. Callow, B. V. Ford-Lloyd and H. J. Newbury, eds*)*, pp. 163–174. CAB International, Newbury.

Akagi, H., Yokozeki, Y., Inagaki, A. and Fujimura, T. (1997). Highly polymorphic microsatellites of rice consist of AT repeats, and a classification of closely related cultivars with these microsatellite loci. *Theoretical and Applied Genetics* **94**, 61–67.

Appels, R. and Lagudah, E. (1990). Manipulation of chromosomal segments from wild wheat for the improvement of bread wheat. *Australian Journal Plant Physiology* **17**, 253–266.

Barbosaneto, J. F., Sorrells, M. E. and Cisar, G. (1996). Prediction of heterosis in wheat using coefficient of parentage and RFLP-based estimates of genetic relationship. *Genome* **39**, 1142–1149.

Barrett, B. A. and Kidwell, K. K. (1998). AFLP-based genetic diversity assessment among wheat cultivars from the Pacific Northwest. *Crop Science* **38**, 1261–1271.

Birch, R. G. (1997). Problems and strategies for practical application. *Plant Transformation* **48**, 297–326.

Blair, M. W., Panaud, O. and McCouch, S. R. (1999). Inter-simple sequence repeat (ISSR) amplification for analysis of microsatellite motif frequency and fingerprinting in rice (*Oryza sativa* L.). *Theoretical and Applied Genetics* **98**, 780–792.

Boivin, K., Deu, M., Rami, J. F., Trouche, G. and Hamon, P. (1999). Towards a saturated sorghum map using RFLP and AFLP markers. *Theoretical and Applied Genetics* **98**, 320–328.

Bothmer, R. V., Jacobsen, N., Baden, C., Jorgensen, R. B. and Linde-Laursen, L. B. (1995). "An Ecogeographical Study of the Genus *Hordeum*". IPGRI, Rome, 15–17.

Brar, D. S. and Khush, G. S. (1997). Alien introgression in rice. *Plant Molecular Biology* **35**, 35–47.

Burkhamer, R. L., Lanning, S. P., Martin, R. J. and Talbert, L. E. (1998). Predicting progeny variance from parental divergence in hard red spring wheat. *Crop Science* **38**, 243–248.

Burstin, J., Charcosset, A., Barriere, Y., Hebert, Y., Devienne, D. and Damerval, C. (1995). Molecular markers and protein quantities as genetic descriptors in maize. 2. Prediction of performance of hybrids for forage traits. *Plant Breeding* **114**, 427–433.

Caetano-Anolles, G., Bassam, B. J. and Gresshoff, P. M. (1992). DNA amplification fingerprinting using very short arbitrary oligonucleotide primers. *Bio Technology* **9**, 553–557.

Castagna, R., Gnocchi, S., Perenzin, M. and Heun, M. (1997). Genetic variability of the wild diploid wheat *Triticum urartu* revealed by RFLP and RAPD markers. *Theoretical and Applied Genetics* **94**, 424–430.

Chen, X., Temnykh, S., Xu, Y., Gho, Y. G. and McCouch, S. R. (1997). Development of a microsatellite framework map providing genome wide coverage in rice (*Oryza sativa* L.). *Theoretical and Applied Genetics* **95**, 553–567.

Chin, H. (1994). Seedbanks: conserving the past for the future. *Seed Science and Technology* **22**, 385–400.

Cho, Y. C., Chung, T. Y. and Suh, H. S. (1995). Genetic characteristics of Korean weedy rice (*Oryza sativa* L.) by RFLP analysis. *Euphytica* **86**, 103–110.

Christou, P. (1996). Transformation technology. *Trends in Plant Science* **1**, 423–431.

Cregan, P. B., Mudge, J., Fickus, E. W., Marek, L. F., Danesh, D., Denny, R., Shoemaker, R., Mathews, B. F., Jarvik, T. and Young, N. D. (1999). Targeted isolation of simple sequence repeat markers through the use of bacterial artificial chromosomes. *Theoretical and Applied Genetics* **98**, 919–928.

Datta, K., Velazhahan, R., Oliva, N., Ona, I., Mew, T., Khush, G. S., Muthukrishnan, S. and Datta, S. K. (1999). Over-expression of the cloned rice thaumatin-like protein (PR-5) gene in transgenic rice plants enhances environmental friendly resistance to *Rhizoctonia solani* causing sheath blight disease. *Theoretical and Applied Genetics* **98**, 1138–1145.

Davila, J. A., Loarce, Y. and Ferrer, E. (1999). Molecular characterization and genetic mapping of random amplified microsatellite polymorphism in barley. *Theoretical and Applied Genetics* **98**, 265–273.

Devienne, D., Mosse, J. M., Maurice, A., Causse, M., Leonardi, A., Touzet, P., Krejci, E., Gouesnard, B., Sanou, J., Panouille, A., Boyat, A., Dubreil, P., Dufour, P., Gallais, A., Lefort, M., Charcosset, A. and Damerval, C. (1994). Molecular markers as a tool for analysing genetic diversity and genome expression in maize (French). *Genetics Selection Evolution* **26**, S21-S34.

Devos, K. M. and Gale, M. D. (1997). Comparative genetics in the grasses. *Plant Molecular Biology* **35**, 3–15.

Dillmann, C. and Guerin, D. (1998). Comparison between maize inbred lines: genetic distances in the expert's eye. *Agronomie* **18**, 659–667.

Dvorak, J., McGuire, P. and Cassidy, B. (1988). Apparent sources of the A genomes of wheats inferred from polymorphism in abundance and restriction fragment length of repeated nucleotide sequences. *Genome* **30**, 680–689.

Dvorak, J., Terlizzi, P. D., Zhang, H. B. and Resta, P. (1993). The evolution of polyploid wheats: identification of the A genome donor species. *Genome* **36**, 21–31.

Dvorak, J., Luo, M., Yang, Z. and Zhang, H. (1998). The structure of the *Aegilops tauschii* genepool and the evolution of hexaploid wheat. *Theoretical and Applied Genetics* **97**, 657–670.

Edney, M. J. (1996). Barley. In "Cereal Grain Quality" (R. J. Henry and P. S. Kettlewell, eds), pp. 113–151. Chapman & Hall, London.

Ellis, R. P., McNicol, J. W., Baird, E., Booth, A., Lawrence, P., Thomas, B. and Powell, W. (1997). The use of AFLPs to examine genetic relatedness in barley. *Molecular Breeding* **3**, 359–369.

Gale, M. D. and Devos, K. M. (1998). Plant comparative genetics after 10 years. *Science* **282**, 656–659.

Garland, S. H., Lewin, L., Abedinia, M., Henry, R. J. and Blakeney, A. (1999). The use of microsatellite polymorphisms for the identification of Australian breeding lines of rice (*Oryza sativa* L.). *Euphytica* **108**, 53–63.

Gu, Z. L., Hillier, L. and Kwok, P. Y. (1998). Single nucleotide polymorphism hunting in cyberspace. *Human Mutation* **12**, 221–225.

Han, E., Ullrich, S. E., Kleinhofs, A., Jones, B. L., Hayes, P. M. and Wesenberg, D. M. (1997). Fine structure mapping of the barley chromosome-1 centromere region containing malting-quality QTLs. *Theoretical and Applied Genetics* **95**, 903–910.

Harlan, J. R. and Wet, J. M. J. D. (1971). Towards a rational classification of cultivated plants. *Taxon* **20**, 509–517.

Hayes, P. M., Chen, F. Q., Kleinhofs, A., Kilian, A. and Mather, D. E. (1996). Barley genome mapping and its applications. In "Methods of Genome Analysis in Plants" (P. P. Jauhar, ed), pp. 229–249. CRC Press, Boca Raton, FL.

Henry, R. J. (1995). Boitechnology applications in the cereal industry: procedures and prospects. *American Association of Cereal Chemists, Inc* **40**, 370–373.

Henry, R. J. (1997). "Practical Applications of Plant Molecular Biology". London, Chapman & Hall.

Henry, R. J. (2000). Cereals with added value. *In* "Cereal Biotechnology" (P. Morris and J. Bryce, eds), pp. 91–106. Woodhead Publishing, Cambridge.

Henry, R. J. and Brown, H. D. (1987). Variation in the carbohydrate composition of wild barley *(Hordeum spontaneum)* grain. *Plant Breeding* **98**, 97–103.

Henry, R. J., McKinnon, G. E., Haak, I. C. and Brennan, P. S. (1993). Use of alpha-amylase inhibitors to control sprouting. *In:* "Pre-Harvest Sprouting in Cereals 1992" (M. K. Walker-Simmons and J. L. Reid, eds), pp. 232–235. American Association of Cereal Chemists, St Paul.

Henry, R. J., Ko, H. L. and Weining, S. (1997). Identification of cereals using DNA-based technology. *Cereal Foods World* **42**, 26–29.

Hernandez, P., Hemmat, M., Weeden, N. F., Dorado, G. and Martin, A. (1999). Development and characterization of *Hordeum chilense* chromosome-specific STS markers suitable for wheat introgression and marker-assisted selection. *Theoretical and Applied Genetics* **98**, 721–727.

Hilu, K. W. (1993). Polyploidy and the evolution of domesticated plants. *American Journal of Botany* **80**, 1494–1499.

Howes, N. K., Woods, S. M. and Townley-Smith, T. F. (1998). Simulations and practical problems of applying multiple marker assisted selection and doubled haploids to wheat breeding programs. *Euphytica* **100**, 225–230.

Ivandic, V., Malyshev, S., Korzum, V., Graner, A. and Borner, A. (1999). Comparative mapping of a gibberellic acid-insensitive dwarfing gene (Dwf2) on chromosome 4HS in barley. *Theoretical and Applied Genetics* **98**, 728–731.

Izawa, T., Ohnishi, T., Nakano, T., Ishida, N., Enoki, H., Hashimoto, H., Itoh, K., Terada, R., Wu, C., Miyazaki, C., Endo, T., Iida, S. and Shimamoto, K. (1997). Transposon tagging in rice. *Plant Molecular Biology* **35**, 219–229.

Jelodar, N. B., Blackhall, N. W., Hartman, T. P. V., Brar, D. S., Khush, G., Davey, M. R., Cocking, E. C. and Power, J. B. (1999). Intergeneric somatic hybrids of rice *(Oryza sativa* L. (+) *Porteresia coarctata* (Roxb.) Tateoka). *Theoretical and Applied Genetics* **99**, 570–577.

Jordan, D. R., Tao, Y. Z., Godwin, I. D., Henzell, R. G., Cooper, M. and McIntyre, C. L. (1998). Loss of genetic diversity associated with selection for resistance to sorghum midge in Australian sorghum. *Euphytica* **102**, 1–7.

Kent, N. L. and Evers, A. D. (1994). "Kent's Technology of Cereals". Pergamon, Oxford.

Ko, H. L. and Henry, R. J. (1994). Rapid cereal genotype analysis. *In* "Improvement of Cereal Quality by Genetic Engineering" (R. J. Henry and J. A. Ronalds, eds), pp. 153–157. Plenum Press, New York.

Ko, H. L., Weining, S. and Henry, R. J. (1996). Application of primers derived from barley alpha-amylase genes to the identification of cereals by PCR. *Plant Varieties and Seeds* **9**, 53–62.

Lanza, L. L. B., Desouza, C. L., Ottoboni, L. M. M., Vieira, M. L. C. and Desouza, A. P. (1997). Genetic distance of inbred lines and prediction of maize single-cross performance using RAPD markers. *Theoretical and Applied Genetics* **94**, 1023–1030.

Law, C. N. (1995). Genetic manipulation in plant breeding – prospects and limitations. *Euphytica* **85**, 1–12.

LeCorre, V. and Bernard, M. (1995). Assessment of the type and degree of restriction fragment length polymorphism (RFLP) in diploid species of the genus Triticum. *Theoretical and Applied Genetics* **90**, 1063–1067.

Lee, S. J. and Penner, G. A. (1997). The conversion of RFLP markers to allele specific amplicons linked to QTLs governing malting quality in barley. *Molecular Breeding* **3**, 457–462.

Lemieux, B., Aharoni, A. and Schena, M. (1998). Overview of DNA chip technology. *Molecular Breeding* **4**, 277–289.

Li, Z. (1998). "Molecular Dissection of Complex Traits". CRC Press, New York.

Liu, B., Liu, Z. and Li, X. W. (1999). Production of a highly asymmetric somatic hybrid between rice and *Zizania latifolio* (Griseb): evidence for inter-genomic exchange. *Theoretical and Applied Genetics* **98**, 1099–1103.

Liu, X. C. and Wu, J. L. (1998). SSR heterogenic patterns of parents for marking and predicting heterosis in rice breeding. *Molecular Breeding* **4**, 263–268.

Liu, Z. Q., Pei, Y. and Pu, Z. J. (1999). Relationship between hybrid performance and genetic diversity based on RAPD markers in wheat, *Triticum aestivum* L. *Plant Breeding* **118**, 119–123.

Louette, D., Charrier, A. and Berthaud, J. (1997). *In situ* conservation of maize in Mexico: genetic diversity and maize seed management in a traditional community. *Economic Botany* **51**, 20–38.

Malik, V. S. (1999). Biotechnology: multibillion dollar industry. *In* "Applied Plant Biotechnology" (V. L. Chopra, V. S. Malik and S. R. Bhat, eds), pp. 1–69. Science Publishers, Enfield.

Mano, Y., Sayed-Tabatabaei, B. E., Graner, A., Blake, T., Takaiwa, F., Oka, S. and Komatsuda, T. (1999). Map construction of sequence-tagged sites (STSs) in barley (*Hordeum vulgare* L). *Theoretical and Applied Genetics* **98**, 937–946.

Maroof, M. A. S., Yang, G. P., Zhang, Q. F. and Gravois, K. A. (1997). Correlation between molecular marker distance and hybrid performance in US Southern long grain rice. *Crop Science* **37**, 145–150.

Maroof, M. S., Zhang, Q. and Chojecki, J. (1994). RFLPs in cultivated barley and their application in the evaluation of malting quality cultivars. *Hereditas* **121**, 21–29.

Martin, J. M., Talbert, L. E., Lanning, S. P. and Blake, N. K. (1995). Hybrid performance in wheat as related to parental diversity. *Crop Science* **35**, 104–108.

McCouch, S. (1998). Toward a plant genomics initiative: thoughts on the value of cross-species and cross-genera comparisons in the grasses. *Proceedings of the National Academy Sciences (USA)* **95**, 1983–1985.

McKinnon, G. E. and Henry, R. J. (1995). Control of gene expression for the genetic engineering of cereal quality. *Journal of Cereal Science* **22**, 203–210.

Merezhko, A. E. (1998). Impact of plant genetic resources on wheat breeding. *Euphytica* **100**, 295–303.

Michelmore, R. W., Paran, I. and Kessel, R. K. (1991). Identification of markers linked to disease resistance genes by bulked segregant analysis: a rapid method to detect markers in specific genomic regions by using segregating populations. *Proceedings of the National Academy of Sciences (USA)* **88**, 9828–9832.

Mingeot, D. and Jacquemin, J. M. (1999). Mapping of RFLP probes characterized for their polymorphism on wheat. *Theoretical and Applied Genetics* **98**, 1132–1137.

Molina-Cano, J. L., Sopena, A., Swanston, J. S., Casas, A. M., Moralejo, M. A., Ubieto, A., Lara, I., Perez-Vendrell, A. M. and Romagosa, I. (1999). A mutant induced in the malting barley cv Triumph with reduced dormancy and ABA response. *Theoretical and Applied Genetics* **98**, 347–355.

Nair, S., Rao, U. P., Bennett, J. and Mohan, M. (1995). Detection of a highly heterozygous locus in recombinant inbred lines of rice and its possible involvement in heterosis. *Theoretical and Applied Genetics* **91**, 978–986.

Nevo, E. (1998). Genetic diversity in wild cereals: regional and local studies and their bearing on conservation *ex situ* and *in situ*. *Genetic Resources and Crop Evolution* **45**, 355–370.

O'Ovidio, R., Tanzarella, O. and Porceddu, E. (1990). Rapid and efficient detection of genetic polymorphism in wheat through amplification by polymerase chain reaction. *Plant Molecular Biology* **15**, 169–171.

Ordon, F., Wenzel, W. and Friedt, W. (1997). Genetics cell biology and physiology ecology and vegetation science. *In* "Progress in Botany" (H. D. B. Heidelberg, J. W. Kadereit, U. Luthge Darmstadl and M. Runge Gottingen, eds), pp. 49–79. Springer-Verlag, Berlin.

Parsons, B. J., Newbury, H. J., Jackson, M. T. and Ford-Lloyd, B. V. (1997). Contrasting genetic diversity relationships are revealed in rice (*Oryza sativa* L.) using different marker types. *Molecular Breeding* **3**, 115–125.

Paterson, A. H., Lin, Y. R., Li, Z. K., Schertz, K. E., Doebley, J. F., Pinson, S. R. M., Liu, S. C., Stansel, J. W. and Irvine, J. E. (1995). Convergent domestication of cereal crops by independent mutations at corresponding genetic loci. *Science* **269**, 1714–1718.

Pereira, A. (1998). Heterologous transposon tagging. *In* "Transgenic Plant Research" (K. Lindsey, ed.), pp. 91–108. Harwood Academic Publishers, Amsterdam.

Perenzin, M., Corbellini, M., Accerbi, M., Vaccino, P. and Borghi, B. (1998). Bread wheat -F-1 hybrid performance and parental diversity estimates using molecular markers. *Euphytica* **100**, 273–279.

Pimentel, D., Harvey, C., Resosudarmo, P., Sinclair, K., Kurz, K., McNair, M., Crist, S., Shpritz, L., Fitton, L., Saffouri, R. and Blair, R. (1995). Environmental and economic costs of soil erosion and conservation benefits. *Science* **267**, 1117–1123.

Poulsen, D. M. E., Henry, R. J., Johnston, R. P., Irwin, J. A. G. and Rees, R. G. (1995). The use of bulk segregant analysis to identify a RAPD marker linked to leaf rust resistance in barley. *Theoretical and Applied Genetics* **91**, 270–273.

Poulsen, D. M. E., Ko, H. L., Meer, J. G. V. D., Putte, P. M. V. D. and Henry, R. J. (1996). Fast resolution of identification problems in seed production and plant breeding using molecular markers. *Australian Journal of Experimental Agriculture* **36**, 571–576.

Rabinovich, S. (1998). Importance of wheat–rye translocations for breeding modern cultivars of *Triticum aestivum* L. *Euphytica* **100**, 323–340.

Ragot, M., Sisco, P. H., Hoisington, D. A. and Stuber, C. W. (1995). Molecular-marker-mediated characterisation of favorable exotic alleles at quantitative trait loci in maize. *Crop Science* **35**, 1306–1315.

Risi, J. L. D., Iyer, V. R. and Brown, P. O. (1997). Exploring the metabolic and genetic control of gene expression on a genomic scale. *Science* **278**, 680–686.

Rooney, L. W. (1996). Sorghum and millets. *In* "Cereal Grain Quality" (R. J. Henry and P. S. Kettlewell, eds), pp. 153–177. Chapman & Hall, London

Scott, K. D., Eggler, P., Seaton, G., Rossetto, M., Ablett, E. M., Lee, L. S. and Henry, R. J. (2000). Analysis of SSRs derived from grape ESTs. *Theoretical and Applied Genetics* **100**, 723–726.

Shan, X., Blake, T. K. and Talbert, L. E. (1999). Conversion of AFLP markers to sequence-specific PCR markers in barley and wheat. *Theoretical and Applied Genetics* **98**, 1072–1078.

Shepherd, M. and Henry, R. J. (1988). Monitoring of fluorescence during DNA melting as a method for discrimination products in variety identification. *Molecular Breeding* **4**, 509–517.

Shewry, P. S. and Lucas, J. A. (1997). Plant proteins that confer resistance to pests and pathogens. *Advances in Botanical Research* **26**, 136–192.

Sorrells, M. E. and Wilson, W. A. (1997). Direct classification and selection of superior alleles for crop improvement. *Crop Science* **37**, 691–697.

Struss, D. and Plieske, J. (1998). The use of microsatellite markers for detection of genetic diversity in barley populations. *Theoretical and Applied Genetics* **97**, 308–315.

Suh, H. S., Sato, Y. L. and Morishima, H. (1997). Genetic characterization of weedy rice (*Oryza sativa* L) based on morpho-physiology isozymes and RAPD markers. *Theoretical and Applied Genetics* **94**, 316–321.

Sun, O. X., Ni, Z. E. and Liu, Z. Y. (1999). Differential gene expression between wheat hybrids and their parental inbreds in seedling leaves. *Euphytica* **106**, 117–123.

Sun, Y., Skinner, D. Z., Liang, G. H. and Hulbert, S. H. (1994). Phylogenetic analysis of sorghum and related taxa using internal transcribed spacers of nuclear ribosomal DNA. *Theoretical and Applied Genetics* **89**, 26–32.

Taketa, S., Harrison, G. E. and Heslop-Harrison, J. S. (1999). Comparative physical mapping of the 5S and 18S-25S rDNA in nine wild *Hordeum* species and cytotypes. *Theoretical and Applied Genetics* **98**, 1–9.

Takumi, S., Murai, K., Mori, N. and Nakamura, C. (1999). Trans-activation of a maize Ds transposable element in transgenic wheat plants expressing the Ac transposase gene. *Theoretical and Applied Genetics* **98**, 947–953.

Tallury, S. P. and Goodman, M. M. (1999). Experimental evaluation of the potential of tropical germplasm for temperate maize improvement. *Theoretical and Applied Genetics* **98**, 54–61.

Taramino, G., Tarchini, R., Ferrario, S., Lee, M. and Pe, M. E. (1997). Characterization and mapping of simple sequence repeats (SSRs) in Sorghum bicolor. *Theoretical and Applied Genetics* **95**, 66–72.

Tsaftaris, S. A. (1995). Molecular aspects of heterosis in plants. *Physiologia Plantarum* **94**, 362–370.

Varshney, R., Sharma, P., Gupta, P., Balyan, H., Ramesh, B., Roy, J., Kumar, A. and Sen, A. (1998). Low level of polymorphism detected by SSR probes in bread wheat. *Plant Breeding* **117**, 182–184.

Vaughan, D. A. (1994). "The Wild Relatives of Rice". International Rice Research Institute, Manila.

Vickers, J. E., Graham, G. C. and Henry, R. J. (1996). A protocol for the efficient screening of putatively transformed plants for *bar*, the selectable marker gene, using the polymerase chain reaction. *Plant Molecular Biology Reporter* **14**, 363–368.

Wang, J., Liu, K. D., Xu, C. G., Li, X. H. and Zhang, Q. F. (1998). The high level of wide compatibility of variety dular has a complex genetic basis. *Theoretical and Applied Genetics* **97**, 407–412.

Wang, R.-L., Stec, A., Hey, J., Lukens, L. and Dooebley, J. (1999). The limits of selection during maize domestication. *Nature* **398**, 236–239.

Weining, S. and Henry, R. J. (1995). Molecular analysis of the DNA polymorphism of wild barley (*Hordeum spontaneum*) germplasm using the polymerase chain reaction. *Genetic Resources and Crop Evolution* **42**, 273–281.

Weining, S. and Langridge, P. (1991). Identification and mapping of polymorphisms in cereals based on the polymerase chain reaction. *Theoretical and Applied Genetics* **82**, 209–216.

Wrigley, C. W. and Morris, C. F. (1996). Breeding cereals for quality improvement. *In* "Cereal Grain Quality" (R. J. Henry and P. S. Kettlewell, eds), pp. 321–369. Chapman & Hall, London.

Xiao, J., Li, J., Yuan, L., McCouch, S. R. and Tanksley, S. D. (1996). Genetic diversity and its relationship to hybrid performance and heterosis in rice as revealed by PCR-based markers. *Theoretical and Applied Genetics* **92,** 637–643.

Xiao, J. H., Li, J. M., Yuan, L. P. and Tanksley, S. D. (1995). Dominance is the major genetic basis of heterosis in rice as revealed by QTL analysis using molecular markers. *Genetics* **140,** 745–754.

Xiong, L. Z., Yang, G. P., Zhang, Q. F. and Maroof, M. A. S. (1998). Relationships of differential gene expression in leaves with heterosis and heterozygosity in a rice diallel cross. *Molecular Breeding* **4,** 129–136.

Yu, S. B., Li, J. X., Tan, Y. E., Gao, Y. J., Li, X. H., Zhang, Q. F. and Maroof, M. A. S. (1997). Importance of epistasis as the genetic basis of heterosis in an elite rice hybrid. *Proceedings of the National Academy of Sciences (USA)* **94,** 9226–9231.

Yu, S. B., Li, J. X., Xu, C. G., Tan, Y. E., Gao, Y. J., Li, X. H., Zhang, Q. F. and Maroof, M. A. S. (1998). Epistasis plays an important role as genetic basis of heterosis in rice. *Science in China, Series C, Life Sciences* **4,** 293–302.

Zhang, O. F., Gao, Y. J., Maroof, M. A. S., Yang, S. H. and Li, J. X. (1995). Molecular divergence and hybrid performance in rice. *Molecular Breeding* **1,** 133–142.

Zhang, Q. F., Zhou, Z. Q., Yang, G. P., Xu, C. G., Liu, K. D. and Maroof, M. A. S. (1996). Molecular marker heterozygosity and hybrid performance in indica and japonica rice. *Theoretical and Applied Genetics* **93,** 1218–1224.

Zhang, Q. F., Liu, K. D., Yang, G. P., Maroof, M. A. S., Xu, C. G. and Zhou, Z. Q. (1997). Molecular marker diversity and hybrid sterility in indica-japonica rice crosses. *Theoretical and Applied Genetics* **95,** 112–118.

Zhao, M. E., Li, X. H., Yang, J. B., Xu, C., Hu, R. Y., Liu, D. J. and Zhang, Q. (1999). Relationship between molecular marker heterozygosity and hybrid performance in intra-and inter-subspecific crosses of rice. *Plant Breeding* **118,** 139–144.

Zhong, G. Y., Peterson, D., Delaney, D., Bailey, M., Witcher, D., Register, J., Bond, D., Li, C. P., Marshall, L., Kulisek, E., Ritland, D., Meyer, T., Hood, E. and Howard, J. (1999). Commercial production of aprotinin in transgenic maize seeds. *Molecular Breeding* **5,** 345–356.

Zhu, H., Briceno, G., Dovel, R., Hayes, P. M., Liu, B. H., Liu, C. T. and Ullrich, S. E. (1999). Molecular breeding for grain yield in barley: an evaluation of QTL effects in a spring barley cross. *Theoretical and Applied Genetics* **98,** 772–779.

Transformation and Gene Expression

PILAR BARCELO, SONRIZA RASCO-GAUNT,
CATHERINE THORPE and PAUL A. LAZZERI

*DuPont Wheat Transformation Laboratory, c/o Rothamsted
Experimental Station, Harpenden, Hertfordshire AL5 2JQ, UK*

I. INTRODUCTION

In order to improve crops by the use of genetic modification (GM) technology, we need to understand the traits we seek to modify at the molecular level, we need cloned genes whose expression will effect the desired modification of these traits, and we need the ability to insert and express these "effect" genes in the target plant. The amenability of crop plants to transformation varies widely between species, this variability largely reflecting how readily the subject may be cultured and regenerated *in vitro*. This is an area of plant biology in which we still have poor understanding of the underlying mechanisms so that in certain crop groups, such as the grain legumes and the cereals, it has taken considerable empirical research to develop efficient culture procedures.

In the cereals in particular, the development of routine transformation technology was also delayed by the fact that *Agrobacterium*-mediated transformation procedures developed for broad-leaved plant species, which are typically based on adventitious shoot regeneration from organ cultures, could not be directly applied to embryogenic cereal culture systems. This difficulty actually provided the major impetus for the development of alternative direct gene transfer transformation techniques, but ironically, as these methods have been refined to give robust and efficient procedures, it has been shown that under correct conditions the cereal × *Agrobacterium* interaction can be efficient and *Agrobacterium*-mediated transformation has now been applied in all the major cereal crops (Komari *et al.*, 1998; Hansen and Chilton, 1999).

While all the economically important cereals can now be routinely transformed, in each of the species there are still remain bottlenecks and technical improvements to be made. For example, despite enormous collective effort in maize transformation, there are still strong genotypic differences in transformation efficiency between selected amenable genotypes and the majority of agronomically elite inbred lines. In wheat, a broad range of germplasm can be transformed by particle bombardment (Rasco-Gaunt *et al.*, 2001), but *Agrobacterium*-mediated transformation is restricted to a few varieties and in barley the majority of transformation research is focused on a single specialist malting variety (Golden Promise), which shows exceptional response *in vitro* (Tingay *et al.*, 1997) and most elite barley varieties are transformed at much lower efficiencies.

However, our experience in crop plant GM over the last decade suggests that we can expect cereal transformation efficiency to be improved steadily by modifications made in practice, so that beyond transgenic plant production *per se*, the larger questions in transformation technology concern the control of transgene integration, expression and stability/inheritance. These are factors that influence the quality of transgenic plants, in terms of whether a particular integration event/transgenic line is useful or not. In all but very basic proof-of-concept transformation experiments (where the question to be answered is whether a transgene has a particular function or not), there is always a degree of wastage of transformation events because of inactive or weakly-active integrations, integrations which become silenced, or integrations which are not stably inherited. These problems will not be addressed by improvements in transformation efficiency (although high-efficiency systems produce large numbers of events, allowing for wastage), but call for the development of new technologies for controlling the insertion and expression of transgenes. This is an area in which primary research to understand integration mechanisms and control of transgene expression is most appropriately done in model plants, followed quickly by system development in crop species where the technology needs to be deployed.

In the following sections of this chapter we will review the current status of cereal transformation, examining methodology for the production of transgenic cereal plants, transgene integration and expression, and three specific areas of technology; the down-regulation of gene expression by antisense and co-suppression techniques, transgene targeting and transgene removal systems, which we consider to be particularly important topics.

II. TRANSGENE DELIVERY

To produce transgenic plants that transmit the introduced trait to progeny generations, a transgene must be delivered to a cell that will eventually give rise to floral tissue and thus produce transgenic gametes. There are two possible approaches that may be taken, first, targeting DNA delivery to meristematic cells of an intact plant (or seed), and second, delivering DNA to cultured cells *in vitro* followed by regeneration of plants. The former approach is intrinsically preferable because tissue culture induction and regeneration are avoided and there is no risk of culture-induced variation among transformants. Consequently, a wide range of *in planta* transformation methods has been explored, including injection of plasmid DNA or *Agrobacterium* into meristems or developing floral organs, application of DNA or *Agrobacterium* at pollination, seed imbibition in DNA or *Agrobacterium* solutions and vacuum infiltration of developing flowers (see Songstad *et al.*, 1995; Birch, 1997; Barcelo and Lazzeri, 1998; Gheysen *et al.*, 1998; and Hansen and Wright 1999, for reviews). However, while a number of these techniques can reproducibly lead to transient transgene expression in a number of plant species, including cereals, it is only in *Arabidopsis* that *in planta* transformation, by infiltration of flowers (Bechtold *et al.*, 1993; Clough and Bent 1998), reliably produces stable transformants. Thus, in all the major crop species, including cereals, established transformation techniques deliver transgenes to cultured tissues.

Two different approaches are in use for transgene delivery to cereals; direct gene transfer (DGT) techniques, in which chemical or physical methods are used to introduce DNA into cells, and *Agrobacterium*-mediated transformation.

2. Cell/Tissue Electroporation

In this technique, high-voltage electrical pulses are used to porate the cell membranes of tissue pieces or cell aggregates to allow DNA uptake from a surrounding buffer solution (Foerster and Neumann, 1989). The technique has been shown to function in all the major cereal species (see Barcelo and Lazzeri, 1998), but it is only in rice and maize (Laursen et al., 1994; Xu and Li, 1994) that transformation efficiency levels are sufficiently high for the method to be considered for the routine production of transgenic lines. The method has advantages over protoplast transformation in that callus cultures or primary explants such as immature embryos or inflorescence tissues may be targeted (D'Halluin et al., 1992; He et al., 2001), but it is technically more demanding than particle bombardment as target tissue preparation is critical, DNA uptake is only efficient in a limited range of cell/tissue types and the post-transformation handling of tissues is complicated. Transient expression studies using cell electroporation suggest that the amount of DNA delivered to targeted cells is less than with particle bombardment, and the observation that transformants recovered from electroporation tend to have low-to-moderate transgene copy numbers (D'Halluin et al., 1992) supports this hypothesis. In comparison with particle bombardment, less research effort has been expended on the development of cell electroporation and it is very likely that the technique could be improved in terms of efficiency and breadth of application, but it is inherently a more complicated method and is unlikely to become competitive with bombardment for cereal biotechnology applications.

3. Silicon Carbon Fibre Vortexing

The use of silicon carbide fibres to mediate DNA uptake into plant cells was the last of the four major DGT techniques to be developed (Kaeppler et al., 1990, 1992). The principle of the method is that plant cells are suspended in a buffer containing plasmid DNA and microscopic silicon carbide fibres (ca 0.3–0.6 μm in diameter and 10–100 μm long) and vortexed, causing the fibres to penetrate cell walls carrying DNA into the cells or allowing it to enter via the holes created (Wang et al., 1995b). In plants, the technique was first applied to suspension cultures (Asano et al., 1991; Kaeppler et al., 1992; Frame et al., 1994) but has subsequently been applied to organized tissues such as embryos (Serik et al., 1996) and embryo-derived callus (Matsushita et al., 1999).

Despite the simplicity of the SiC fibre transformation technique and the fact that little specialized equipment is needed, the method is not widely used to generate transgenic cereal plants for two reasons. Most important, it is currently only efficient when used to target certain suspension culture cells and, as explained above (Section II. A.1), in most cereals the production of embryogenic suspensions is difficult and time consuming. Second, the microscopic SiC fibres are potentially hazardous to human health and need careful handling, which has probably discouraged wider uptake and development.

4. Particle Bombardment

Particle bombardment (biolistic transformation) is the most widely used plant DGT method and is the technology which first made the genetic modification of the major cereal crops possible. The principle of the process is simple: dense microscopic particles are accelerated through plant cell walls, allowing DNA that has been precipitated on to them to be released and to integrate into the target cell genome (Klein *et al.*, 1987; Klein and Fitzpatrick-McElligott, 1993). The first particle gun developed (Sanford *et al.*, 1987) used an explosive charge to accelerate tungsten, but this was superseded by a helium-driven gun (Kikkert, 1993).

In the early years of biolistic transformation research, a number of alternative particle acceleration systems were developed, using compressed nitrogen or air discharge (Morikawa *et al.*, 1989; Iida *et al.*, 1990; Oard, 1993; Sautter, 1993), flowing helium (Takeuchi *et al.*, 1992; Vain *et al.*, 1993) or electrostatic discharge (Rech *et al.*, 1991; McCabe and Christou, 1993). All of these devices were capable of delivering particles to plant cells leading to transient gene expression, but three instruments have proved practicable for plant GM: the electrostatic (ACCELL) gun (McCabe and Christou, 1993); the particle inflow gun (PIG) (Vain *et al.*, 1993a); and the commercial helium-driven (PDS-1000/He) gun (Kikkert, 1993). The latter is the most widely used and is today the standard particle bombardment device.

As DNA delivery via particle bombardment is essentially independent of plant species and tissue and requires minimal preparation of target tissues, it is a powerful research tool for the analysis of gene expression in cells of intact tissues via transient expression assays. This has become a standard technique in plant molecular biology. The only limitation is that the tissues/organs must be detached from the parent plant and be of a size to fit into the particle gun vacuum chamber, but this still allows a range of plant organs to be targeted. Recently, the target size limitations for bombardment have been overcome by the development of a hand-held particle gun, which allows transgene delivery to tissues of intact plants, allowing the analysis of effects of transgene expression *in planta* with minimal disturbance (Chaure *et al.*, 2000).

The stable transformation of cereal crops is, however, the application in which particle bombardment has had its most important impact. The transformation of maize (Gordon-Kamm *et al.*, 1990), wheat (Vasil *et al.*, 1992), barley (Wan and Lemaux, 1994), rye (Castillo *et al.*, 1994) and oats (Somers *et al.*, 1992) was made viable by bombardment technology and the first commercial GM cereal varieties were produced by this method (e.g. maize protected from infestation by the European Corn Borer through the expression of *Bacillus thuringiensis* Bt toxin; Koziel *et al.*, 1993).

There are two major requirements for efficient transformation by particle bombardment. First, efficient delivery of particles to large numbers of target cells coupled with minimal cell damage and, second, high levels of division and subsequent regeneration of targeted cells.

The two main variables of DNA delivery by bombardment are particle type/ preparation and particle acceleration (Klein and Fitzpatrick-McElligott, 1993; Sanford *et al.*, 1993). As these are chemical and physical processes, respectively, they are relatively amenable to optimization. A series of studies has examined the influence of different particle types (e.g. tungsten *vs* gold), sizes and different procedures for the precipitation of DNA on to particles (e.g. Rasco-Gaunt *et al.*, 1999a; Harwood *et al.*, 2000, and references therein). In parallel, particle acceleration parameters, such as propellant force, vacuum pressure, target distance and bombardment number, have been examined to define optimal conditions in a range of species. The majority of studies have been performed using the PDS 1000/He gun, which has meant that, while there are clearly interactions between target tissue/plant species and delivery parameters, it has generally been possible to extrapolate between studies to derive broadly applicable bombardment parameters.

While improved particle delivery is generally expected to improve transformation efficiency, there may actually be a negative aspect to individual cells receiving multiple particles and large amounts of plasmid DNA, as there appears to be evidence for the recovery of increased numbers of high copy-number transformants in "high-efficiency" transformation systems (see Section IV below).

The cell biology components of bombardment-mediated transformation have proved more difficult to optimize than DNA delivery parameters and, in the majority of systems, cell biology factors limit transformation efficiency. In cereals, two types of target tissues are used for bombardment. First, primary explants, which are bombarded immediately or soon after isolation and subsequently induced to become embryogenic and regenerate (e.g. rice, Christou *et al.*, 1991; wheat, Barro *et al.*, 1997) and, second, pre-established proliferating embryogenic cultures, which are bombarded, followed by further proliferation and regeneration (e.g. oats, Somers *et al.*, 1992; wheat, Vasil *et al.*, 1992; maize, Gordon-Kamm *et al.*, 1990). In the former approach, each transformation event is generally represented by one or a few plants, while in the latter approach proliferation after bombardment can lead to the recovery of numbers of clonal plants per transformation event. The disadvantage of the former system is that a constant supply of target explants must be available. In the latter system a single embryogenic line may yield large numbers of events, but there is the possibility that any culture-induced (somaclonal) variation event (Karp, 1991) can be present in a number of transgenic lines.

The major cell biology approach to the improvement of bombardment transformation efficiency has been a gradual improvement of cereal culture and regeneration systems, increasing the frequency of conversion of targeted cells into plants (Lazzeri and Shewry, 1993; Birch, 1997; Hansen and Wright, 1999). In addition, modifications to culture procedures that allow targeted tissues to better tolerate the damage/stress of the bombardment process have had dramatic effects in several species. Such treatments include plasmolysis of tissues prior to bombardment, or culture on high-osmoticum media (Finer *et al.*, 1999) and the incorporation of ethylene antagonists in the culture medium (authors' observations).

B. *AGROBACTERIUM*-MEDIATED TRANSFORMATION

Transformation of the important cereals by *Agrobacterium* was a holy grail of cereal biotechnology during the later eighties and much of the nineties because it was clear that facile and efficient transformation technology would have a major impact on crop breeding and production.

Early work on the interaction between monocotyledonous plants and *Agrobacterium* showed that they typically did not produce tumours in response to inoculation with oncogenic bacteria (see review by Smith and Hood, 1995), but from 1984 onwards a series of experiments on monocots demonstrated opine production (Hooykaas *et al.*, 1984), T-DNA integration (Schaefer *et al.*, 1987) and the recovery of transgenic plants (Bytebier *et al.*, 1987) from *Agrobacterium* transformation. This work stimulated research on *Agrobacterium* transformation of cereal crops and subsequent publications presented indicative evidence for T-DNA transfer to cereal cells, based either on the production of opines in transformed tissues (Graves and Goldman, 1987) or the use of replicative virus molecules cloned into the T-DNA (Agroinfection) as an assay for gene transfer (Grimsley *et al.*, 1988; Boulton *et al.*, 1989). It was shown that T-DNA transfer in cereals, as in dicots, was dependent on the activity of the *Agrobacterium vir* genes (Graves and Goldman, 1987; Grimsley *et al.*, 1989) and that broad host range *Agrobacterium* strains, such as C58, were able to transfer T-DNA to a range of cereals (Boulton *et al.*, 1989). Evidence for T-DNA integration and the expression of reporter and selectable marker genes in transformed tissues was reported first for rice (Raineri *et al.*, 1990) with transgenic maize plants being reported shortly after (Gould *et al.*, 1991). However, it was the publication by Hiei and co-workers (Hiei *et al.*, 1994) of the transformation of several japonica rice cultivars at efficiencies up to 30% that confirmed the utility of *Agrobacterium* transformation for cereal genetic modification.

In the Hiei *et al.* (1994) study on japonica rice, the best transformation frequencies were obtained from *Agrobacterium* co-cultivation of scutellum-derived calluses, using bacterial strains carrying a "super-virulent" Ti plasmid (Komari, 1990). A similar approach has subsequently been applied with success in the other major cereals: javanica and indica rice (Dong *et al.*, 1996; Rashid *et al.*, 1996), maize (Ishida *et al.*, 1996), barley (Tingay *et al.*,

1997) and wheat (Cheng *et al.*, 1997). This has been a development of major significance for crop plant biotechnology, although the current situation is that the "real-world" efficiency of application of *Agrobacterium* transformation varies very considerably between the major cereal species. The technology is most efficient and shows least genotypic limitation in rice (Hiei *et al.*, 1997), but shows decreasing efficiency and increased genotype dependence as one moves to maize and then barley and wheat. In rice, transformation efficiencies in the range of 5–20% are regularly obtained (e.g. Hiei *et al.*, 1994; Toki, 1997; Cheng *et al.*, 1998); in maize, efficiencies in the range of 5–20% are obtained in tissue culture model genotypes (i.e. genotypes selected for good response to *in vitro* culture) (Ishida *et al.*, 1996), but much lower efficiencies are seen in typical inbred lines. In barley and wheat, mean efficiencies of *ca* 4% (Tingay *et al.*, 1997) and 1.6% (Cheng *et al.*, 1997) have been reported, respectively, but in each case from a single tissue culture model genotype. It is not coincidence that these two genotypes, cvs Golden Promise and Bob White, are also the "standard", most productive varieties for particle bombardment transformation. It is clear that the present amenability of the major cereals to *Agrobacterium* transformation closely reflects the efficiency and genotypic limitations of tissue culture and regeneration systems in these four species.

The optimization of *Agrobacterium*-mediated transformation of cereals is complex as a wide range of factors influence the bacteria and the plant tissues involved in the process, in addition to the complicated interactions between the *Agrobacterium* and plant cells (Gheysen *et al.*, 1998; Hansen and Chilton, 1999). The body of work on rice has identified a number of important components of the procedure including: the type and developmental stage of the tissue targeted; the bacterial strain and Ti plasmid combination; bacterial growth and co-cultivation conditions such as pH and temperature; *vir* gene induction by the addition of phenolic compounds; and plant tissue culture media and conditions (Hiei *et al.*, 1997). These primary components will be important in other cereals, but need to be analysed and optimized for each species and to some extent at the cultivar level within species.

A primary requirement for *Agrobacterium* transformation is that plant cells co-cultivated with bacteria should be in active division. As these cells must also be regenerable, the preferred tissue source in cereals is the scutella of immature embryos (Ishida *et al.*, 1996; Tingay *et al.*, 1997; Cheng *et al.*, 1997), or embryogenic cultures derived from them (Hiei *et al.*, 1994). Actively dividing cells are expected to be more effective in producing compounds that induce *Agrobacterium vir* gene expression, essential for the T-DNA transfer process, and DNA replication is thought to be required for foreign DNA integration (see Section IV.A below).

The use of bacterial strains harbouring "super-virulent" Ti plasmids has been important in the development of *Agrobacterium* transformation

TABLE II

Advantages and disadvantages of particle bombardment versus Agrobacterium *transformation methods*

Particle bombardment	*Agrobacterium* transformation
Advantages	**Advantages**
Species- and tissue-independent DNA delivery, same procedures for all targets, readily transferable technology	Facile process, efficiency can be very high under optimal conditions
Can deliver vector "fragments" containing only sequences desired to be inserted into recipient genome	Tendency for simple, low-copy number insertions with good frequency of transgene expression
Facile construction of transformation vectors, cloning and vector multiplication in *E. coli*	Possibility of transferring trait gene and selection marker on separate T-DNAs to facilitate segregation-apart in progeny
Disadvantages	**Disadvantages**
Tendency for multiple-copy complex insertion patterns, which may lead to unstable transgene expression	Transfer of large DNA fragments is possible
Transfer of large constructs, intact and functional, is difficult	High frequency of integration of undesired, extra T-DNA Ti plasmid sequences into recipient genome
Specialized equipment (particle gun) is required	Construction of transformation vectors is more complicated, need to transfer from cloning host to *Agrobacterium*

methodology for cereals. In rice and maize, the use of the "super-binary" vector pTOK233 (Komari, 1990) was suggested to be a central factor for efficient transformation (Hiei *et al.*, 1994; Ishida *et al.*, 1996) and the bacterial strain used by Tingay *et al.* (1997) for barley transformation was also super-virulent. In contrast, in wheat Cheng *et al.* (1997), used a standard *Agrobacterium* strain, C58, containing a normal binary vector. It may be that the need for super-virulent bacterial strains will be reduced as optimal conditions for bacterial × plant cell interactions are defined.

The current status of *Agrobacterium* transformation of cereals can be summarized as follows: in rice it is a reliable and efficient procedure, applicable in a broad range of germplasm and suitable for the generation of large populations of transgenic lines (e.g. Cheng *et al.*, 1998). In maize, it can be highly efficient in germplasm amenable to tissue culture and in industry is routinely applied to produce thousands of commercial events (J. P. Ranch, Pioneer Hi-Bred International, personal communication). There is still, however, significant genotypic limitation to application and elite inbred lines not highly amenable in tissue culture are transformed less

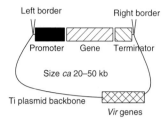

Fig. 2. Structure of a typical *Agrobacterium* transformation vector.

efficiently. In wheat and barley, the methodology is still under development. Currently, efficient application of *Agrobacterium* transformation (i.e. at routine efficiencies >1%) is confined to a very limited range of model genotypes, which have exceptional response in tissue culture. These genotypes, typified by cv Golden Promise in barley and cv Bob White in wheat, show a much more rapid proliferative response in culture than is seen in typical agronomically elite germplasm, which is probably associated with their amenability to *Agrobacterium* infection (authors' observations).

There are, however, a number of advantages of *Agrobacterium* transformation over particle bombardment that will drive the further development of the methodology in the more recalcitrant cereals (see Table II). These advantages are broadly in the areas of simplicity, efficiency and the production of plants with simpler integration patterns. If *Agrobacterium* transformation follows the pattern of development of other cell and transformation methodologies in the cereals we should see routine procedures for the recalcitrant cereal species within the next 5 years.

The structure of a typical *Agrobacterium* transformation vector is shown in Fig. 2.

III. TISSUE CULTURE AND SELECTION

A. TISSUE CULTURE METHODOLOGY

Tissue culture is an essential component of virtually all cereal genetic transformation systems. Indeed, success in obtaining transgenic cereals has been repeatedly emphasized to correlate with the efficiency of plant regeneration procedures. Nevertheless, it is not trivial to establish regenerable cultures in cereals since only a limited range of tissues can be induced to produce plants at good efficiency.

Plant regeneration in cereals is primarily via somatic embryogenesis. Although a variety of explants have been tested for establishing embryogenic callus cultures (see Table I), the most reliable results have been restricted to cultures initiated from immature zygotic embryos, particularly with respect to obtaining transgenic plants. However, the isolation of immature embryos, more precisely the scutellar tissue, is labour intensive. Hence, alternative systems are continually sought or re-evaluated to widen the selection of explants able to be used as targets for transformation.

In recent years, a number of laboratories reported experiments testing immature inflorescences, shoot meristems and mature embryos for transformation. Presently, these tissues seem to provide the best alternative to immature embryos. For example, transgenic barley (Zhang *et al.*, 1999), maize (Lowe *et al.*, 1995) and oats (Zhang *et al.*, 1999) can be recovered using shoot meristems. Similarly, transgenic wheat (Rasco-Gaunt and Barcelo, 1999), sorghum (Casas *et al.*, 1993) and tritordeum (Barcelo *et al.*, 1994) can be recovered using immature inflorescences. Mature seeds or embryos have been used successfully for rice transformation (e.g. Chen *et al.*, 1998b; Matsushita *et al.*, 1999), and recently, mature seed cultures were used to produce transgenic oats (Torbert *et al.*, 1998; Cho *et al.*, 1999). As transformation target explants, these three tissue types have several advantages over the use of immature embryos. For instance, immature inflorescences are easier and quicker to isolate, are less variable in response to culture, and allow more efficient use of greenhouse/growth chamber space due to faster turnover of donor plants. The main limitation, however, is that inflorescences from only a narrow range of genotypes are responsive to both tissue culture and transformation (e.g. Rasco-Gaunt and Barcelo, 1999). In contrast, shoot meristem cultures can be obtained in a wide range of genotypes (Lowe *et al.*, 1995; Lemaux *et al.*, 1999; Zhang *et al.*, 1999), they do not require growth of donor plants as seedlings are used as source of explants, and recovery of morphologically normal and fertile plants can be expected as the occurrence of somaclonal variation is reduced by comparison with somatic tissue explants (Lemaux *et al.*, 1999). The drawback of meristem transformation systems is that culture periods can be extensive as it takes longer to establish meristem cultures relative to embryo cultures, and the rate of chimerism can be high. In addition, transformation efficiencies are to date lower than in embryo-derived systems (Lowe *et al.*, 1995; Lemaux *et al.*, 1999). Mature embryos are a more convenient and less costly tissue source for transformation as there is no necessity to grow donor plants. However, much work still needs to be done to improve culture response and to prove applicability across cereal species. Table III presents the various tissues used successfully to establish cultures for producing transgenic cereal plants.

In transformation targeting somatic explants, transgenes are usually delivered into cells before the process of differentiation (morphogenesis) begins. The normal pathway of morphogenesis is regeneration via somatic embryo formation rather than organogenesis (adventitious shoot production), which is common in dicotyledonous plants. Typically, gene

TABLE III
Explants used for achieving stable transformation in cereals

Cereal crop	Explant	Example reference
Barley	Protoplast	Funatsuki *et al.* (1995)
	Microspore	Jaehne *et al.* (1994)
	Immature embryo	Wan and Lemaux (1994)
	Shoot meristem	Zhang *et al.* (1999)
Maize	Protoplast	Golovkin *et al.* (1993)
	Cell suspension	Gordon-Kamm *et al.* (1990)
	Immature embryo	Brettschneider *et al.* (1997)
	Shoot meristem	Lowe *et al.* (1995)
Oat	Cell suspension	Torbert *et al.* (1995)
	Immature embryo	Somers *et al.* (1992)
	Mature embryo	Torbert *et al.* (1998)
	Seedling leaf base	Gless *et al.* (1998)
	Shoot meristem	Zhang *et al.* (1999)
Rice	Protoplast	Shimamoto *et al.* (1989)
	Cell suspension	Nandadeva *et al.* (1999)
	Immature embryo	Christou *et al.* (1991)
	Mature embryo	Matsushita *et al.* (1999)
Rye	Immature embryo	Castillo *et al.* (1994)
Sorghum	Immature embryo	Casas *et al.* (1993)
Tritordeum	Immature inflorescence	Castillo and Casas (1999)
	Immature inflorescence	Barcelo *et al.* (1994)
Wheat	Immature embryo	Vasil *et al.* (1992)
	Immature inflorescence	Rasco-Gaunt and Barcelo (1999)

delivery is made either directly into undifferentiated cells of primary explants before callus initiation or into proliferating embryogenic tissue. Embryogenic cells in the targeted tissue divide and grow to produce either one or several somatic embryos, which can germinate to give plants. Thus, the system leads to the production either of unique transgenic plants or groups of clonal plants derived from a single transformation event. In either case, the production of non-chimeric transformants is usual, although when two adjacent cells in a target explant are transformed independently a plant chimeric for two different events may result, as in some cases more than one cell may be involved in the development of a single somatic embryo. The production of chimeric plants containing transgenic and non-transgenic sectors is generally inhibited by the application of selection pressure (see Section III.B) inhibiting the growth of non-transformed cells, but may occur under "light" selection regimes.

A number of laboratories also use long-term embryogenic callus cultures as target tissues. This is possible in rice, but long-term culture establishment, however, is more problematic in most other cereals because of loss of regenerability, increase in albinism, loss of fertility and frequent occurrence of abnormal phenotypes among regenerants (e.g. Gordon-Kamm *et al.*, 1999; Lemaux *et al.*, 1999).

Apart from the importance of choice of explant for successful cereal transformation, there are other important factors that influence culture establishment and transformation. Genotype is a primary factor. During the early period of cereal transformation, "model" genotypes known to be amenable in culture were used almost exclusively. More recently, however, efforts are being made to extend the technology to elite genotypes, which are either agronomically important breeding lines or current commercial varieties (Iser *et al.*, 1999; Tang *et al.*, 1999; Zhang *et al.*, 1999).

While it is recognized that *in vitro* response is to large extent genetically controlled and that culture requirements vary between genotypes, broadly applicable culture systems can contribute to extending transformation capability across a range of germplasm, as the development of separate systems for individual genotypes is generally impractical. Recent work towards developing broadly applicable systems has focused on the examination of culture medium components (Perl *et al.*, 1992; Zhang *et al.*, 1999; Rasco-Gaunt *et al.*, 2000) and culture conditions (Takumi and Shimada, 1996), evaluation of explant tissues (Lowe *et al.*, 1995; Lemaux *et al.*, 1999) and examination of the influence of DNA delivery methods on regeneration (Becker *et al.*, 1994; Koprek *et al.*, 1996; Brettschneider *et al.*, 1997; Gless *et al.*, 1998; Rasco-Gaunt *et al.*, 1999a). In addition to identifying general conditions that improve regeneration and transformation across cereal species, the importance of optimal donor plant material is stressed, particularly for elite genotypes, which are less robust in culture (e.g. Sivamani et al., 1996).

The manipulation of media components also has proven fruitful for improving transformation. Generally, MS (Murashige and Skoog, 1962) is the main salt formulation used for cereal species from the Triticeae, while N6 (Chu *et al.*, 1975) is preferred for rice and maize cultures. 2,4-Dichloroacetic acid (2,4-D), with or without low concentration of cytokinins, such as zeatin and benzyadenine purine (BAP), is the auxin commonly used for inducing callus, while for shoot regeneration, higher concentrations of cytokinins are used, with or without auxin supplementation. A number of other culture medium components have been found to enhance callus proliferation and/or regeneration, and the recovery of transformants:

1. osmotic treatment utilizing high sugar or sugar alcohols (Vain *et al.*, 1993b; Brettschneider *et al.*, 1997; Chen *et al.*, 1998b; Rasco-Gaunt *et al.*, 2000);
2. carbon source (Zhang *et al.*, 1999);
3. auxin regime (Barro *et al.*, 1998; Rasco-Gaunt *et al.*, 2000);
4. metal ions such as silver and more copper (Perl *et al.*, 1992; Zhang *et al.*, 1999; Cho *et al.*, 1999; Rasco-Gaunt *et al.*, 2001).

B.   SELECTION SYSTEMS

Establishing the preferential growth of transgenic cells over non-transgenic cells *in vitro*, termed "selection", is a key component of most genetic engineering systems. The selection process is particularly important in the transformation of cereal crops in which typically a great majority of non-transgenic cells need elimination due to low transformation frequencies.

Current cereal transformation systems make use of marker genes, which are typically resistance genes that confer tolerance to toxic substances such as antibiotics and herbicides, to inhibit or kill wild-type cells while allowing transgenic cells to proliferate. The transgenic cells and eventually plants expressing the chosen marker gene are recovered, in which a large percentage are likely to carry the trait gene.

1.   Antibiotic Selection

A variety of selection methods and marker genes for plants have evolved through the years, most of which have been applied in cereal transformation (Table IV). The first generation of selectable marker genes conferred resistance to antibiotics such as the aminoglycosides, kanamycin, geneticin (G418) and paromomycin, which were shown to function in a range of species, e.g. barley (Ritala *et al.*, 1994; Funatsuki *et al.*, 1995; Hagio *et al.*, 1995), maize (Gordon-Kamm *et al.*, 1990), wheat (Nehra *et al.*, 1994; Ortiz *et al.*, 1996; Cheng *et al.*, 1997), rice (Li *et al.*, 1993; Christou and Ford, 1995; Sivamani *et al.*, 1996; Chen *et al.*, 1998c) and oat (Torbert *et al.*, 1995; Cho *et al.*, 1999). Of this group of compounds, kanamycin is less favoured because it is generally least toxic to cereal cells (Dekeyser *et al.*, 1989; Hauptmann *et al.*, 1988). Among the first generation of selection markers, hygromycin-mediated selection has been used successfully in rice transformation (Datta, 1999) but antibiotic selection is generally not favoured for maize transformation (Gordon-Kamm *et al.*, 1999). However, there are also new antibiotic resistance genes currently being evaluated for plant transformation. For example, the *sat*3 gene, which confers resistance to streptothricins was recently introduced by Jelenska *et al.* (2000). This marker gene system is potentially usable for cereal transformation as the mechanism of protein inhibition has similarities with the aminoglycoside antibiotics.

The second generation of marker genes for cereals evolved because of the need for alternative systems and for more efficient markers, as the effectiveness of aminoglycoside antibiotics was seen to be less satisfactory in cereals than in dicots. Two marker systems were developed almost simultaneously: the selectable herbicide resistance genes and visual (scorable) selection markers.

2. Herbicide Selection

Genes conferring resistance to herbicidal compounds provided an alternative to antibiotics for selection. In some cases, the herbicide resistance traits also offer an opportunity to improve crop agronomy. Herbicide resistance genes generally code for modified target proteins insensitive to the herbicide or for an enzyme that degrades or detoxifies the herbicide in the plant.

The most commonly used herbicide resistance gene *bar* or *pat* (phosphinothricin acetyl transferase; De Block *et al.*, 1995) confers tolerance to formulations based on phosphinothricin (PPT), such as glufosinate, bialaphos, Basta and Liberty by detoxification of these compounds. Groups who have tested different formulations of PPT have shown that there are differences in their effectiveness in maize and in wheat (Dennehey *et al.*, 1994). Nevertheless, PPT resistance-based selection has been shown to be effective in a range of plant species for which selection procedures/parameters have been analysed in detail and optimized to improve efficiency (De Block *et al.*, 1995; Haensch *et al.*, 1998). Further, convenient *in vitro* and *in vivo* screening assays have been developed for PPT-resistant lines in a number of species, including cereals (Kramer *et al.*, 1993; De Block *et al.*, 1995; Rasco-Gaunt *et al.*, 1999c). Consequently, PPT-based selection has been used in all transformable cereals, e.g. wheat (e.g. Becker *et al.*, 1994; Nehra *et al.*, 1994; Altpeter *et al.*, 1996; Barro *et al.*, 1998; Iser *et al.*, 1999), barley (e.g. Jaehne *et al.*, 1994; Wan and Lemaux, 1994; Brinch-Pedersen *et al.*, 1996; Jensen *et al.*, 1996), rice (e.g. Oard *et al.*, 1996; Kim *et al.*, 1999; Kohli *et al.*, 1999a), maize (e.g. Gordon-Kamm *et al.*, 1990; Dennehey *et al.*, 1994; Brettschneider *et al.*, 1997), oat (e.g. Somers *et al.*, 1992; Gless *et al.*, 1998; Zhang *et al.*, 1999) and rye (Castillo *et al.*, 1994). Although widely used and efficient, this selection system may have side effects, as indicated recently by Bregitzer *et al.* (1998). This study suggested that PPT had mutagenic effects in barley following the observation that the proportion of albino plantlets regenerated from transgenic tissues subjected to selection was higher than from transgenic tissues not subjected to selection.

Less commonly used herbicide selection systems in cereals are those which exploit tolerance to glyphosate or sulfonylurea compounds. Resistance to these herbicides has been obtained by using genes encoding the mutant target enzymes EPSPS/GOX and ALS (see Table IV), respectively. These two systems have been applied in the transformation of rice (Dekeyser *et al.*, 1989), maize (Gordon-Kamm *et al.*, 1990; Fromm *et al.*, 1990; Register *et al.*, 1994), wheat (Zhou *et al.*, 1995; Zhang *et al.*, 2000), barley (Wan and Lemaux, 1994) and oats (Somers *et al.*, 1992).

A further herbicide marker, the cyanamide-based selection system used by Troy Weeks in wheat (US Department of Agricultural, 1998), makes use of cyanamide hydratase as a marker gene, which when expressed in cells

TABLE IV

Selection systems applied for the stable transformation of cereals

Mode of selection	Marker gene	Source of gene	Selection agent/substrate	Cereal crop
Antibiotic	*neo, npt* II (neomycin phosphotransferase)	*Escherichia coli*	Kanamycin Geneticin G418 Paromomycin	Barley Maize Oat Rice Wheat
	hpt, hph, aph-IV (hygromycin phosphotransferase)	*Escherichia coli*	Hygromycin	Barley Maize Oat Rice Wheat
	Mutant *dhfr* (dihydrofolate reductase)	Mouse	Methotrexate	Maize
	aadA (spectinomycin resistance)		Streptomycin	Maize
Herbicide	*bar, pat* (phosphinothricin acetyltransferase)	*Streptomyces hygroscopicus, Streptomyces viridochromogenes*	Glufosinate Bialaphos Basta Herbiace	Barley Maize Oat Rice Rye Sorghum Tritordeum Wheat
	epsps/gox (enolphyruvylshikimate phosphate synthase/glyphosate oxidoreductase)	*Agrobacterium* sp. strain CP4	Glyphosate	Wheat

Type	Gene	Source organism	Selective agent/substrate	Crop species
Reporter	*als* (acetolactate synthase)	*Zea mays*	Sulfonylureas	Maize
	cah (cyanamide hydratase)	*Myrothecium verrucaria*	Cyanamide	Wheat
	cat (chloramphenicol acetyltransferase)		None	Barley
	uidA (β-glucuronidase)	*Escherichia coli*	Glucuronide	Barley Maize Oat Rice Rye Sorghum Wheat
	gfp (green fluorescence protein)	*Aequorea victoria* (jellyfish)	None	Barley Maize Wheat
	luc (luciferase)	*Photinus pyralis* (firefly)	Luciferin	Rice
	B, C1 (anthocyanin regulatory element)	*Zea mays*	None	Sorghum Wheat
Positive	*manA* (phosphomannose isomerase)		Mannose	Maize Wheat
	xylA (xylose isomerase)	*Thermoanaerobacterium thermosulfurogenes*	Xylose	Maize
	lysC, AK (lysine-threonine aspartokinase)	*Escherichia coli*	Lysine, threonine	Barley
	lec (leafy cotyledon)	*Zea mays*	None	Maize Wheat

enables the cells to convert cyanamide into urea. Cyanamide is used both as a fertilizer and herbicide in agriculture. However, the fact that cyanamide is a highly toxic substance makes the system less appealing for routine use in the laboratory.

Finally, although the use of herbicide resistance genes for selection of transgenic plants in culture is effective and attractive from the practical point of view, increasing environmental concerns about the possible transfer of the resistance traits from transgenic crops to sexually compatible wild grass/weed relatives and the difficulties of using the same resistance marker in cereal and broad-leaved crops in rotation systems has motivated the development of alternative selection technology.

3. Scorable Gene-mediated Selection

Scorable marker genes are used typically as visual markers for transient expression studies for evaluating DNA delivery, but scorable genes are generally not used as selectable markers for transformation as their use is more complicated and typically relies on repeated visual sampling over the selection period. However, they can potentially alleviate reliance on a small group of selectable markers.

In the early years of plant transformation, the β-glucuronidase marker gene (uidA/GUS; Jefferson, 1987), was been used extensively as an indicator of DNA transfer into plant cells and tissues. As it requires destructive assays, GUS is not a viable selection marker, but there are several alternative visual markers, such as the anthocyanin and luciferase systems, which have been used cereal transformation. The anthocyanin marker genes result simply in accumulation of pigment visible under a light microscope, but may have detrimental effects on transformed cells, impairing their development (Chawla et al., 1999; Mentewals et al., 1999). In contrast, the luciferase (luc) gene, which confers luciferase (LUC) activity to transformed cells, does not seem to have toxic effects on cereal cells (Lonsdale et al., 1998; Baruah-Wolff et al., 1999), although limitations to the use of this system are the costs of the equipment required to measure LUC activity and the fact that the substrate must be supplied to expressing cells.

The green fluorescent protein (GFP) system is a more recently available scorable marker that is increasingly applied in plant transformation. This marker allows the non-destructive, visual identification of transgenic cells by standard fluorescence microscopy. The system is tissue- and genotype-independent, shows low toxicity and is cell autonomous. Recent reports on GFP indicate that its use for cereal genetic transformation has considerable promise (Pang et al., 1996; Vain et al., 1998; Ahlandesberg et al., 2000).

4. Positive Selection

Human health and environmental safety concerns about with the use of antibiotic and herbicide resistance have stimulated the development of new "benign" marker systems designed to be neutral in terms of consumer (animal or human) and environmental impact. Positive selection (Haldrup *et al.*, 1998), a strategy that gives transgenic cells a metabolic and hence growth advantage over non-transgenic cells is particularly suitable for the deployment of benign markers.

The most advanced positive selection system to date deployed in cereals is the mannose selection system, which has recently been used successfully for wheat transformation by Reed *et al.* (1999). This system uses the phosphomannose isomerase (*man*A or *pmi*) gene as selectable marker and mannose as selective agent. PMI-expressing cells acquire the ability to convert mannose-6-phosphate (produced by endogenous plant hexokinase from mannose) to fructose-6-phosphate, while non-transgenic cells accumulate mannose-6-phosphate to cytotoxic levels. A similar system makes use of the xylose isomerase gene (*xyl*A) and xylose as selective agent. The gene enables the transgenic cells to utilize xylose as a carbohydrate source (Brinch-Pedersen *et al.*, 1999). These two carbohydrate-based selection systems constitute a significant step forward in the development of effective technology with minimal environmental safety and human/animal health concerns.

A second positive selection system proposed for use in cereals targets amino acid metabolism (Brinch-Pedersen *et al.*, 1996), The basis of the technology is that the cultivation of wild-type cells on media containing lysine and threonine results in exhaustion of the cellular pools of methionine, eventually causing cell death. However, the expression of a feedback-insensitive aspartate kinase (*lys*C) gene maintains methionine synthesis allowing transgenic cells to survive. This system has been demonstrated in barley but has yet to be applied in other species.

An alternative approach to positive selection of transgenic cereal lines, based on modification of *in vitro* culture performance, has recently been introduced by Lowe *et al.* (2000). The method is based on the overexpression of the maize homologue to the *Arabidopsis* "leafy cotyledon" gene (*lec*), which in culture imparts a growth advantage to transgenic tissues by enhancement of somatic embryogenesis and regeneration. The system has been shown to allow the visual selection of maize and wheat transformants (Lowe *et al.*, 2000).

Several other positive selection strategies that have been shown to function in dicots have yet to be tested in cereal systems. For example, one system exploits the growth-promoting effects of cytokinins (CK). Isopentenyl transferase (*ipt*) catalyses the production of isopentenyl AMP, a precursor of several cytokinins. It was shown that cells transformed with the *ipt* gene would produce elevated levels of CK, stimulating regeneration of shoots from calluses or explants grown in hormone-free media (Ebinuma *et al.*, 1997; Kunkel *et al.*, 1999; Sugita *et al.*, 1999). A second strategy, developed by Joersbo and Okkels (1996), uses the GUS gene in combination with a CK-based selection system. The method exploits the fact that exogenous CK is needed by explants for shoot regeneration. CK is added as an inactive glucuronide derivative to the culture medium and cells which express GUS are able to convert the CK glucuronide to active CK, which promotes shoot development, while non-transformed cells do not regenerate (Joersbo and Okkels, 1996).

The selection of transgenic cereal cells *in vitro* is not trivial as cereal cells are in general more tolerant of the common selection agents than are dicot cells. This leads to "leaky" selection regimes and higher frequencies of "escapes", i.e. non-transgenic plants surviving selection. However, simply increasing the levels of selection agents often results in transformants with low to moderate levels of expression of the resistance gene being lost.

Different approaches have been taken to increase cereal selection efficiency, including: modifying culture procedures and timing of selection, i.e. application of selecting agents pre-callus, post-callus or post-shoot initiation; selecting highly active promoters to drive resistance genes; and engineering nucleotide sequences of marker genes to match the codon-usage of the target tissues to maximum expression. In cases where relatively high escape frequencies cannot be avoided without reducing transformation efficiency, facile *in planta* screens for marker gene activity allow non-transformants to be discarded within days of transfer to soil (e.g. Rasco-Gaunt *et al.*, 1999b).

As the presence of selectable marker genes is not always desired in transformants, attention has been devoted to the development of technology for the selective elimination of marker sequences from the transgenic plant. This may be achieved by normal segregation (e.g. Komari *et al.*, 1996) but more predictable systems use site-specific recombination or gene replacement mechanisms. These technologies are discussed in Section VII.

IV.   TRANSGENE INTEGRATION

A.   TRANSGENE INTEGRATION VIA *AGROBACTERIUM* AND DGT METHODS

Transgenes can be stably introduced into the cereal plant genome by two different delivery methods, *Agrobacterium*-mediated transformation and direct gene transfer. In both methods, two major steps are required: firstly, foreign DNA has to be introduced into the plant cell nucleus; and, secondly, the DNA must be integrated into the plant genome (Fig. 3). The major difference between the two transformation methods lies in the way in which foreign DNA is prepared and introduced into the nucleus of the plant cell. In the case of *Agrobacterium*-mediated gene transfer, the bacterium recognizes a susceptible plant cell and attaches to it. Then a T-complex, consisting of a single-stranded T-DNA covered with Vir (virulence) proteins is produced,

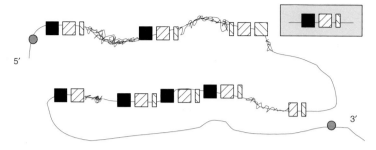

Fig. 3. Schematic of transgene integration. Transgenes mainly integrate by illegitimate recombination at one locus. Plants typically contain between 1 and 5 transgene insertions, but rearrangements and/or concatemers are common. These are due to: (1) Prior to integration – plasmid molecules recombining/ligating, mechanical shearing, or host endonuclease activity (concatemers may be formed by HR between plasmid molecules or via blunt-end ligation); and (2) during integration – DSB repair at the integration point causing deletions and synthesis of filler DNA around the transgenes. Triplets of boxes (■-▧-▷) represent transformation vector components promoter–gene–terminator (see also Figs 1 and 2).

which is transported from the bacterium into the plant cell through a Vir-assembled channel. After that, the T-complex is targeted to the nucleus for integration into the plant genome (for detail, see Gheysen *et al.*, 1998).

In contrast, in the case of DGT methods, foreign DNA is introduced into the plant cell nucleus by the action of physical or chemical stimulus as a naked double-stranded DNA molecule (see Barcelo and Lazzeri, 1998).

Once transgenes have entered the plant cell nucleus by either of the two delivery methods, integration into the plant genome may occur. This process appears to be similar in the two transformation methods. Genetic analysis and *in situ* hybridization to T-DNA insertions suggests that transgenes integrate into the plant genome by illegitimate recombination typically at one or two loci, sited at random over the genome (Ambros *et al.*, 1986; Thomas *et al.*, 1994; Wang *et al.*, 1995a). However, although integration sites appear randomly distributed over chromosomes, several experiments with promoter-less marker genes have indicated that transgenes integrate predominantly in transcriptionally active regions. This holds true for transgenes introduced by either *Agrobacterium*-mediated transformation (Koncz *et al.*; 1989, Topping *et al.*, 1991) or DGT methods (Salgueiro *et al.*, 1998). Although integration occurs by illegitimate recombination, it is also believed that partial, short homologies (not full homologous recombination) between the introduced foreign DNA and the recipient genomic DNA are responsible for initiating integration (Matsumoto *et al.*, 1990; Gheysen *et al.*, 1991 (*Agrobacterium*); Hamada *et al.*, 1993 (DGT)), which is then completed by DNA repair processes at the transgene-genomic DNA junctions (see Section IV. B.2 for more details).

B.   TRANSGENE INTEGRATION PATTERN

1.   Locus and Insertion Number
Independent of the DNA delivery method and integration process, transgenes are usually integrated into the transgenic plant genome at one or two loci. At the transgenic locus/loci, the number of insertions of the transgene can be highly variable between transformation methods, experiments and among plant species. In general, protoplast-based methods tend to produce transformants with larger numbers of insertions than other DGT methods, such as particle bombardment.

Agrobacterium-mediated transformation tends to produce lower insertion numbers. This fact is clearly illustrated in rice, in which large populations of transgenics have been produced from all three methods (Christou *et al.*, 1991; Hiei *et al.*, 1994; Datta, 1999). In other cereals, however, most transgenic plants have been produced by particle bombardment and, only more recently by *Agrobacterium* transformation. In some species, present data suggest that there are broad similarities between plants produced by either method and that variation between species and experiments may be as great as variation between transformation methods. For example, in wheat, the majority of transgenic plants produced by particle bombardment (Weeks *et al.*, 1993; Nehra *et al.*, 1994; Becker *et al.*, 1994; Barro *et al.*, 1998; Cannell *et al.*, 1999; Stoger *et al.*, 1998; Rasco-Gaunt *et al.*, 2001) and *Agrobacterium* (Cheng *et al.*, 1997) appear to contain between 1 and 5 transgene insertions, although with particle bombardment, plants with a larger number of insertions are also recovered. Other sets of data in wheat again show similarity between transgenic plants produced by different methods. In work by Cheng *et al.* (1997), *Agrobacterium* gave rise to 35% of transgenics with single copy and 44% single locus insertions, while Rasco-Gaunt *et al.* (2001) recovered 24% single copy and ~50% single locus transformants from particle bombardment. However, results obtained in wheat may be rather different from those observed in other cereals. In maize and rice, particle bombardment tends to produce plants with larger numbers of insertions than it does in wheat (e.g. Gordon-Kamm *et al.*, 1999; Vain *et al.*, 1999), whereas *Agrobacterium*-produced plants in rice generally contain fewer transgene insertions (Hiei *et al.*, 1994; Aldemita and Hodges, 1996; Ishida *et al.*, 1996). However, differences not only emerge when transformation methods are compared but also when plants produced by the same method in different experiments are compared. For example, Vain *et al.* (1999) reported obtaining transgenic rice plants by particle bombardment with more than 80 transgene insertions, whereas Kohli *et al.* (1999a) reported only plants with less than five insertions. The striking difference between the plants obtained in these two experiments, even though they were made in the same species and by the same laboratory, reflects the complexity of the transformation process and the difficulty of making comparative studies.

Differences between experiments holds true also for *Agrobacterium* transformation, particularly in relation to the complexity of integration pattern (see below).

2. Factors Influencing the Complexity of Transgene Integration Patterns

As reviewed above, transgene insertion number is variable between cereal species and experiments, but this need not represent a problem *per se* for stable expression of transgenes in cereal crops. The phenomenon that is of more concern for stable gene expression and which is relatively common in plants produced with either transformation method, is that a significant proportion of the transformants contain rearranged/truncated and/or concatenized transgene insertions. These complex patterns of integration are a problem because they tend to lead to instability and silencing of transgene expression (see Section V).

There are many factors than may influence the complexity of the integration pattern a transgenic plant finally has, some that operate during the transformation procedure itself and others that affect the fate of transgenes once they are inside the plant cell nucleus or after integration into the host chromosome.

Factors related to the transformation method can be considered using particle bombardment, the major DGT method for cereals, as an example. The method of preparation and physical status of plasmid DNA prior to entering the nucleus (i.e. supercoiled, circular or linear molecules), the degree of degradation, the concentration and amount of DNA delivered per shot, the method of coating particles and the degree of plasmid shearing during delivery are some of the steps of the procedure that can influence the integrity of plasmid molecules and therefore the pattern of integration. In practice, these components of the procedure are difficult to standardize and monitor and they may account for a significant proportion of the variability observed between experiments. Most of these factors have not been examined in comparative studies, although they potentially have a major effect on integration pattern complexity. One factor, however, that has been studied in some detail is the influence of the physical status of plasmid DNA before it is delivered into the plant cell.

Two recent studies have investigated whether linearized or supercoiled, single (ss) or double (ds) stranded plasmid DNA is most effective for transformation. Uze *et al.* (1999) found that they could obtain transgenic wheat plants with plasmids delivered in any of the forms mention above, although linear forms in general were suggested to be slightly more efficient than supercoiled forms in integrating into the plant genome and therefore in producing stable transformants. When comparing ss and ds forms, they found that ds forms were more efficient in integrating into the plant genome than ss forms, although the difference was probably not significant. To obtain firm conclusions from this experiment is difficult since not only is there a difficulty in preparing "clean" ds and ss forms, but also conversion from the ss form to the ds form occurs naturally inside the cell (Furner *et al*., 1989), making it impossible to know the status of the plasmid DNA prior to integration. The same laboratory performed the same experiment in rice (Nandadeva *et al*., 1999), which gave similar results in that delivery of ds plasmids in linear form was the most effective plasmid form to produce transformants, leading to an overall two-fold increase in transformation efficiency. These data were, however, based on hygromycin-resistant colonies and not on confirmed transgenic calli or plants. Moreover, the work did not address to what extent plasmid linearization/degradation occurs before delivery and how this factor influences the final pattern in which transgenes are integrated.

Moving to factors that can affect the fate of transgenes once inside the plant cell nucleus, these are relatively better characterized and factors operating both prior to integration and during the integration process have been investigated. Once the foreign DNA is inside the nucleus of the plant cell, the introduced sequences can either be transcribed, resulting in transient expression, or integrated into the genome resulting in stable transformation. At the stage prior to integration, there is good evidence that plasmid molecules can recombine, ligate and concatenize, either by homologous recombination between common plasmid sequences or by blunt-end ligation of linearized molecules (De Neve *et al*., 1997; Gorbunova and Levy 1997). The plasmid molecules may also be digested by nucleases. Nuclease activity inside the cell is likely to linearize a proportion of the introduced plasmid molecules and exonuclease activity to "nibble" the ends of linearized molecules deleting parts of them and/or creating overhangs of ssDNA able to be re-ligated with themselves or with other molecules to form a novel rearranged integration patterns (Gorbunova and Levy, 1997). This process of recombination/degradation prior to integration is likely to contribute significantly to the complexity of transgene integration patterns. In addition, further rearrangements/deletions, etc. can happen at the time of integration.

For integration of foreign DNA to be possible, a double-strand break (DSB) in the plant genome has to occur. At this break point, digestion may also start to occur and therefore deletions at these points are common. The break and digestion process activates DNA repair mechanisms and the synthesis of filler DNA and subsequently the re-ligation of the two ends via a non-homologous end-joining repairs the damage. It is postulated that T-DNA sequences are integrated into the genome by a breakage/repair process (Gheysen *et al*., 1991; Salomon and Puchta, 1998), and essentially the same mechanism is seen to operate for integration of foreign DNA when introduced by DGT methods (Gorbunova and Levy, 1997).

In cereals, there have been relatively few studies analysing transgene insertion loci in detail. The results of experiments published to date (Takano *et al.*, 1997; Kohli *et al.*, 1998; Pawlowski and Somers, 1998), however, demonstrate that deletions, sequence reshuffling and the synthesis of filler DNA commonly occur.

Takano *et al.* (1997) analysed transgene integration sites in three transgenic rice plants produced by protoplast transformation. They found large-scale rearrangements in the recipient genome at the integration site, occurrence of filler DNA and microhomologies of five base pairs between plasmid and target DNA. Interestingly, they also found a common AT-rich repetitive sequence structure in all three junctions of the target genome. This adenine and thymine (AT)-rich sequence had several characteristics of a SAR (scaffold-attachment region) and preliminary analysis revealed SAR activity of this fragment. Previous work has suggested that AT-rich sequences are preferred targets for DNA integration (Gheysen *et al.*, 1987).

In oats, Pawlowski and Somers (1998) analysed the structure of transgene loci in 13 lines produced by microprojectile bombardment and found that transgene sequences had integrated into the oat genome as a mixture of intact and rearranged, and/or truncated transgene fragments interspersed with host genomic DNA. The number of insertions of transgenic DNA within the transgene loci varied from 2 to 12 among the 13 lines. Although the lengths of interspersing host DNA within the transgene loci were unknown, the sizes of the restriction fragments that hybridized to transgene sequences (ranging from 3.6 to *ca* 60 kb), as revealed by restriction digests with endonucleases that did not cleave the introduced plasmids, indicated that the entire transgene loci ranged from 35 to 280 kb. The observation that all transgenic lines analysed exhibited genomic interspersion of multiple clustered transgenes suggests a predominating integration mechanism. The authors proposed that transgene integration at multiple clustered DNA replication forks could account for the integration structures observed.

Contrasting results to these were published by Kohli *et al.* (1998) in rice. They found plasmid sequences intervening between transgene sequences at the integration locus in seven out of eight lines analysed. This pattern was attributed to multiple reorganizations between plasmid molecules prior to integration. The remaining line appeared to have an intervening sequence with no homology to rice sequences in the database. These sequences are likely to be the result of reshuffling in the template used for the synthesis of "filler" DNA, as Gorbunova and Levy (1997) postulated for a similar finding in transgenic tobacco lines. The authors, however, hypothesized that the initiation of integration at an specific site in the genome makes the site more receptive for further integration events becoming a "temporary hot spot" for integration.

In general, it is difficult to conclude whether a particular integration pattern results from events occurring in the plant cell nucleus prior to integration, from events happening during integration, or both. For example, it seems that the interspersing "filler" DNA in the majority of transgenic lines is of plasmid origin with some exceptions of plant or unknown origin (see references above). In some studies "filler" DNA is believed to result from non-homologous end-joining of plasmid molecules prior to integration, whereas in other cases it is postulated that "filler" DNA is newly synthesized to repair DSBs (Gorbunova and Levy, 1997). It seems that plasmid backbone sequences are significant contributors to the complexity of integration patterns. If this is true, current experiments in which plants are being produced with DNA molecules without backbone sequences (vector "fragments" or "inserts"), with the aim of excluding non-essential sequences from transgenic plants, should give simpler patterns of integration.

Finally, other factors that have been seen to influence transgene insertion number and integration pattern complexity in species other than cereals are the tissue used as target for DNA delivery (Grevelding *et al.*, 1993) and the stage of the cell cycle at which transformation occurred (Kartzke *et al.*, 1990). Grevelding *et al.* (1993) found that in *Arabidopsis* single copy T-DNA insertions were the predominant form of integration pattern in root-derived transformants (64%), whereas in leaf disc-derived transgenics, multiple insertions were more common (89%). In an independent experiment, Kartzke *et al.* (1990) showed that regenerants obtained from transformations of non-synchronized tobacco protoplasts predominantly contained one non-rearranged insertion of the introduced DNA, whereas the integration pattern of S- or M-phase transformants contained multiple insertions. In the case of the S-phase transformants, rearrangements were also common. These two experiments have not been reproduced in other systems, so it is difficult to say whether these findings are broadly applicable or are isolated occurrences. However, it is hard to believe that cell-cycle stage, which profoundly influences chromatin organization, does not influence the transformation/integration process.

There is clearly much yet to be understood about transgene integration, and thence how to direct integration to particular areas of the genome by either homologous recombination or gene targeting, and how to control the complexity of transgene integration patterns. This information will be necessary for biotechnologists to be able reliably to produce transgenic lines with predictable and stable transgene expression.

V. TRANSGENE EXPRESSION

A. PROMOTERS

Three major classes of promoters are currently used in plant biotechnology: those able to drive expression constitutively, those driving expression specifically in certain tissues and at certain developmental stages, and those only able to drive expression under inductive conditions.

1. Constitutive Promoters

These are mainly used to drive expression of selectable marker genes for the identification of transgenic tissues *in vitro*. The aim is to protect the transformed cell from the action of the selection agent by expressing the selectable marker gene highly in all cell types; this can be achieved by using promoters with strong and ubiquitous activity.

Constitutive promoters are also used for driving the expression of genes at high levels in all tissues and stages of development of the whole plant, such as in the application of engineered herbicide resistance.

One of the first promoters to be used in cereal transformation was the 35S promoter isolated from the tobacco cauliflower mosaic virus (Odell *et al.*, 1985). This promoter, although able to drive gene expression in a broad range of dicots at high levels, was less efficient when tested in cereals. It was only when monocot intron sequences, for example, the maize alcohol dehydrogenase *adh* I intron 1, were cloned behind the promoter sequence that expression of marker genes in cereals routinely reached high enough levels for either visual detection (scorable markers) or efficient selection (selectable markers) (e.g. Callis *et al.*, 1987; Chibbar *et al.*, 1993; Barro *et al.*, 1998). Other introns were also tested and gave similar results in enhancing expression, as in the case of the rice actin *Act* 1 intron 1 (McElroy *et al.*, 1991) (see also Section V.C).

Meanwhile, other constitutive promoter sequences have been isolated from cereals, such as the rice *Act* 1 (McElroy *et al.*, 1991) and the maize polyubiquitin *Ubi* 1 sequences (Christensen *et al.*, 1992). Both these promoters are also used in conjunction with their respective first intron sequences and, together with the chimeric 35S-*adh*I sequence mentioned above, are the promoters most widely used in cereal transformation.

The rice *Act* 1 promoter cassette includes a 1.3 kb 5′ upstream region, the 5′ non-coding exon 1, intron 1 and the 5′ part of the first coding exon of the *Act* 1 gene (McElroy *et al.*, 1990). The intron sequence is a crucial part of the regulatory cassette since its removal abolishes expression completely. There are examples of the use of this promoter to drive expression of genes in transgenic wheat (Nehra *et al.*, 1994; Barro *et al.*, 1998), barley (Funatsuki *et al.*, 1995), rice (Zhang *et al.*, 1991) and maize (Zhong *et al.*, 1996).

The maize ubiquitin *Ubi* 1 promoter cassette has a similar organization to that of the *Act* 1 cassette in that it contains exon and intron sequences. It is a strong promoter in young, actively dividing cells but expression decreases significantly as the plant matures, although it shows good activity in floral tissues and the developing embryo and endosperm (authors' data). The *Ubi* 1 promoter cassette has been used to drive expression of marker or trait genes in rice (Cornejo *et al.*, 1993), wheat (Weeks *et al.*, 1993; Barro *et al.*, 1998;

Stoger et al., 1999b; Rasco-Gaunt et al., 2000), barley (Wan and Lemaux, 1994) and rye (Castillo et al., 1994). Again, in this case the intron sequence is an integral part of the promoter, since its removal greatly reduces the activity of the sequence.

In a different approach to achieve high levels of transgene expression in cereals, the chimeric pEmu promoter was constructed by adding a set of enhancer elements to the 5′ end of a truncated Adh I promoter and first intron (Last et al., 1991). In transient assays in cereals it proved to have up to 400-fold higher activity than the CaMV35S promoter. However, in stable transformation it was much less effective and the recovery of transgenic plants was difficult. There are a few reports of the production of transgenic rice plants using the pEmu promoter to drive the selectable marker gene (Chamberlain et al., 1994; Li et al., 1997), but this promoter is not widely used. It is suspected that recombination in the plasmid molecule due to the presence of the duplicated enhancer sequences (four copies of the 40 bp octopine synthase enhancer) was one reason for the problems with this promoter construct. Recombinogenic spots have also been detected in the CaMV35S promoter sequence (see Section V.C).

2.   Tissue-specific and Developmentally Regulated Promoters

These are classes of promoter that allow the expression of transgenes only in specific tissues or under certain developmental conditions, leaving the rest of the tissues in the plant unmodified by transgene expression. For engineering crops, there are many circumstances in which such type of regulation is required for expressing transgenes; e.g. genes involved in grain quality may only need to be expressed in endosperm cells. In cereals, promoters have been characterized for a number of tissues: endosperm, embryo, anther tapetum, meristem, mesophyll and phloem, and some sequences regulated by stress and light stimuli have been described (see Baga et al., 1999, for a list of promoters tested in cereals). In this section we will discuss some of the promoters that have been more widely used in cereal biotechnology.

Improving grain quality is a primary target in cereal GM, and for engineering seed quality traits, endosperm- or embryo-specific promoters are required. Several storage proteins and starch biosynthetic enzymes are coded for by well-characterized genes, whose promoters have also been isolated and tested in cereals. Starting with the storage protein genes, wheat HMW (high molecular weight) glutenin subunit promoters (e.g. those from the HMW subunit 1Ax1 and 1Dx5 genes) have been proven to drive endosperm-specific expression in transgenic bread wheat (Blechl and Anderson, 1996; Barro et al., 1997), and durum wheat (He et al., 1999; Lamacchia et al., 2001). However, the strict endosperm-specificity of these promoters may be lost when they are used to express genes in other cereals. Experiments have been made in which HMW subunit promoter sequences have been used to express GUS in transgenic oats and barley. In oats, GUS expression was observed in the aleurone layer and in floral tissues, in addition to the endosperm (P. Morris, personal communication), while in barley, expression was also seen in roots, leaves and pollen grains (Y. Zhang, personal communication). This is a good example of how promoters can behave unpredictably when used in heterologous systems, even in closely related species. Other storage protein promoters with endosperm-specific expression that have been used for transgene expression are the barley B- and D-hordein (Cho *et al.*, 1998), rice glutelin (Zheng *et al.*, 1993), and maize γ-zein sequences (Russell and Fromm, 1997).

Promoters from genes for starch biosynthetic enzymes are also obvious tools for achieving transgene expression in endosperm tissue. The granule-bound starch synthase (GBSS) promoters from rice and maize have been isolated and tested in the species of origin. In both cases, expression was observed in both endosperm and pollen grains (Itoh *et al.*, 1997; Russell and Fromm, 1997). Another promoter from a starch biosynthetic enzyme, the rice ADP-glucose pyrophosphorylase (AGP) small subunit, showed endosperm-specific expression when tested driving the *uid*A gene in transgenic rice (Russell and Fromm, 1997).

Considering embryo-specific promoters for cereal biotechnology, there is little public information available, although considerable work has been done within the cereal biotechnology industry. Some use has been made of the maize globulin (*Glb*1) promoter. This sequence drives gene expression in the whole embryo and has been successfully used for increasing lysine and methionine levels in maize kernels (C. Falco, personal communication).

Another important group of tissues for cereal biotechnology comprises the anther tapetum and pollen grains, since promoters able to direct expression in these tissues are of interest for creating nuclear male sterile lines for hybrid production (DeBlock *et al.*, 1997). A number of tapetum-specific promoters are available from maize (e.g. Ca 55) and rice (e.g. E1 and T72). These three promoters have been tested in wheat by driving the expression of the ribonuclease gene, *barnase*, giving rise to sterile transgenic plants (De Block *et al.*, 1997). Another tapetum-specific promoter, that of the rice Osg6B gene, has been tested expressing the *uidA* gene in transgenic rice (Yokoi *et al.*, 1997).

Two other tissue-specific promoters characterized in transgenic cereals are the rice sucrose synthase (Rss) and the rice tungro bacilliform virus (RTBV) sequences, which are active in phloem. The Rss promoter has been tested in rice and wheat and appears to be effective in driving lectin expression in this tissue (Rao *et al.*, 1998; Stoger *et al.*, 1999a), while the RTBV promoter has been used to achieve phloem-specific transgene expression in rice (Bhattacharyya-Pakrasi *et al.*, 1993). Such promoters are very useful for engineering resistance to sucking insects (e.g. aphids) via the expression of protectant molecules in the phloem stream.

mechanism by which eukaryotic genomes defend themselves from the invasion of prokaryotic DNA sequences to which they are constantly exposed. This process has not yet been shown to operate in plants, but a common conclusion from studies such as that of Scrable and Stambrook (1999) is that the more native or closely related a sequence is to the recipient genome and the simpler the integration pattern is, the higher the chance is of achieving proper expression. However, a negative aspect of the use of native or closely related transgene sequences is that they may trigger unintended co-suppression or sense suppression. In this process, the use of homologous transgene sequences not only gives rise to lack of expression of the transgene but also of the endogenous homologue. Co-suppression will be discussed in more detail in Section VI.

Apart from the origin of the introduced foreign sequence, the structural integrity of the transgene in terms of the presence or absence of introns, and of 5'- and 3'-sequences may also influence expression stability (Gutierrez *et al.*, 1999). Most transgenic plants produced contain cDNA clones instead of genomic sequences, which implies that introns and probably other 3' gene sequences are absent. In consequence, the mRNA produced may not accumulate at the same level as that produced by sequences containing introns and native 3' sequences. This effect was examined by Lugones *et al.* (1999) in experiments performed in the aerial hyphae of the fungus *Schizophyllum commune*, in which transformation with cDNA sequences led to lines with no expression. In contrast, when genomic sequences were used, normal expression was detected. The authors observed that, although introns were not necessary for transcription initiation, but they were required for mRNA accumulation. Equally, when artificial introns were introduced in the correct orientation into the intron-less transcriptional unit, accumulation of mRNA was detected. It is clear that, in this system, splicing is required for the normal processing of primary transcripts. To date, a similar study addressing the stability of expression of transgenic lines containing cDNA clones versus those containing genomic clones has not been performed in transgenic plants.

The reasons most transgenic plants contain cDNA clones are: first, because the availability of cDNA sequences from EST and gene isolation programmes is much greater than that of genomic sequences; and, secondly, because cDNA clones readily permit the use of different promoter sequences to direct expression of transgenes at levels and in patterns different from those of the native promoter. Thus, there is more flexibility with the use of cDNA clones, but the practice may also be responsible for a proportion of the transgene expression problems observed in genetically modified plants.

Another feature of the foreign sequences used for transformation that can also influence transgene expression stability is the presence of recombinogenic sequences in structural genes or promoters. A well-documented case is that of the CaMV 35S promoter, which contains recombinogenic sequences (Geldreich *et al.*, 1986; Vaden and Melcher, 1990), which may have been responsible for the loss of expression in some transgenic rice lines (Kohli *et al.*, 1999b).

Finally, one of the major factors affecting transgene expression stability is the pattern in which transgenes are integrated (see Section IV). Complex patterns of integration with either inverted repeats, tandem arrays, and/or rearranged insertions are more likely to trigger silencing than simple integration patterns.

C. IMPROVING TRANSGENE EXPRESSION AND STABILITY

After reviewing some of the factors that can affect transgene expression stability, an obvious question is how to minimize silencing of transgenes and to ensure improved and predictable transgene expression.

A first consideration is that the recombinant foreign DNA sequence matches the isochore composition of the host genome as much as possible. Very often this is not the case when heterologous genes of either bacterial, viral or even non-related plant origin are used. For instance, a 44–70% guanine and cytosine (GC) value has been reported for monocot genes, of which the majority are in the 60–70% range (Matassi *et al.*, 1989). Genes from dicots have, however, a more narrow distribution of GC residues, with values ranging from 40 to 56% (Matassi *et al.*, 1989). The main difference in base content between monocot and dicot coding regions is because an overall higher proportion of GC appears in the third position of the codon (Murray *et al.*, 1989; Fennoy and Bailey-Serres, 1993). This difference in codon usage between monocot and dicot genes is even seen when homologous sequences are compared (Matassi *et al.*, 1989). The importance of using coding regions with the right codon usage and GC content for optimal gene expression in plants has been demonstrated by several examples; *Bacillus thuringiensis* gamma-endotoxin (*Bt* toxin) gene was modified for expression in dicots by raising the content of GC in the coding region but without altering the encoded protein. That modification led to an overall 50-fold increase in expression (Ohme-Takagi *et al.*, 1993). Similar types of modifications to other *Bt* toxin genes have also been made to allow efficient expression in monocots (Fujimoto *et al.*, 1993; Koziel *et al.*, 1993). Other examples of improvement of gene expression in an heterologous host genome by codon usage modification are the bacterial heat-stable (1,3–1,4)-β-glucanase gene and the jelly fish green fluorescence protein. The (1,3–1,4)-β-glucanase gene was successfully expressed during germination in transgenic barley only after modification of the codon usage of the original bacterial gene (Jensen *et al.*, 1996), and GFP expression was improved in cereals (e.g. rice; Vain *et al.*, 1998) by removing a cryptic intron and changing the third position of the codons to increase GC content from 32 to 60% (Rouwendal *et al.*, 1997) and by replacing the Ser_{65} codon with a threonine or cysteine codon (Pang *et al.*, 1996).

Moving to the design of transformation vectors, the use of duplicated sequences, e.g. several copies of the same promoter or terminator sequences likely to recombine and form secondary structures should be avoided. This practice is rather common in transformation vectors as in the case of one of the most widely used vectors in cereal biotechnology, pAHC25 (Christensen and Quail, 1996), which contains two copies of the ubiquitin promoter and two copies of the *nos* terminator. Further, the use of vector backbone sequences likely to be recognized as non-host and trigger silencing should also be excluded from the transformation vector. This is relatively easy in transformation vectors used for direct gene transfer as one can cut the fragment containing the promoter-structural gene-terminator complex out and use that for transformation. In the case of *Agrobacterium* transformation, gaining control over the sequences that are finally integrated into the recipient genome is less straightforward. Although normally the foreign DNA transferred and integrated into the plant genome is that contained between the right and left borders (T-DNA), backbone sequences also can be co-integrated alongside the T-DNA. At present, this integration has been observed to occur in as many as 75% (Kononov *et al.*, 1997) to 85% (Hanson *et al.*, 1999) of transformants. Reducing this high frequency of integration of backbone sequences is not easy to control experimentally. One possible approach, published recently by Hanson *et al.* (1999), involves the use of a lethal gene (*barnase*) incorporated into the non-T-DNA portion of the binary vector, along with the scorable marker for luciferase expression. In the experiment, a population of 50 tobacco transgenic plants was produced with the novel vector and only one plant was found to contain backbone sequences, compared with an 85% frequency of integration of backbone sequences in the control population. Using the barnase vector reduced transformation frequency by about 30%, which in the case of tobacco is not a problem, but would be a serious difficulty in many crop species.

A further option for DNA sequences used for transformation vectors that can improve transgene expression stability is the use of genomic clones instead of cDNA sequences. It is now well documented that introns (Callis *et al.*, 1987; Luehrsen and Walbot 1991; Mascarenhas *et al.*, 1990; Tanaka *et al.*, 1990), 5′-untranslated regions (De Loose *et al.*, 1995) and specific sequences downstream of polyadenylation sites (MacDonald *et al.*, 1991) have a significant influence on expression levels or mRNA stability (see Section V.B). In the case of introns, one view is that intron-mediated enhancement of transgene expression does not increase, but rather restores expression of genes with impaired expression (Rethmeier *et al.*, 1998). This suggests that the use of coding sequences without the original genomic conformation could be a common cause of expression level and stability problems in transgenic plants (see Baga *et al.*, 1999, for review).

A group of DNA elements that are able to insulate transgene expression from the effect of the surrounding genome and/or increase the levels of transgene expression are specific DNA sequences that can attach themselves to the nuclear matrix. These elements are called MARs (matrix attachment regions) or SARs, which interact with the nuclear matrix to form chromatin into loops. After transgene integration the native chromatin organization is likely to be disrupted and it is believed that the inclusion of MAR elements into the transformation vector may help to mimic native chromatin organization after transgene integration (Holmes-Davis and Comai, 1998). MAR or SAR elements have been used in a number of experiments and, although the results are still somewhat confusing, there seem to be effects of these elements on transgene expression levels and stability (Holmes-Davis and Comai, 1998).

There are, however, several reasons why a number of the experiments performed with MAR elements are inconclusive. The most common problem is that the constructs used for the experiments contain the selectable marker linked to the MAR effect reporter gene. This may prevent the retrieval of plants from insertion events with loci giving low levels of expression (which do not survive selection), biasing the population of plants towards high expressers. Another problem is the lack of standard control vectors containing "stuffer" DNA fragments that preserve the distance between the reporter gene and other elements, and that the "stuffer" DNA should closely resemble the active MAR sequence. A secondary problem is that most experiments do not take into account that the sizes of the loops generated by the different MAR-containing constructs are different, which may influence expression. A factor not taken account of in several experiments is the fact that transgenic plants often contain complex integration patterns with rearrangements/deletions/inverted repeats/concatemers, etc., and that these complex integration patterns are likely to cause expression instability, which may confound the analysis of MAR activity. Also, any incomplete insertions of the complex (MAR–promoter–gene–terminator–MAR) should not be included as functional. Finally, with very few exceptions (Allen *et al.*, 1996) most experiments have been conducted with MAR sequences heterologous to the recipient genome and few data are available on the comparative activity of endogenous MAR sequences.

Although extensive analysis of MAR element function in cereals has been performed by industry laboratories, the only published study in cereals is that by Vain *et al.* (1999). In this paper, two MAR elements, the tobacco RB7 and the yeast ARS1, are examined for their effect on GUS expression in rice plants. The constructs used were generated by Allen *et al.* (1996) and contained the *uid*A gene driven by the 35S promoter and terminated by the *nos* terminator. For this work, 83 callus lines were generated from which plants were obtained. The authors observed a reduction in the number of non-expressing lines in the MAR-containing population compared with the population of plants without MARs. They observed also that, in the presence of MARs, GUS activity increased in proportion to transgene insertion number up to 20 copies, but was generally reduced in lines carrying a higher insertion number. The authors suggested that the RB7 MAR significantly improved the stability of transgene expression levels over two generations and therefore appeared to offer protection against transgene silencing. A point of concern in this study, however, is that the population of plants generated in the experiment had an abnormally high number of transgene insertions (up to 80 *uid*A copies) and showed high levels of complexity of integration (the majority of lines containing large numbers of rearranged inserts). As a consequence, the plants were estimated as having a certain number of insertions by densitometry and complete and incomplete transgene insertions were counted as equivalent. As explained above, it is difficult to draw clear conclusions on the effect of MAR elements on transgene stability when the plants being compared do not have simple and comparable integration patterns, and it will be of interest to see whether similar results will emerge from analysis of MAR function in cereal lines with standardized transgene insertion numbers and patterns.

Future experiments on assessing the stabilizing effect of MAR elements will hopefully ensure not only that the constructs are built correctly to allow direct comparisons of MAR+ and MAR– events, but also that the populations of plants analysed contain unlinked non-selected events with similar insertion numbers free of rearrangements and deletions, and, finally, that experiments use homologous MAR elements with strong matrix binding properties. Recently, new MAR elements from plants are being isolated and characterized (e.g. Avramova *et al.*, 1995, 1998; Michalowski *et al.*, 1999) and populations of transgenic cereals with simple integrations of MAR elements are being generated via the use of transposable elements to separate the MAR/reporter gene sequences from the selection cassette and other transgene sequences (M. Cannell, personal communication).

VI. DOWN-REGULATION OF GENE EXPRESSION

The use of plant transformation technology to down-regulate the expression of endogenous genes is a powerful tool for both plant science research and the development of improved crops. In the field of plant science the silencing of individual genes in "knockout" experiments can be used to study gene function. In the development of improved crop species, targeted silencing of genes offers potential for the modification of agronomic and quality traits to produce varieties with increased commercial value. For example, the first GM food product to be introduced on to the UK market employed the down-regulation of gene expression. In this case the down-regulation of poly-galacturonase activity in tomato was used to develop a slower ripening variety. This variety could remain on the vine longer, and was used to produce a more flavoursome tomato paste in a more energy-efficient manner (Dale, 2000).

Early plant transformation experiments aimed at the down-regulation of gene expression took an antisense approach, with the first example being the down-regulation of nopaline synthase activity in tobacco (Rothstein *et al.,* 1987). The antisense technique involves introduction of a construct containing a copy of all or part of the target gene in an antisense orientation. The mechanism behind this approach is that the complementary RNA transcribed from the introduced antisense DNA hybridizes to the RNA transcribed from the endogenous, sense target gene to form an RNA duplex. The formation of this RNA duplex between sense and antisense transcripts prevents expression of the target gene. Subsequent studies have indicated that antisense down-regulation is not entirely mediated by duplex formation, but that additional activities are triggered, leading to the degradation of the target gene RNA transcript (Stam *et al.*, 2000). Mechanisms proposed for this effect will be discussed in more detail below.

Subsequent to the first antisense experiments, unexpected incidences of endogenous gene silencing were observed during the introduction into plants of extra copies of genes in the sense orientation. Although the original aim of these studies was to increase gene expression levels, in some cases a silencing of both the transgene and the homologous endogenous gene was observed. This phenomenon is termed co-suppression (Montgomeri and Fire, 1998).

A.   MECHANISMS OF DOWN-REGULATION OF GENES

The exact mechanisms responsible for the down-regulation of gene expression in plants via antisense and co-suppression are as yet not fully understood. However, several models have recently been put forward. Because these models of gene silencing involve interactions between homologous or complementary nucleic acid sequences they have been termed homology-dependent gene silencing (HDGS) (Cogoni and Macino, 1999).

Homology-dependent gene silencing effects have been divided into two categories, based on the level at which silencing occurs. Transcriptional gene silencing (TGS) occurs at the DNA level and prevents transcription from occurring, whilst in post-transcriptional gene silencing (PTGS) transcription of the gene is unaffected, but subsequent degradation of the RNA transcript occurs.

1.   *Transcriptional Gene Silencing*
Transcriptional gene silencing generally involves the interaction of genes that share homology in promoter regions and is associated with increased promoter methylation that can be meiotically heritable (Park *et al.*, 1996). Such increases in promoter methylation have been shown to be accompanied

by alterations to chromatin structure in transgenic lines of *Petunia hybrida* (van Blockland *et al.*, 1997). Studies such as this in plants and others in animal systems (Jones *et al.*, 1998) indicate that it is probable that promoter methylation does not directly prevent gene transcription, but rather DNA methylation is associated with the targeting of transcriptional repressive protein complexes to the transgene region. These protein complexes bind to the methylated DNA, cause chromatin condensation and thus prevent transcription of the gene. There is recent evidence for links between DNA methylation, chromatin remodelling and TGS in plants. In *Arabidopsis* a chromomethylase (CHMET) protein has been found, containing both a chromodomain and a DNA methyltransferase domain (Henikoff and Comai, 1998). In a second study in *Arabidopsis,* the loss of function of a gene involved in chromatin structure (DDM1) was found to cause a reduction in DNA methylation (Jeddeloh *et al.*, 1999). A range of factors are known to be involved in the methylation and silencing of transgenes in plants, as discussed in Section V. These factors include the site of insertion, transgene copy number and the conformation of the inserted DNA sequences.

In cases of TGS where co-suppression occurs so that both the inserted transgene and homologous chromosomal genes are down-regulated via methylation, the transcriptional gene silencing mechanism must involve *trans*-acting methylation signals. To date there is limited knowledge of the targeting of methylation by different DNA methyltransferases in plants (Finnegan *et al.*, 1998), but in the case of TGS recent evidence points towards methylation signals acting through DNA–DNA pairing or an RNA–DNA interaction. In *Petunia* a chalcone synthase (*Chs*) transgene containing an inverted repeat has been suggested to cause methylation and silencing of an unlinked homologous sequence due to DNA–DNA pairing (Stam *et al.*, 1998). A study of TGS in tobacco has demonstrated *trans*-methylation due to the transcription of aberrant, possibly double-stranded, RNA from a promoter sequence in an inverted repeat arrangement (Mette *et al.*, 1999).

2. Post-transcriptional Gene Silencing

The second category of homology-dependent gene silencing event that causes the co-suppression of transgenes and homologous endogenous genes has been termed post-transcriptional gene silencing (PTGS). The main characteristic of this phenomenon is the absence of RNA accumulation despite transcription levels being unaffected. In cases of co-suppression the degradation of RNA transcripts from both the transgene and the homologous chromosomal genes occurs.

Several models have been put forward to explain the triggering of PTGS by transgenic RNA and the mechanism by which the homology-dependent RNA degradation process takes place. Since under normal circumstances gene expression does not trigger silencing, there is an implication that the triggering of PTGS must occur via a mechanism that recognizes the transgenic RNA molecule as being distinct from normal cellular RNA. It has been postulated that the presence of the RNA in a double-stranded form may lead to the triggering of PTGS. Evidence for double-stranded RNA (dsRNA) acting as a trigger of PTGS in plants has been obtained in studies where reproducible PTGS is induced by the introduction of transgene constructs specifically designed to produce dsRNA molecules. The introduction of a marker gene construct with self-complementarity (forming an RNA hairpin) into a rice plant already containing the same marker gene resulted in marker gene silencing in almost 90% of transformants, while transformation with constructs with single sense or antisense marker genes led to silencing in only a few cases (Waterhouse *et al.*, 1998). In the same study, constructs containing viral genes in various configurations were introduced into tobacco plants that were subsequently tested for viral immunity. Results showed that viral immunity was conferred much more effectively by transgene configurations allowing the formation of dsRNA than by constructs in which only simple sense or antisense transgenes were present. These results suggest that the underlying biological role of PTGS in plants may be as an antiviral defence mechanism, where the production of dsRNA by viruses triggers the degradation of these molecules (Cogoni and Macino, 1999; Waterhouse *et al.*, 1999).

A more recent study involving *Petunia* transformed with antisense chalcone synthase (*Chs*) genes in various configurations showed that the strongest silencing of the endogenous *Chs* gene is induced by antisense *Chs* genes, which are arranged as inverted repeats and therefore able to form dsRNA structures. Weaker silencing was found in transformants containing a single copy antisense *Chs* gene, despite it being transcriptionally more active than the inverted repeat antisense *Chs* form (Stam *et al.*, 2000). The measurement of transcription levels of the antisense genes showed that the amount of antisense *Chs* RNA produced was much lower than the amount of sense mRNA present. The differences in abundance indicate that there were insufficient transgene-derived antisense RNA molecules to hybridize to every sense *Chs* mRNA to form a duplex. Given the growing evidence that dsRNA acts as a trigger for PTGS, it is possible that gene silencing produced by the introduction of antisense transgenes generally occurs via the triggering of PTGS by dsRNA molecules formed between the transcript of the antisense transgene and mRNAs transcribed from endogenous copies of the gene.

If the down-regulation of endogenous gene expression via antisense is indeed mediated by PTGS it seems relatively simple to explain that the dsRNA acting as the trigger for the process is formed from hybridization between transgene-derived antisense RNA molecules and the endogenous sense mRNA transcripts. However, in co-suppression, where the introduction of sense transgenes causes silencing of the both transgenes and homologous endogenous genes, there must be alternative sources of the dsRNA trigger for PTGS.

There are several situations in which such dsRNA could be formed (Montgomery and Fire, 1998). In one example, if the sense transgene is inserted proximal to a host promoter on the opposite DNA strand, then a low level of antisense RNA could be transcribed in addition to the transcription of sense RNA from the transgene promoter. Hybridization of these two transcripts would form dsRNA. In a second example, where the insertion of two copies of the transgene into the genome in opposing orientations occurs (i.e. an inverted repeat configuration), occasional read-through transcription of the terminators of these genes would produce RNA molecules that could undergo intramolecular hybridization to produce hairpin structures containing areas of dsRNA. A further possibility for the production of dsRNA from sense transgenes is via the activity of an RNA-dependent RNA polymerase (RdRP) (Fire, 1999). Such an enzyme has recently been identified in tomato (Schiebel *et al.*, 1998) and shown to convert ssRNA to dsRNA *in vitro*. It is obvious that the activity of such an enzyme must be tightly regulated to prevent dsRNA production from standard, cellular RNAs. Models that account for the specific activity of RdRP on introduced transgenes have been proposed (Wassenegger and Pélissier, 1998) and relate to the production of aberrant RNA (abRNA) and the recognition of this by the RdRP enzyme. Such aberrant RNA could be produced in a number of ways, including high RNA turnover rates from transcriptionally active promoters and multiple transgene copies, irregular termination of transcription and irregular RNA processing. It has been observed that PTGS-inducing transgenes are frequently methylated within their coding regions (Ingelbrecht *et al.*, 1994; English *et al.*, 1996). Methylation could lead to premature termination of transcription and therefore the production of abRNA species.

Recent evidence supporting the involvement of an RdRP in the production of antisense RNA (asRNA) molecules as part of the PTGS triggering mechanism comes from the discovery of 25 nucleotide RNA fragments with antisense polarity to the targeted host gene in transgenic plants where PTGS has occurred (Hamilton and Baulcombe, 1999). These 25 nucleotide asRNAs could be made directly from an abRNA target or could be processed from a longer initial product.

Whatever the source of the dsRNA, whether it is directly transgene-derived, or the result of RdRP action on an abRNA transcribed from a transgene, most models of PTGS propose that dsRNA is the trigger for the PTGS process that finally leads to the degradation of the RNA transcripts from both the transgene and homologous chromosomal gene and therefore the down-regulation of gene expression.

The exact mechanism involved in the homology-dependent degradation of chromosomally derived RNA transcripts is still unknown, although several models have been proposed. Some models involve the presence of a dsRNA-dependent RNA polymerase (dsRdRP) (Dougherty and Parks,

1995; Wassenegger and Pélissier, 1998; Waterhouse *et al.*, 1998; Fire, 1999), which uses the dsRNA triggers to synthesize large numbers of copy RNA (cRNA) molecules of both a sense and antisense form. These cRNAs can then hybridize with each other to form new dsRNA triggers and amplify the PTGS response. More importantly, the antisense cRNA molecules can hybridize with homologous chromosomally derived mRNAs to form dsRNA. As an alternative to the presence of a dsRdRP, the cRNA synthesis could be carried out by an RdRP enzyme assuming the presence of additional enzymes, which are capable of acting on the dsRNA to produce partially single-stranded forms as templates for cRNA production.

A second model of the PTGS mechanism developed during studies in the nematode *Caenorhabditis elegans* involves the activity of a multi-round enzyme, which unwinds part of the triggering dsRNA, producing a single-stranded portion, which can then hybridize directly with the homologous chromosomally derived mRNA (Montgomery *et al.*, 1998).

Once a RNA duplex has been formed with the target mRNA, this molecule then becomes the target for degradation. Studies in tobacco indicate that the initial step in this degradation process may be an endonucleolytic cleavage involving a dsRNA-specific nuclease (van Eldik *et al.*, 1998). Alternatively, a ssRNA-specific nuclease could be present as a complex together with the dsRdRP during cRNA synthesis, becoming attached to the cRNA and cleaving the single-stranded areas of the target mRNA adjacent to the RNA duplex formed after cRNA/mRNA hybridization (Waterhouse *et al.*, 1998). A further possibility is that modifications of the RNA duplex, for example, the de-amination of adenosines to inosines, may occur, leading to the "tagging" of the duplex as a target for degradation (Fire, 1999).

Once the target RNA duplex has been cleaved or modified, it is then rapidly degraded via an as yet uncharacterized mechanism that may involve PTGS-specific degradation pathways and/or normal decay mechanisms.

A further aspect of PTGS is the observation that PTGS can spread systematically through a plant (Palauqui *et al.*, 1997; Wassenegger and Pélissier, 1999). The silencing signalling molecule must be of a form capable of travelling from cell to cell through plasmodesmata and over long distances via the phloem (Kooter *et al.*, 1999). One possible component of this systemic signalling mechanism could be the recently discovered PTGS-associated 25 nucleotide RNA fragments (Hamilton and Baulcombe, 1999) discussed above.

B. APPLICATION OF TRANSGENE-MEDIATED DOWN-REGULATION OF GENES IN CEREALS

While the application of techniques for down-regulation of endogenous gene expression via the introduction of transgenes in cereals lags behind the state of the art in dicot crops, there are now several examples (see Table V).

TABLE V

Examples of transgene-mediated down-regulation of endogenous genes in cereals

Cereal crop	Transgene type	Target gene	Reference
Rice	Antisense	14–16 kDa allergenic	Tada *et al.* (1996)
	Sense	proteins	Itoh *et al.* (1997)
	Antisense	GBSS (waxy)	Fujisawa *et al.*
		Heterotrimeric G protein	(1999)
Wheat	Sense	HMW glutenin	Blechl *et al.* (1998)
	Sense	HMW glutenin	Alvarez *et al.* (2000)
	Antisense	S-Adenosyl methionine	Rasco-Gaunt *et al.*
		decarboxylase	(1999a)
Barley	Antisense	Snf1-kinase	Y. Zhang (personal communication)
Maize	Sense	Starch synthesis genes	K. Broglie (personal communication)

An antisense approach was employed for the production of transgenic rice plants with reduced levels of the 14–16 kDa allergenic proteins (Tada *et al.*, 1996). In this experiment, antisense constructs of the gene coding for the 16 KDa allergen protein under the control of four different rice seed-specific promoters were introduced into rice protoplasts. From a population of 120 polymerase chain reaction (PCR)-positive lines, 11 were fertile and set seed. From these, several lines were selected by immunoblot screening with antibodies to the allergenic protein and shown to have strongly reduced levels of the allergen in comparison to non-transformed control lines. The high levels of reduction were observed to be inherited stably over three generations.

In a second gene down-regulation experiment in rice, a sense *waxy* gene (GBSS) was introduced into protoplasts and transgenic lines showing co-suppression of the transgene and endogenous gene were produced. These lines, however, showed down-regulation of the GBSS expression in pollen, but not in endosperm (although a decrease in GBSS protein was observed in a subset of the transgenic lines) (Itoh *et al.*, 1997). Again, this effect was transmitted to progeny, showing normal Mendelian segregation.

In a further experiment in rice, transgenic plants containing an antisense cDNA for the α-subunit of rice heterotrimeric G protein were seen to produce little or no mRNA for the subunit and exhibited abnormal morphology, including dwarf morphology and the setting of small seeds (Fujisawa *et al.*, 1999).

In wheat, co-suppression of endogenous genes has been observed, as in the case of wheat lines containing a chimaeric sense transgene containing portions of two HMW glutenin subunits (Blechl and Anderson 1996; Blechl *et al.*, 1998). A similar phenomenon was also observed by another group in a number of transgenic wheat lines containing multiple copies of a single HMW glutenin subunit in sense orientation (Alvarez *et al.*, 2000). In contrast, down-regulation was not observed in transgenic wheat by a third group working with the same set of genes (Barro *et al.*, 1997), even when the transgene insertion number of the plants recovered varied from one to more than fifteen.

The use of an antisense strategy to down-regulate endogenous gene expression has also been demonstrated in wheat, in a study in which lines contained an antisense version of the S-adenosyl methionine decarboxylase (*samdc*) gene from tritordeum (Dresselhaus *et al.*, 1996). In this experiment, two lines containing the samdc antisense construct showed reduced polyamine levels, the effect predicted to be the result of reduced SAMDC activity (Rasco-Gaunt *et al.*, 1999).

An antisense approach has also been used in barley to achieve down-regulation of the activity of barley SnRK1 genes that are homologues of the yeast Snf1-kinase, which is involved in global regulation of carbohydrate metabolism (Y. Zhang, personal communication).

Maize is the cereal crop that receives most attention in terms of applied biotechnology and transgene-mediated down-regulation of gene expression routinely used in a number of AgBiotech companies and in several public laboratories. However, there are to date no published experiments despite the large volume of work that has been done in areas of economic importance, for example, the down-regulation of a number of genes involved in starch synthesis has been achieved, leading to the generation of maize starches with altered structure (K. Broglie, personal communication). One reason for this is that much of the work is commercially sensitive and thus will not be published until results are at a stage to allow patent applications to be made.

C.   STRATEGIES FOR INCREASING EFFICIENCY OF GENE DOWN-REGULATION IN CEREALS

An understanding of the mechanisms involved in the triggering of HDGS phenomena is essential for the design of experiments aimed at the efficient silencing of specific genes in cereals. Based on current knowledge of the gene silencing mechanisms, it appears that optimal triggering of the gene silencing pathway should be a key objective.

In cases of TGS, it is postulated that silencing of the inserted transgene and the endogenous target gene occurs via promoter methylation and chromatin condensation. TGS is thought to occur in cases where homology is present between the promoter regions of the transgene and the target gene (Park *et al.*, 1996), thus experiments aimed at the silencing gene expression via TGS must introduce transgenes under the control of promoter sequences homologous to those controlling expression of the target gene.

Unfortunately, the circumstances under which the triggering of TGS occurs remain largely unclear, although insertion of the transgene in or next to silent hypermethylated genomic sequences is thought to be one likely cause (Pröls and Meyer, 1992; Vaucheret *et al.*, 1998). Since it is currently impossible to target transgene insertion efficiently to specific chromosomal locations, proximity-based triggering of TGS cannot yet be used for directed gene silencing studies. It has also been observed that transgenes frequently undergo TGS when present as multiple copies (Assaad *et al.*, 1993), although the reason for this remains unclear. This trigger of TGS could potentially be exploited via the design of multi-transgene constructs for introduction into cereals. Again, however, it is likely that the insertion site of these transgenes will probably have a large influence on their ability to trigger TGS.

In cases of PTGS, the presence of dsRNA seems to be the critical trigger of the pathway. The nature of this trigger readily suggests strategies for the design of transformation constructs with the potential to trigger PTGS more effectively.

To date most experiments undertaken with the aim of down-regulation of a specific gene have used an antisense approach. This approach is based on the theory that the antisense transcripts produced hybridize directly with mRNA expressed from the target endogenous gene and thus block translation. However, as discussed above, recent evidence suggests that gene silencing via antisense transgenes, in some cases at least, is only partly due to the blocking of mRNA translation by duplex formation and that complete gene silencing occurs via the dsRNA triggering of the PTGS pathway (Stam *et al.*, 2000). If, as the authors of this study suggest, both antisense- and sense-mediated gene silencing operate via PTGS, the distinction between the two approaches to gene silencing becomes largely irrelevant and the design of experiments for down-regulation of genes in the future should be focussed on the effective production of dsRNA.

With increased understanding of gene silencing mechanisms, particularly the fact that dsRNA acts as a trigger of PTGS, there are good prospects for increasing the frequency of gene silencing events by improved design of transgene constructs. To ensure the production of dsRNA, transgene sequences in both sense and antisense orientation should be introduced into plants and expressed from strong promoters to ensure high levels of transcription. These sense and antisense transgenes could be introduced on separate constructs under the control of separate promoters. An alternative approach, which has already been demonstrated to be effective in silencing of marker genes in transgenic rice plants, is to design constructs where sense and antisense transgene sequence will be produced in a single transcript, under the control of a single promoter (Waterhouse *et al.*, 1998). Having internal regions of complementarity, these transcripts are able to form duplexes in a hairpin structure. Introduction of these transgene constructs were found to produce silencing of the marker gene in 90% of cases.

Further experiments are now needed to show if such a high level of gene silencing can be achieved using such hairpin constructs targeted at the silencing of endogenous genes in cereal species. In order to produce transgenic lines demonstrating a stable down-regulation phenotype, the ideal situation would be a small number of sites of transgene insertion containing highly transcribed sequences, producing transcripts able to form dsRNA directly.

Once mechanisms of HDGS are better understood, new strategies for increasing the efficiency of down-regulation of gene expression will be formulated. These may involve the production of RNA molecules in a form to be the optimal target for the RdRP enzyme, ensuring production of high levels of dsRNA triggers. Additionally, constructs could be designed that will produce dsRNA in a form that will trigger PTGS more effectively. Further studies are needed to understand the exact nature of the homologous interaction between transgene and target gene needed for gene silencing to occur. This will be important in cases where the target genes are members of gene families, with many related genes with varying levels of homology being present in the genome. Better knowledge of the influence of homology levels in gene silencing phenomena will also be important in studies involving polyploid cereals such as oats and wheat, where the silencing of certain alleles, whilst maintaining expression from other alleles, will allow the development of interesting partial phenotypes for some traits.

Although gene silencing was initially viewed as an obstacle to the development of transgenic cereals where overexpression of gene activity was required, many potential applications of the specific down-regulation of gene expression can now be identified. The role of genes involved in metabolic pathways and plant development can be elucidated by the production of "knockout" mutants via the introduction of transgene constructs. Targeted gene silencing studies will also allow the development of new cereal lines with novel agronomic and quality traits. Such traits could include novel disease resistances and alterations to protein, starch and lipid profiles to provide new end-use characteristics.

VII. METHODS FOR GENERATING MARKER-FREE TRANSGENICS AND FOR THE RESOLUTION OF INTEGRATION PATTERNS

Despite progress in the production of transgenic plants from crop species considered in the past to be recalcitrant to transformation, a contentious issue in the commercialization of genetically modified plants is the fact that most contain marker genes. Marker genes are used in the transformation procedure for the selection of transformed cells and tissues (see Section III). Historically, genes coding for resistance to antibiotics or herbicides have been used, but regulatory authorities and environmental groups have seen them as potentially harmful to the environment and human health. Irrespective of the presence or absence of good scientific evidence for this contention, the opposition to these group of genes has motivated the development of two avoidance strategies; one towards the development of new selectable markers with neutral impact on the environment and potential consumers (see Section III), and a second strategy towards the removal of the marker genes from the recipient genome after transformation.

The simplest way of eliminating marker genes from the recipient genome would be to rely on the natural occurrence of segregation, in those cases in which the marker gene integrates at a different locus from that in which the trait gene is integrated. After the first seed generation, it should then be possible to recover transgenic lines that only contain the "trait gene" locus. Unfortunately, for plants generated by DGT methods, integration at more that one locus is rare (approximately 25% depending on experiment and species) and even more rare is the situation in which all insertions of the marker gene fall in just one of the loci rather than being distributed between two loci. However, in the case of *Agrobacterium* transformation it has been shown in tobacco and rice that the use of two T-DNA vectors can result in the recovery of marker-free transgenic lines at acceptable frequencies (65% for rice) (Komari *et al.*, 1996).

Artificial systems for marker removal have been proposed that either involve transposition or site-specific recombination (Yoder and Goldsbrough, 1994). Intragenomic relocation of transgenes via transposable elements involves a cut-and-paste mechanism that results in the excision of transgenes or any other sequence flanked by the appropriate elements from one locus, and the subsequent reinsertion of this material into a second locus. The transgenic loci may then segregate apart as the result of crossing-over during meiosis resulting in marker-free segregants still containing the trait gene appearing among the next generation. Transposition has also been used in combination with positive selectable markers, as it is the case of the multi-auto-transformation (MAT) system. This system combines transposition with a positive selectable marker, *ipt* (isopentenyltransferase) to generate marker-free transgenic tobacco plants (Ebinuma *et al.*, 1997). The MAT system relies on the overexpression of the *ipt* gene, resulting on an increase in endogenous cytokinins and on the subsequent production of an extreme shooty phenotype (ipt-shooty). Upon excision of the *ipt* gene by transposition, normal shoots develop from ipt-shooty phenotypes. However, the low efficiency of this system stimulated the development of a similar MAT system combining positive selectable markers with site-directed recombination (the yeast R/RS system) (Sugita *et al.*, 1999).

There are at present two other recombination/excision systems that have been shown to work in cereals; the bacteriophage P1 *cre/lox* (Srivastava *et al.*, 1999) and the yeast *Flp/Frt* (Lyznik *et al.*, 1993, 1996). Both, as the yeast

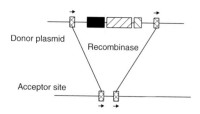

Fig. 4. The use of site-specific recombination to direct transgene insertion to a site flanked by two previously inserted acceptor sites.

R/RS system, are simple two-component recombination system (recombinase and its target site). All these three site-specific recombination systems rely on the excision of sequences flanked by *lox*, *Frt* or RS sites upon exposure to their corresponding recombinases *cre*, *Flp* or R.

Similar recombination methods can also be used to resolve integration patterns from multiple to single insertions. As discussed above, the integration of transgenes in complex multiple-insertion patterns is a significant factor in transgene expression instability. While it is true that single insertion transgenics do not always stably express the transgene, overall they show lower incidence of transgene inactivation than complex-insertion transformants.

There are different strategies for improving transgene stability (see Section V), but an important approach is to reduce the number of transgene insertions. Two experiments performed in cereals suggest that multiple insertions can be reduced either by influencing integration complexity by using inhibitors of DNA repair enzymes (DeBlock *et al.*, 1997) or by reducing insertion numbers by excising them from the genome after integration, using site-specific recombination (Srivastava *et al.*, 1999) (Fig. 4).

In the first experiment, the authors showed that they could increase the production of single-copy transgenic wheat plants by using niacinamide in the culture medium. The logic behind this experiment is that niacinamide is an inhibitor of the action of DNA repair enzymes like the nuclear enzyme poly (ADP-ribose) polymerase (PARP) (De Murcia and De Murcia, 1994). This enzyme is also involved in cell division (Eki, 1994), apoptosis (Monti *et al.*, 1994), and recombination (Waldman and Waldman, 1990) and in several other processes in animal cells. Although PARP is found in plants (Lepiniec *et al.*, 1995), only few facts about its function are known. However, one observation is that in tobacco and *Arabidopsis* the addition of PARP inhibitors to the culture medium results in an enhancement of intrachromosomal homologous recombination (Puchta *et al.*, 1995). Conversely, the same inhibitor used in the culture medium of animal cells results in a reduction of illegitimate recombination (Waldman and Waldman, 1990). In the experiment published by De Block *et al.* (1997), although the results show an effect of the compound on the proportion of plants recovered with low-insertion numbers, the total number of plants produced is rather low. To our knowledge, no further experiment has been performed on this approach, but it would certainly be interesting to extend the initial experiment by testing the action of other PARP-inhibitor compounds (e.g. nicotinamide, isoquinolinone, 3-aminobenzamide, 8-hydroxy-2-methylquinazolin-4-one, etc.), or by inhibiting other DNA repair enzymes.

In the second experiment, Srivastava and co-workers (Srivastava *et al.*, 1999) reported that they could convert multiple-copy wheat transgenics to single-copy by flanking the transgenes with *lox* sites and crossing the plants with transgenic lines expressing *cre* recombinase. In principle, this is the same strategy as used for marker removal (see above), although if in the same experiment one wants to use the *cre/lox* system for both purposes, then a second set of *lox* sites has to be used. In the published experiment, the authors chose to flank the selectable marker gene with wild-type *lox*P sites and to use the mutant variant *lox*511 for flanking the outermost part of the transgene. The *lox*P sites flanking the selectable marker were placed in the correct orientation to be recombined upon crosses with *cre* recombinase expressing plants, resulting in the removal of all marker insertions that contained intact *lox*P sites. The mutant *lox*511 sites were placed in such an orientation that recombination only happened between the outermost sites, resulting in multiple insertions being converted into single insertions. Again, this recombination system depends on the integrity of the *lox* sites. This idea has been shown to work in cereals but it has a some disadvantages. First, it relies on insertions being intact and, therefore does not resolve complex rearranged patterns of integration, which are those most likely to create problems of expression. Second, it requires either a round of sexual crossing or transient expression of the recombinase. In our opinion, although excision systems can be used to reduce insertion number, a system that controls complexity prior to integration is inherently a better solution. As discussed in Section IV above, better understanding of the mechanism of transgene integration should lead to approaches for the control of integration complexity.

VIII. CONCLUSIONS

In the last decade, cereal GM has moved from a situation in which the focus was the development of basic gene delivery methods and the production of any transgenic plants was a significant achievement, to one in which (at least in maize) commercial laboratories are producing thousands of independent transgenic lines containing genes to modify agronomic traits. However, the various sections in this chapter should make it clear that there are still needs for major transformation technology development in a number of areas and a general observation is that the pace of application of the technology has frequently outstripped progress in understanding the underlying biology of the systems that are being used.

For example, in the area of tissue culture, highly efficient culture systems have been developed in several crops, but we still have little understanding of the cellular and genetic processes involved and these systems are frequently confined to specific explants and limited ranges of germplasm. Gene delivery technology for cereals has received much attention and a range of DGT methodology has been invented and put into application, but recently we have come full circle to the application of *Agrobacterium* techniques in the major cereals, although the full implications of using one or other approach have yet to be resolved. In the area of selection, we have moved from antibiotic markers to herbicides to positive selection systems, but we still need more efficient and benign selection markers. This is an area where new technology could have a major impact.

The field in which we have most to learn and most still to develop, however, is that of transgene integration and expression (including both novel/overexpression of transgenes and the modification of expression of endogenous genes). In this area, we currently have very little control over the site of transgene insertion or of the number and configuration of transgene copies inserted, but we increasingly realize that these factors have major influence on expression level and stability. Currently, in most systems a relatively high proportion of transgenic plants produced are not useful because of problems with transgene expression, stability and heritability, and a better understanding of these processes has the potential radically to reduce the numbers of lines that must be produced per gene construct to be evaluated and thus the effectiveness and cost of transformation programmes.

REFERENCES

Ahlandsberg, S., Sathish, P., Sun, C. X. and Jansson, C. (2000). Green fluorescent protein as a reporter system in the transformation of barley cultivars. *Physiologia Plantarum* **107**, 194–200.

Aldemita, R. R. and Hodges, T. K. (1996). *Agrobacterium tumefaciens*-mediated transformation of *japonica* and *indica* rice varieties. *Planta* **199**, 612–617.

Allen, G. C., Hall, G. E., Michalowski, S., Newman, W., Spiker, S., Weissinger, A. K. and Thompson, W. F. (1996). High-level transgene expression in plant cells: effects of a strong scaffold attachment region from tobacco. *Plant Cell* **8**, 899–913.

Altpeter, F., Vasil, V., Srivastava, V., Stoeger, E. and Vasil., I. K. (1996). Accelerated production of transgenic wheat (*Triticum aestivum* L.) plants. *Plant Cell Reports* **16**, 12–17.

Alvarez, M. L., Guelman, S., Halford, N. G., Lustig, S., Reggiardo, M. I., Ryabushkina, N., Shewry, P., Stein, J. and Vallejos, R. H. (2000). Silencing of

HMW glutenins in transgenic wheat expressing extra HMW subunits. *Theoretical Applied Genetics* **100**, 319–327.

Ambros, P. F., Matzke, A. J. M. and Matzke, M. A. (1986). Localization of *Agrobacterium rhizogenes* T-DNA in plant chromosomes by *in situ* hybridization. *EMBO Journal* **5**, 2073–2077.

Asano, Y., Otsuki, Y. and Ugaki, M. (1991). Electroporation-mediated and silicon carbide fiber-mediated DNA delivery in *Agrostis alba* L. (Redtop). *Plant Science* **79**, 247–252.

Assaad, F. F., Tucker, K. L. and Signer, E. R. (1993). Epigenetic repeat-induced gene silencing (RIGS) in arabidopsis. *Plant Molecular Biology* **22**, 1067–1085.

Avramova, Z., SanMiguel, P., Georgieva, E. and Bennetzen, J. L. (1995). Matrix attachment regions and transcribed sequences within a long chromosomal continium containing maize *Adh1*. *Plant Cell* 1667–1680.

Avramova, Z., Tikhonov, A., Chen, M. and Bennetzen, J. L. (1998). Matrix attachment regions and structural colinearity in the genomes of two grass species. *Nucleic Acids Research* **26**, 761–767.

Baga, M., Chibbar, R. N. and Kartha, K. K. (1999). Expression and regulation of transgenes for selection of transformants and modification of traits in cereals. *In* "Molecular Improvement of Cereal Crops", Advances in Cellular and Molecular Biology of Plants, Vol. 5 (I. K. Vasil, ed.), pp. 83–131. Kluwer Academic Publishers.

Barcelo, P. and Lazzeri, P. A. (1998). Direct gene transfer: Chemical, electrical and physical methods. *In*: "Transgenic Plant Research" (K. Lindsey, ed.), pp. 35–55. Harwood Academic Publishers, Amsterdam.

Barcelo, P., Hagel, C., Becker, D., Martin, A. and Loerz, H. (1994). Transgenic cereal (tritordeum) plants obtained at high efficiency by microprojectile bombardment of inflorescence tissue. *Plant Journal* **5**, 583–592.

Barro, F., Rooke, L., Bekes, F., Gras, P., Tatham, A. S., Fido, R., Lazzeri, P. A., Shewry, P. R. and Barcelo, P. (1997). Transformation of wheat with HMW subunit genes results in improved functional properties. *Nature Biotechnology* **15**, 1295–1299.

Barro, F., Cannell, M. E., Lazzeri, P. A. and Barcelo, P. (1998). The influence of auxins on transformation of wheat and tritordeum and analysis of transgene integration patterns in transformants. *Theoretical Applied Genetics* **97**, 684–695.

Baruah-Wolff, J., Harwood, W. A., Lonsdale, D. A., Harvey, A., Hull, R. and Snape, J. W. (1999). Luciferase as a reporter gene for transformation studies in rice (*Oryza sativa* L.). *Plant Cell Reports* **18**, 715–720.

Bechtold, N., Ellis, J. and Pelletier, G. (1993). *In planta Agrobacterium*-mediated gene transfer by infiltration of adult *Arabidopsis thaliana* plants. *Comptes Rendues Academie Sciences* **316**, 1194–1199.

Becker, D., Brettschneider, R. and Loerz, H. (1994). Fertile transgenic wheat from microprojectile bombardment of scutellar tissue. *Plant Journal* **5**, 299–307.

Bhattacharyya-Pakrasi, M., Peng, J., Elmer, J. S., Laco, G., Shen, P., Kaniewska, M. B., Kononowicz, H., Wen, F., Hodges, T. K. and Beachy, R. N. (1993). Specificity from a promoter from the rice tungro baciliform virus for expression in phloem tissues. *Plant Journal* **4**, 71–79.

Birch, R. G. (1997). Plant transformation: problems and strategies for practical application. *Annual Review of Plant Physiology and Plant Molecular Biology* **48**, 297–326.

Blechl, A. E. and Anderson, O. D. (1996). Expression of a novel high-molecular-weight glutenin subunit in transgenic wheat. *Nature Biotechnology* **14**, 875–879.

Blechl, A. E., Le, H. Q. and Anderson, O. D. (1998). Engineering changes in wheat flour by genetic transformation. *Journal of Plant Physiology* 152, 703–707.

Boulton, M. I., Buchholz, W. G., Marks, M. S., Markham, P. G. and Davies, J. W. (1989). Specificity of *Agrobacterium*-mediated delivery of maize streak virus DNA. *Plant Molecular Biology* 12, 31–40.

Bregitzer, P., Halbert, S. E. and Lemaux, P. G. (1998). Somaclonal variation in the progeny of transgenic barley. *Theoretical Applied Genetics* 96, 421–425.

Brettschneider, R., Becker, D. and Loerz, H. (1997). Efficient transformation of scutellar tissue of immature maize embryos. *Theoretical Applied Genetics* 94, 737–748.

Brinch-Pedersen, J., Galili, G., Knudsen, S. and Holm, P. B. (1996). Engineering of the aspartate family biosynthetic pathway in barley (*Hordeum vulgare* L.) by transformation with heterologous genes encoding feedback insensitive aspartate kinase and dihydrodipicolinate synthase. *Plant Molecular Biology* 32, 611–620.

Brinch-Pedersen, H., Olsen, O., Knudsen, S. and Holm, P. B. (1999). An evaluation of feed-back insensitive aspartate kinase as a selectable marker for barley (*Hordeum vulgare* L.) transformation. *Hereditas* 131, 239–245.

Bytebier, B., Deboek, F., De Greve, H., van Montagu, M. and Hernalsteens, J-P. (1987). T-DNA organisation in tumour cultures and transgenic plants of the monocotyledon *Asparagus officinalis*. *Proceedings of the National Academy of Sciences (USA)* 84, 5345–5349.

Caddick, M. X., Greenland, A. J., Jepson, I., Krause, K. P., Qu, N., Riddell, K. V., Salter, M. G., Schuch, W., Sonnewald, U. and Tomsett, A. B. (1998). An ethanol inducible gene switch for plants used to manipulate carbon metabolism. *Nature Biotechnology* 16, 177–180.

Callis, J., Fromm, M. and Walbot, V. (1987). Introns increase gene expression in cultured maize cells. *Genes and Development* 1, 1183–1200.

Cannell, M. E., Doherty, A., Lazzeri, P. A. and Barcelo, P. (1999). A population of wheat and tritordeum transformants showing a high degree of marker gene stability and heritability. *Theoretical Applied Genetics* 99, 772–784.

Casas, A. M., Kononowicz, A. K., Zehr, U. B., Tomes, D. T., Axtell, J. D., Butler, L. G., Bressan, R. A. and Hasegawa, P. M. (1993). Transgenic sorghum plants via microprojectile bombardment. *Proceedings of the National Academy of Sciences (USA)* 90, 11212–11216.

Castillo, A. M. and Casas, A. M. (1999). Transgenic cereals: *Secale cereale* and *Sorghum bicolor* (rye and sorghum). Transgenic cereals: *Hordeum vulgare* L. (barley). *In* "Molecular Improvement of Cereal Crops" (I. K. Vasil, ed.), pp. 341–360. Kluwer Academic Publishers, Dordrecht.

Castillo, A. M., Vasil, V. and Vasil, I. K. (1994). Rapid production of fertile transgenic plants of rye (*Secale cereale* L.). *Bio/Technology* 12, 1366–1371.

Chamberlain, D. A., Brettell, R. I. S., Last, D. I., Witrzens, B., Mc Elroy, D., Dolferus, R. and Dennis, E. S. (1994). The use of Emu promoter with antibiotic and herbicide resistance genes for the selection of transgenic wheat callus and rice plants. *Australian Journal of Plant Physiology* 21, 95–112.

Chaure, P., Gurr, S. J. and Spanu, P. (2000). Stable transformation of *Erysiphe graminis*, an obligate biotrophic pathogen of barley. *Nature Biotechnology* 18, 205–207.

Chawla, H. S., Cass, L. A. and Simmonds, J. A. (1999). Developmental and environmental regulation of anthocyanin pigmentation in wheat tissues transformed with anthocyanin regulatory genes. *In Vitro Cellular and Developmental Biology – Plant* 35, 403–408.

Chen, L., Marmey, P., Taylor, N. J., Brizard, J. P., Espinoza, C., D'Cruz, P., Huet, H., Zhang, S., de Kochko, A., Beachy, R. N. and Fauquet, C. M. (1998a). Expression and inheritance of multiple transgenes in rice plants. *Nature Biotechnology* **16**, 1060 – 1064.

Chen, L., Zhang, S., Beachy, R. N. and Fauquet, C. M. (1998b). A protocol for consistent, large-scale production of fertile transgenic rice plants. *Plant Cell Reports* **18**, 25–31.

Chen, W. P., Gu, X., Liang, G. H., Muthukrishnan, S., Chen, P. D., Liu, D. J. and Gill, B. S. (1998c). Introduction and constitutive expression of a rice chitinase gene in bread wheat using biolistic bombardment and the bar gene as a selectable marker. *Theoretical and Applied Genetics* **97**, 1296–1306.

Cheng, M., Fry, J. E., Pang, S., Zhou, H., Hironaka, C. M., Duncan, D. R., Conner, T. W. and Wan, Y. (1997). Genetic transformation of wheat mediated by *Agrobacterium tumefaciens. Plant Physiology* **115**, 971–980.

Cheng, X., Sardana, R., Kaplan, H. and Altosaar, I. (1998). *Agrobacterium*-transformed rice plants expressing synthetic *cry*IA(b) and *cry*IA(c) genes are highly toxic to striped stem borer and yellow stem borer. *Proceedings of the National Academy of Sciences (USA)* **95**, 2767–2772.

Chibbar, R. N., Kartha, K. K., Datla, R. S.S., Leung, N., Caswell, K., Mallard, C. S. and Steinhauer, L. (1993). The effect of different promoter-sequences on transient expression of gus reporter gene in cultured barley (*Hordeum vulgare* L.) cells. *Plant Cell Reports* **12**, 506–509.

Cho, M. J., Ha, C. D., Buchana, B. B. and Lemaux, P. G. (1998). Subcellular targeting of barley hordein promoter-*uid*A fusions in transgenic barley seeds. *Congress of* In vitro *Biology*, pp. 1023.

Cho, M. J., Jiang, W. and Lemaux, P. G. (1999). High-frequency transformation of oat via microprojectile bombardment of seed-derived highly regenerative cultures. *Plant Science* **148**, 9–17.

Christensen, A. H. and Quail, P. H. (1996). Ubiquitin promoter-based vectors for high-level expression of selectable and/or screenable marker genes in monocotyledonous plants. *Transgenic Research* 5, **213**–218.

Christensen, A. H., Sharrock, R. A. and Quail, P. (1992). Maize polyubiquitin genes: structure, thermal perturbation of expression and transcript splicing, and promoter activity following transfer to protoplasts by electroporation. *Plant Molecular Biology* **18**, 675–689.

Christou, P. and Ford, T. L. (1995). The impact of selection parameters on the phenotype and genotype of transgenic rice callus and plants. *Transgenic Research* **4**, 44–51.

Christou, P., Ford, T. F. and Kofron, M. (1991). Production of transgenic rice (*Oryza sativa*) plants from agronomically important *indica* and *japonica* varieties via electric discharge particle acceleration of exogenous DNA into immature zygotic embryos. *Bio/Technology* **9**, 957–962.

Chu, C. C., Wang C. C., Sunc, C. S., Hsu, C., Yin, K. C., Chu, C. Y. and Bi, F. Y. (1975). Establishment of an efficient medium for anther culture of rice through comparative experiments on the nitrogen sources. *Scientifica Sinica* **18**, 659–668.

Clough, S. J. and Bent, A. (1998). Floral dip: a simplified method for *Agrobacterium*-mediated transformation of *Arabidopsis thaliana. Plant Journal* **16**, 735–743.

Cogoni, C. and Macino, G. (1999). Homology-dependent gene silencing in plants and fungi: a number of variations on the same theme. *Current Opinions in Microbiology* **2**, 657–662.

Cornejo, M., Luth, D., Blankenship, K. M., Anderson, O. D. and Blechl, A. E. (1993). Activity of the maize ubiquitin promoter in transgenic rice. *Plant Molecular Biology* **23**, 567–581.

Dale, S. (2000). The gene police. *New Scientist* **4**, 28–31.

Datta, S. K. (1999). Transgenic cereals: *Oryza sativa* (rice). *In* "Molecular Improvement of Cereal Crops", Advances in Cellular and Molecular Biology of Plants, Vol. 5 (I. K. Vasil, ed.), pp. 149–187. Kluwer Academic Publishers, Dordrecht.

De Block, M., De Sonville, A. and Debrouwer, D. (1995). The selection mechanism of phosphinothricin is influenced by the metabolic status of the tissue. *Planta* **197**, 619–626.

De Block, M., Debrouwer, D. and Moens, T. (1997). The development of a nuclear male sterility system in wheat. Expression of the *barnase* gene under the control of tapetum specific promoters. *Theoretical Applied Genetics* **95**, 125–131.

De Loose, M., Danthinne, X., van Blockstaele, E., van Montagu, M. and Depicker, A. (1995). Different 5′ leader sequences modulate beta-glucuronidase accumulation levels in transgenic *Nicotiana tabacum* plants. *Euphytica* **85**, 209–216.

De Murcia, G. and De Murcia, M. J. (1994). Poly(ADP-ribose) polymerase: a molecular nick-sensor. *Trends in Biochemical Sciences* **19**, 172–176.

De Neve, M., De Buck, S., Jacobs, A., van Montagu, M. and Depicker, A. (1997). T-DNA integration patterns in co-transformed plant cells suggest that T-DNA repeats originate from ligation of separate T-DNAs. *Plant Journal* **11**, 15–29.

Dekeyser, R., Claes B., Marichal, M., van Montagu, M. and Caplan, A. (1989). Evaluation of selectable markers for rice transformation. *Plant Physiology* **90**, 217–223.

Dennehey, B. K., Petersen, W. L., Ford-Santino, C., Pajeau, M. and Armstrong, C. L. (1994). Comparison of selective agents for use with the selectable marker gene bar in maize transformation. *Plant Cell Tissue and Organ Culture* **36**, 1–7.

D'Halluin, K., Bonne, E., Bossut, M., De Beuckleer, M. and Leemans, J. (1992). Transgenic maize plants by tissue electroporation. *Plant Cell* **4**, 1495–1505.

DiMaio, J. J. and Shillito, R. D. (1989). Cryopreservation technology for plant cell cultures. *Journal of Tissue Culture Methods* **12**, 163–169.

Dong, J., Teng, W., Buchholz, W. G. and Hall, T. C. (1996). *Agrobacterium*-mediated transformation of javanica rice. *Molecular Breeding* **2**, 276–276.

Dougherty, W. G. and Parks, T. D. (1997). Transgenes and gene suppression: telling us something new? *Current Opinions in Cell Biology* **7**, 399–405.

Dresselhaus, T., Barcelo, P., Hagel, C., Loerz, H. and Humbeck K. (1996). Isolation and characterization of a tritordeum cDNA encoding S-adenosylmethionine decarboxylase that is circadian-clock-regulated. *Plant Molecular Biology* **30**, 1021–1033.

Ebinuma, H., Sugita, K., Matsunaga, E. and Yamakado, M. (1997). Selection of marker-free transgenic plants using the isopentenyl transferase gene as a selectable marker. *Proceedings of the National Academy of Sciences (USA)* **94**, 2117–2121.

Eki, T. (1994). Poly(ADP-ribose) polymerase inhibits DNA replication by human replicative DNA polymerase alpha, gamma and epsilon *in vitro*. *FEBS Letters* **356**, 261–266.

English, J. J., Mueller, E. and Baulcombe, D. C. (1996). Suppression of virus accumulation in transgenic plants exhibiting silencing of nuclear genes. *Plant Cell* **8**, 179–188.

Fennoy, S. L. and Bailey-Serres, J. (1993). Synomymous codon usage in *Zea mays* L. nuclear genes is varied by levels of C and G-ending codons. *Nucleic Acids Research* **21**, 5294–5300.

Finer, J. J., Finer, K. R. and Ponappa, T. (1999) Particle bombardment mediated transformation. *Plant Biotechnology* **240**, 59–80.

Finnegan, E. J., Genger, R. K., Peacock, W. J. and Dennis, E. S. (1998). DNA methylation in plants. *Annual Review of Plant Physiology and Plant Molecular Biology* **49**, 223–247.

Fire, A. (1999). RNA-triggered gene silencing. *Trends in Genetics* **15**, 358–363.

Foerster, W. and Neumann, E. (1989). Gene transfer by electroporation: a practical guide. *In* "Electroporation and Electrofusion in Cell Biology" (E. Neumann, A. E. Sowers and C. A. Jordan, eds), pp. 299–318. Plenum Press, New York.

Frame, B. R., Drayton, P. R., Bagnall, S. V., Lewnau, C. J., Bullock, W. P., Wilson, H. M., Dunwell, J. M., Thompson, J. A. and Wang, K. (1994). Production of fertile transgenic maize plants by silicon carbide fiber-mediated transformation. *Plant Journal* **6**, 941–948.

Fromm, M. E., Morrish, F., Armstrong, C., Williams, R., Thomas, J. and Klein, T. M. (1990). Inheritance and expression of chimeric genes in the progeny of transgenic maize plants. *Bio/Technology* **8**, 833–839.

Fujimoto, H., Itoh, K., Yamamoto, M., Kyozuka, J. and Shimamoto, K. (1993). Insect resistant rice generated by introduction of a modified gamma-endotoxin gene of *Bacillus thuringiensis*. *Bio/Technology* **11**, 1151–1155.

Fujisawa, Y., Kato, T., Ohki, S., Ishikawa, A., Kitano, H., Sasaki, T., Asahi, T. and Iwasaki, Y. (1999). Suppression of the heterotrimeric G protein causes abnormal morphology, including dwarfism, in rice. *Proceedings of the National Academy of Sciences(USA)* **96**, 7575–7580.

Funatsuki, H., Kuroda, H., Kihara, M., Lazzeri, P. A., Mueller, E., Loerz, H. and Kishinami, I. (1995). Fertile transgenic barley generated by direct DNA transfer to protoplasts. *Theoretical Applied Genetics* **91**, 707–712.

Furner, I. J., Higgins, E. S. and Berrington, A. W. (1989). Single-stranded DNA transform plant protoplasts. *Molecular and General Genetics* **220**, 65–68.

Geldreich, A., Lebeurier, G. and Hirth, L. (1986). *In vivo* dimerization of cauliflower mosaic virus DNA can explain recombination. *Gene* **48**, 277–286.

Gheysen, G., van Montagu, M. and Zambryski, P. (1987). Integration of *Agrobacterium tumefaciens* transfer DNA (T-DNA) involves rearrangements of target plant DNA sequences. *Proceedings of the National Academy of Sciences (USA)* **84**, 6169–6173.

Gheysen, G., Villarroel, R. and van Montagu, M. (1991). Illegitimate recombination in plants: a model for T-DNA integration. *Genes and Development* **5**, 287–297.

Gheysen, G., Angenon, G. and van Montagu, M. (1998). *Agrobacterium*-mediated plant transformation: a scientifically intriguing story with significant application. *In* "Transgenic Plant Research" (K. Lindsey, ed.), pp. 1–33. Harwood Academic Publishers, Amsterdam.

Gless, C., Loerz, H. and Jaehne-Gaertner, A. (1998). Transgenic oat plants obtained at high efficiency by microprojectile bombardment of leaf base segments. *Journal of Plant Physiology* **152**, 151–157.

Golovkin, M. V., Abraham, M., Morocz, S., Bottka, S., Feher, A. and Dudits, D. (1993). Production of transgenic maize plants by direct DNA uptake into embryonic protoplasts. *Plant Science* **90**, 41–52.

Gorbunova, V. and Levy, A. A. (1997). Non-homologous DNA end joining in plant cells is associated with deletions and filler DNA insertions. *Nucleic Acids Research* **25**, 4650–4657.

Gordon-Kamm, W. J., Baszczynski, C. L., Bruce, W. B. and Tomes, D. T. (1999). Transgenic cereals: *Zea mays* (maize). *In* "Molecular Improvement of Cereal Crops" Advances in Cellular and Molecular Biology of Plants, Vol. 5 (I. K. Vasil, ed.), pp. 189–253. Kluwer Academic Publishers, Dordrecht.

Gordon-Kamm, W. J., Spencer, M. T., Mangano, M.l., Adams, T. R., Daines, R. J., Start, W. G., O'Brien, J. V., Chambers, S. A., Adams, Jr., W. R., Willetts, N. G., Rice, T. B., Mackey, C. J., Krueger, R. W., Kausch, A. P. and Lemaux, P. G. (1990). Transformation of maize cells and regeneration of fertile transgenic plants. *Plant Cell* **2**, 603–618.

Gould, J., Devery, M., Hasegawa, O., Ulian, E. C., Peterson, G. and Smith, R. H. (1991). Transformation of *Zea mays* L. using *Agrobacterium tumefaciens* and the shoot apex. *Plant Physiology* **95**, 426–434.

Graves, A. C.F. and Goldman, S. L. (1987). *Agrobacterium tumefaciens*-mediated transformation of the monocot genus *Gladiolus*: detection of expression of T-DNA encoded genes. *Journal of Bacteriology* **169**, 1745–1746.

Greenland, A. J., Bell, P., Hart, C., Jepson, I., Nevshemal, T., Register, J., III and Wright, S. (1997). Reversible male sterility: a novel system for the production of hybrid corn. *Society of Experimental Biology* **SEB 1044**, 141–147.

Grevelding, C., Fantes, V., Kemper, E., Schell, J. and Masterson, R. (1993). Single-copy T-DNA insertions in *Arabidopsis* are the predominant form of integration in root-derived transgenics, whereas multiple insertions are found in leaf discs. *Plant Molecular Biology* **23**, 847–860.

Grimsley, N. H., Ramos, C., Hein, T. and Hohn, B. (1988). Meristematic tissues of maize plants are most susceptible to agroinfection with maize streak virus. *Bio/Technology* **6**, 185–189.

Grimsley, N. H., Hohn, B., Ramos, C., Kado, C. and Rogowsky, P. (1989). DNA transfer from *Agrobacterium* to *Zea mays* or *Brassica* by agroinfection is dependent on bacterial virulence functions. *Molecular and General Genetics* **217**, 309–316.

Gutierrez, R. A., MacIntosh, G. C. and Green, P. J. (1999). Current perspective on mRNA stability in plants: multiple levels and mechanisms of control. *Trends in Plant Science* **4**, 429–437.

Haensch, R., Mendel, R. R. and Schulze, J. (1998). A rapid and sensitive method to evaluate genotype specific tolerance to phosphinothricin-based selective agents in cereal transformation. *Journal of Plant Physiology* **152**, 145–150.

Hagio, T., Hirabayashi, T., Machii, H. and Tomotsune, H. (1995). Production of fertile transgenic barley (*Hordeum vulgare* L.) plant using the hygromycin-resistance marker. *Plant Cell Reports* **14**, 329–334.

Haldrup, A., Petersen, S. G. and Okkels, F. T. (1998). Positive selection: a plant selection principle based on xylose isomerase, an enzyme used in the food industry. *Plant Cell Reports* **18**, 76–81.

Hamada, T., Sasaki, H., Seki, R. and Sasaki, Y. (1993). Mechanism of chromosomal integration of transgenes in microinjected mouse eggs; sequence analysis of genome-transgene and transgene-transgene junctions at two loci. *Gene* **128**, 197–202.

Hamilton, A. J. and Baulcombe, D. C. (1999). A species of small antisense RNA in posttranscriptional gene silencing in plants. *Science* **286**, 950–952.

Hansen, G. and Chilton M. D. (1999). Lessons in gene transfer to plants by a gifted microbe. *Plant Biotechnology* **240**, 21–57.

Hansen, G. and Wright, M. S. (1999). Recent advances in the transformation of plants. *Trends in Plant Science* **4**, 226–231.

Hanson, B., Engler, D., Moy, J., Newman, B., Ralston, E. and Gutterson, N. (1999). A simple method to enrich an *Agrobacterium*-transformed population for plants containing only T-DNA sequences. *Plant Journal* 19, 727–734.

Harwood, W. A., Ross, S. M., Cilento, P. and Snape, J. W. (2000). The effect of DNA/ gold particle preparation technique, and particle bombardment device, on the transformation of barley (*Hordeum vulgare*). *Euphytica* 111, 67–76.

Hauptmann, R. M., Ozias-Akins, P., Tabaeizadeh, Z., Rogers, S. G., Fraley, R. T., Horsch, R. B. and Vasil, I. K. (1988). Evaluation of selectable markers for obtaining stable transformants in the *Gramineae*. *Plant Physiology* 86, 602–606.

He, G. Y., Rooke, L., Steele, S., Bekes, F., Gras, P., Tatham, A. S., Fido, R., Barcelo, P., Shewry, P. R. and Lazzeri, P. A. (1999). Transformation of pasta wheat (*Triticum turgidum* L. var. *durum*) with high-molecular-weight glutenin subunit genes and modification of dough functionality. *Molecular Breeding* 5, 377–386.

He, G. Y., Lazzeri, P. A. and Cannell, M. E. (2001). Fertile transgenic plants obtained from tritordeum inflorescences by tissue electroporation. *Plant Cell Reports* (in press).

Henikoff, S. and Comai, L. (1998). A DNA methyltransferase homolog with a chromodomain exists in multiple polymorphic forms in *Arabidopsis*. *Genetics* 149, 307–318.

Hiei, Y., Ohta, S., Komari, T. and Kumashiro, T. (1994). Efficient transformation of rice (*Oryza sativa* L.) mediated by *Agrobacterium* and sequence analysis of the boundaries of the T-DNA. *Plant Journal* 6, 271–282.

Hiei, Y., Komari, T. and Kubo, T. (1997). Transformation of rice mediated by *Agrobacterium tumefaciens*. *Plant Molecular Biology* 35, 205–218.

Holmes-Davis, R. and Comai, L. (1998). Nuclear matrix attachment regions and plant gene expression. *Trends in Plant Science* 3, 91–97.

Hooykaas van Slogteren, G. M.S., Hooykaas P. J.J. and Schilperoort, R. A. (1984). Expression of Ti-plasmid genes in monocotyledonous plants infected with *Agrobacterium tumefaciens*. *Nature* 311, 763–764.

Iida, A., Seki, M., Kamada, M., Yamada, Y. and Morikawa, K. (1990). Gene delivery into cultured plant cells by DNA-coated gold particles accelerated by a pneumatic particle gun. *Theoretical Applied Genetics* 80, 813–816.

Ingelbrecht, I., van Houdt, H., van Montague, M. and Depicker, A. (1994) Posttranscriptional silencing of reporter transgenes in tobacco correlates with DNA methylation. *Proceedings of the National Academy of Sciences (USA)* 91, 10502–10506.

Iser, M., Fettig, S., Scheyhing, F., Viertel, K. and Hess, D. (1999). Genotype-dependent stable genetic transformation in German spring wheat varieties selected for high regeneration potential. *Journal of Plant Physiology* 154, 509–516.

Ishida, Y., Saito, H., Ohta, S., Hiei, Y., Komari, T. and Kumashiro, T. (1996). High efficiency transformation of maize (*Zea mays* L.) mediated by *Agrobacterium tumefaciens*. *Nature Biotechnology* 14, 745–750.

Itoh, K., Nakajima, M. and Shimamoto, K. (1997). Silencing of *waxy* genes in rice containing *Wx* transgenes. *Molecular and General Genetics* 255, 351–358.

Jaehne, A., Becker, D., Brettschneider, R. and Loerz, H. (1994). Regeneration of transgenic microspore-derived fertile barley. *Theoretical Applied Genetics* 89, 525–533.

Jefferson, R. (1987) Assaying chimeric genes in plants: the GUS gene fusion system. *Plant Molecular Biology Reporter* 5, 387–405.

Jeddeloh, J. A., Stokes, T. L. and Richards, E. J. (1999). Maintenance of genomic methylation requires a SW12/SNF2-like protein. *Nature Genetics* 22, 94–97.

Jelenska, J., Tietze, E., Tempe, J. and Brevet, J. (2000). Streptothricin resistance as a novel selectable marker for transgenic plant cells. *Plant Cell Reports* **19**, 298–303.

Jensen, L. G., Olsen, O., Kops, O., Wolf, N., Thomsen, K. K. and Von Wettstein, D. (1996). Transgenic barley expressing a protein-engineered, thermostable (1,3–1,4)-beta-glucanase during germination. *Proceedings of the National Academy of Sciences (USA)* **93**, 3487–3491.

Joersbo, M. and Okkels, F. T. (1996). A novel principle for selection of transgenic plant cells: positive selection. *Plant Cell Reports* **16**, 219–221.

Jones, P. L., Veenstra, G. J.C., Wade, P. A., Vermaak, D., Kass, S. U., Landsberger, N., Strouboulis, J. and Wolffe, A. P. (1998). Methylated DNA and MeCP2 recui histone deacetylase to repress transcription. *Nature Genetics* **19**, 187–191.

Kaeppler, H. F., Gu, W., Somers, D. A., Rines, H. W., Cockburn, A. F. (1990). Silicon carbide fiber-mediated DNA delivery into plant cells. *Plant Cell Reports* **9**, 415–418.

Kaeppler, H. F., Somers, D. A., Rines, H. W. and Cockburn, A. F. (1992). Silicon carbide fibre-mediated stable transformation of plant cells. *Theoretical Applied Genetics* **84**, 560–566.

Karp, A. (1991). On the current understanding of somaclonal variation. *Oxford Surveys of Plant Molecular and Cell Biology* **7**, 1–58.

Kartzke, S., Saedler, H. and Meyer, P. (1990). Molecular analysis of transgenic plants derived from transformations of protoplasts at various stages of the cell cycle. *Plant Science* **67**, 63–72.

Kikkert, J. R. (1993). The Biolistic PDS-1000/He device. *Plant Cell Tissue and Organ Culture* **33**, 221–226.

Kim, J. K., Duan, X. L., Wu, R., Seok, S. J., Boston, R. S., Jang, I. C., Eun, M. Y. and Nahm, B. H. (1999). Molecular and genetic analysis of transgenic rice plants expressing the maize ribosome-inactivating protein b-32 gene and the herbicide resistance bar gene. *Molecular Breeding* **5**, 85–94.

Klein, T. M. and Fitzpatrick-McElligott, S. (1993). Particle bombardment: a universal approach for gene transfer to cells and tissues. *Current Opinion in Biotechnology* **4**, 583–590.

Klein, T. M., Wolf. E. D., Wu, R. and Sanford, J. C. (1987). High-velocity microprojectiles for delivery of nucleic acids into living cells. *Nature* **327**, 70–73.

Kohli, A., Leech, M., Vain, P., Laurie, D. A. and Christou, P. (1998). Transgene organization in rice engineered through direct DNA transfer supports a two-phase integration mechanism mediated by the establishment of integration hot spots. *Proceedings of the National Academy of Sciences (USA)* **95**, 7203–7208.

Kohli, A., Gahakwa, D., Vain, P., Laurie, D. A. and Christou, P. (1999a). Transgene expression in rice engineered through particle bombardment: Molecular factors controlling stable expression and transgene silencing. *Planta* **208**, 88–97.

Kohli, A., Griffiths, S., Palacios, N., Twyman R. M., Vain, P., Laurie, D. A. and Christou, P. (1999b). Molecular characterization of transforming plasmid rearrangements in transgenic rice reveals a recombination hotspot in the CaMV 35S promoter and confirms the predominance of microhomology mediated recombination. *Plant Journal* **17**, 591–601.

Komari, T. (1990). Transformation of cultured cells *Chenopodium quinoa* by binary vectors that carry a fragment of DNA from the virulence region of pTiBo542. *Plant Cell Reports* **9**, 303–306.

Komari, T., Hiei, Y., Saito, Y., Murai, N. and Kumashiro, T. (1996). Vectors carrying two separate T-DNAs for co-transformation of higher plants mediated by *Agrobacterium tumefaciens* and segregation of transformants free from selection markers. *Plant Journal* **10**, 165–174.

Komari, T., Hiei, Y., Ishida, Y., Kumashiro, T. and Kubo, T. (1998). Advances in cereal gene transfer. *Current Opinion in Plant Biology* **1**, 161–165.

Koncz, C., Martini, N., Mayerhofer, R., Koncz-Kalman, Z., Koerber, H., Redei, G. P. and Schell, J. (1989). High-frequency T-DNA-mediated gene tagging in plants. *Proceedings of the National Academy of Sciences (USA)* **86**, 8467–8471.

Kononov, M. E., Bassuner, B. and Gelvin, S. V. (1997). Integration of T-DNA binary vector "backbone" sequences into the tobacco genome: evidence for multiple complex patterns of integration. *Plant Journal* **11**, 945–957.

Kooter, J. M., Matzke, M. A. and Meyer, P. (1999). Listening to the silent genes, transgene silencing, gene regulation and pathogen control. *Trends in Plant Science* **4**, 340–347.

Koprek, T. M., Haensch, R., Nerlich, A., Mendel, R. R. and Schulze, J. (1996). Fertile transgenic barley of different cultivars obtained by adjustment of bombardment conditions to tissue response. *Plant Science* **119**, 79–91.

Koziel, M. G., Beland, G. L., Bowman, C., Carozzi, N. B., Crenshaw, R., Crossland L., Dawson, J., Desai, N., Hill, M., Kadwell, S., Launis, K., Lewis, K., Maddox, D., McPherson, K., Meghji, M. R., Merlin, E., Rhodes, R., Warren, G. W., Wright, M. and Evola, S. V. (1993). Field performance of elite transgenic maize plants expressing an insecticidal protein derived from *Bacillus thuringiensis*. *Bio/Technology* **11**, 194–200.

Kramer, C., DiMaio, J., Carswell, G. K. and Shillito, R. D. (1993). Selection of transformed protoplast-derived *Zea mays* colonies with phosphinothricin and a novel assay using the pH indicator chlorophenol red. *Planta* **190**, 454–458.

Kumpatla, S. P. and Hall, T. C. (1999). Organizational complexity of a rice transgene locus susceptible to methylation-based silencing. *IUBMB Life* **48**, 459–467.

Kunkel, T., Niu, Q. W., Chan, Y. S. and Chua, N. H. (1999). Inducible isopentenyl transferase as a high-efficiency marker for plant transformation. *Nature Biotechnology* **17**, 916–919.

Lamacchia, C., Shewry, P. R., Di Fonzo, N., Harris, N., Lazzeri, P. A., Napier, J. A., Halford, N. G. and Barcelo, P. (2001). Tissue-specific activity of a seed protein gene promoter in transgenic wheat endosperms – characterization of an important tool for wheat biotechnology. *Journal of Experimental Botany* (in press).

Last, D. I., Brettell, R. I.S., Chamberlain, D. A., Chaudhury, A. M., Larkin, P. J., Marsh, E. L., Peacock, W. J. and Dennis, E. S. (1991). pEmu: an improved promoter for gene expression in cereal cells. *Theoretical Applied Genetics* **81**, 581–588.

Laursen, C. M., Krzyzek, R. A., Flick, C. E., Anderson, P. C. and Spencer, T. M. (1994). Production of fertile transgenic maize by electroporation of suspension culture cells. *Plant Molecular Biology* **24**, 51–61.

Lazzeri, P. A. and Shewry, P. R. (1993) Biotechnology of cereals. *In* "Biotechnology and Genetic Engineering Reviews" (M. Tombs, ed.), pp. 79–146. Intercept Ltd, Andover.

Leckband, G. and Loerz, H. (1998). Transformation and expression of a stilbene synthase gene of *Vitis vinifera* L. in barley and wheat for increased fungal resistance. *Theoretical Applied Genetics* **96**, 1004–1012.

Lemaux, P. G., Cho, M. J., Zhang, S. and Bregitzer, P. (1999). Transgenic cereals: *Hordeum vulgare* L. (barley). *In* "Molecular Improvement of Cereal Crops" (I. K. Vasil, ed.), pp. 255–316. Kluwer Academic Publishers, Dordrecht.

Lepiniec, L., Babiychuk, E., Kushnir, S., van Montagu, M. and Inze, D. (1995). Characterization of an *Arabidopsis thaliana* cDNA homologue to animal poly(ADP-ribose) polymerase. *FEBS Letters* **364**, 103–108.

Li, L., Rongda, Q., Kochko, de A., Fauguet, C. M. and Beachy, R. N. (1993). An improved rice transformation system using the biolistic method. *Plant Cell Reports* **12**, 250–255.

Li, Z., Upadhyaya, N. M., Meena, S., Gibbs, A. J. and Waterhouse, P. M. (1997). Comparison of promoters and selectable marker genes for use in *Indica* rice transformation. *Molecular Breeding* **3**, 1–14.

Lonsdale, D. M., Lindup, S., Moisan, L. J., Harvey, A. J. (1998). Using firefly luciferase to identify the transition from transient to stable expression in bombarded wheat scutellar tissue. *Physiologia Plantarum* **102**, 447–453.

Lowe, K., Bowen, B., Hoerster, G., Ross, M., Bond, D., Pierce, D. and Gordon-Kamm, W. (1995). Germline transformation of maize following manipulation of chimeric shoot meristems. *Bio/Technology* **13**, 677–682.

Lowe, K., Abbitt, S., Glassman, K., Gregory, C., Hoerster, G., Sun, X., Rasco-Gaunt, S., Lazzeri, P. and Gordon-Kamm, B. (2000). Use of maize LEC1 to improve transformation. "2000 World Congress on *In Vitro* Biology", San Diego, California, USA, 10–15 June 2000.

Luehrsen, K. R. and Walbot, V. (1991). Intron enhancement of gene expression and the splicing efficiency of introns in maize cells. *Molecular and General Genetics* **225**, 81–93.

Lugones, L. G., Scholtmeijer K., Klootwijk, R. and Wessels, J. G.H. (1999). Introns are necessary for mRNA accumulation in *Schizophyllum commune*. *Molecular Microbiology* **32**, 681–689.

Lyznik, L. A., Mitchell, J. C., Hiriyama, L. and Hodges, T. K. (1993). Activity of yeast FLP recombinase in maize and rice protoplasts. *Nucleic Acids Research* **21**, 969–975.

Lyznik, L. A., Rao, K. V. and Hodges, T. K. (1996). FLP-mediated recombination of FRT sites in the maize genome. *Nucleic Acids Research* **24**, 3784–3789.

MacDonald, M. H., Mogen, B. D. and Hunt, A. G. (1991). Characterization of the polyadenylation signals from the T-DNA-encoded octopine synthase gene. *Nucleic Acids Research* **19**, 5575–5581.

Maqbool, S. B. and Christou, P. (1999). Multiple traits of agronomic importance in transgenic indica rice plants: analysis of transgene integration patterns, expression levels and stability. *Molecular Breeding* **5**, 471–480.

Mascarenhas, D., Mettler, I. J., Pierce, D. A. and Lowe, H. W. (1990). Intron-mediated enhancement of heterologous gene expression in maize. *Plant Molecular Biology* **15**, 913–920.

Matassi, G., Montero, L. M., Salinas, J. and Bernardi, G. (1989). The isochore organization and the compositional distribution of homologous coding sequences in the nuclear genome of plants. *Nucleic Acids Research* **17**, 5273–5290.

Matsumoto, S., Ito, Y., Hosoi, T., Takahashi, Y. and Machida, Y. (1990). Integration of *Agrobacterium* T-DNA into a tobacco chromosome; possible involvement of DNA homology between T-DNA and plant DNA. *Molecular and General Genetics* **224**, 309–316.

Matsushita, J., Otani, M., Wakita, Y., Tanaka, O. and Shimada, T. (1999). Transgenic plant regeneration through silicon carbide whisker-mediated transformation of rice (*Oryza sativa* L.). *Breeding Science* **49**, 21–26.

McCabe, D. and Christou, P. (1993) Direct DNA transfer using electric discharge particle acceleration (ACCELL technology). *Plant Cell Tissue and Organ Culture* **33**, 227–236.

McElroy, D., Zhang, W., Ca, J. and Wu, R. (1990) Isolation of an efficient Actin promoter for use in rice transformation. *Plant Cell* **2**, 163–171.

McElroy, D., Blowers, A. D., Jenes, B. and Wu, R. (1991). Construction of expression vectors based on the rice actin 1 (*Act1*) 5′ region for use in monocot transformation. *Molecular and General Genetics* **231**, 150–160.

Mentewab, A., Letellier, V., Marque, C. and Sarrafi, A. (1999). Use of anthocyanin biosynthesis stimulatory genes as markers for the genetic transformation of haploid embryos and isolated microspores in wheat. *Cereal Research Communication* **27**, 17–24.

Mette, M. F., van der Winden, J., Matzke, M. A. and Matzke, A. J.M. (1999). Production of aberrant promoter transcripts contributes to methylation and silencing of unlinked homologous promoters *in trans*. *EMBO Journal* **18**, 241–248.

Michalowski, S. M., Allen, G. C., Hall, G. E., Thompson, W. F. and Spiker, S. (1999). Characterization of randomly-obtained matrix attachment regions (MARs) from higher plants. *Biochemistry* **38**, 12795–12804.

Montgomery, M. K. and Fire, A. (1998). Double-stranded RNA as a mediator in sequence-specific genetic silencing and co-suppression. *Trends in Genetics* **14**, 255–258.

Montgomery, M. K., Xu, S. and Fire, A. (1998). RNA as a target of double-stranded RNA-mediated genetic interference in *Caenorhabditis elegans*. *Proceedings of the National Academy of Sciences (USA)* **95**, 15502–15507.

Monti, D., Cossarizza, A., Salvioli, S., Francesqui, C., Rainaldi, G., Straface, E., Rivabene, R. and Malorni, W. (1994). Cell-death protection by 3-amino-benzamide and other poly(ADP-ribose) polymerase inhibitors: different effects of human natural killer and lymphokine-activated killer cell activities. *Biochemistry and Biophysics Research Communications* **199**, 525–530.

Morikawa, H., Iida, A., Yamada, Y. (1989). Transient expression of foreign genes in plant cells and tissues obtained by a simple biolistic device (particle gun). *Applied Microbiology and Biotechnology* **31**, 320–322.

Murashige, T. and Skoog, F. (1962). A revised medium for rapid growth and bioassays with tobacco tissue cultures. *Plant Physiology* **15**, 473–497.

Murray, E. E., Lotzer, J. and Eberle, M. (1989). Codon usage in plant genes. *Nucleic Acids Research* **17**, 477–493.

Nandadeva, Y. L., Lupi, C. G., Meyer, C. S., Devi, P. S., Potrykus, I. and Bilang, R. (1999). Microprojectile-mediated transient and integrative transformation of rice embryogenic suspension cells: effects of osmotic cell conditioning and of the physical configuration of plasmid DNA. *Plant Cell Reports* **18**, 500–504.

Nehra, N. S., Chibbar, R. N., Leung, N., Caswell, K., Mallard, C., Steinhauer, L., Bagga, M. and Kartha, K. K. (1994). Self-fertile transgenic wheat plants regenerated from isolated scutellar tissues following microprojectile bombardment with two distinct gene constructs. *Plant Journal* **5**, 285–297.

Oard, J. (1993). Development of an airgun device for particle bombardment. *Plant Cell Tissue and Organ Culture* **33**, 247–250.

Oard, J. H., Linscombe, S. D., Braverman, M. P., Jodari, F., Blouin, D. C., Leech, M., Kohli, A., Vain, P., Cooley, J. C., and Christou, P. (1996). Development of field evaluation, and agronomic performance of transgenic herbicide resistant rice. *Molecular Breeding* **2**, 359–368.

Odell, J. T., Nagy, F. and Chua, N. (1985). Identification of DNA sequences required for activity of the cauliflower mosaic virus 35S promoter. *Nature* **313**, 810–812.

Ohme-Takagi, M., Taylor, C. B., Newman, T. C. and Green, P. J. (1993). The effect of sequences with high AU content on mRNA stability in tobacco. *Proceedings of the National Academy of Sciences (USA)* **90**, 11811–11815.

Ortiz, J. P.A., Reggiardo, M. I., Ravizzini, R. A., Altabe, S. G., Cervigni, G. D.L., Spitteler, M. A., Morata, M. M., Elias, F. E. and Vallejos, R. H. (1996). Hygromycin resistance as an efficient selectable marker for wheat stable transformation. *Plant Cell Reports* **15**, 877–881.

Palauqui, J. C., Elmayan, T., Pollien, J.-M. and Vaucheret, H. (1997). Systemic acquired silencing: transgene-specific post transcriptional silencing is transmitted by grafting from silenced stocks to non-silenced scions. *EMBO Journal* **16**, 4738–4745.

Pang, S., De Boer, D. L., Wan, Y., Ye, G., layton, J. G., Neher, M. K., Armstrong, C. L., Fry, J. E., Hinchee, M. A.W. and Fromm, M. E. (1996). An improved green fluorescent protein gene as a vital marker in plants. *Plant Physiology* **112**, 893–900.

Park, Y-D., Papp. I., Moscone E. A., Iglesias V. A., Vaucheret H., Matzke A. J. and Matzke M. A. (1996). Gene silencing mediated by promoter homology occurs at the level of transcription and results in meiotically heritable alterations in methylation and gene activity. *Plant Journal* **9**, 183–194.

Paszkowski, J., Shillito, R. D., Saul, M., Mandak, V., Hohn, T., Hohn B. and Potrykus, I. (1984). Direct gene transfer to plants. *EMBO Journal* **3**, 2717–2722.

Pawlowski, W. P. and Somers, D. A. (1998).Transgenic DNA integrated into the oat genome is frequently interspersed by host DNA. *Proceedings of the National Academy of Sciences (USA)* **95**, 12106–12110.

Perl, A., Kless, H., Blumenthal, A., Galili, G. and Galun, E. (1992). Improvement of plant regeneration and GUS expression in scutellar wheat calli by optimization of culture conditions and DNA-microprojectile delivery procedures. *Molecular and General Genetics* **235**, 279–284.

Pröls, F. and Meyer, P. (1992). The methylation patterns of chromosomal integration regions influence gene activity of transferred DNA in *Petunia hybrida*. *Plant Journal* **9**, 183–194.

Puchta, H., Swoboda, P. and Hohn, B. (1995). Induction of intrachromosomal recombination in whole plants. *Plant Journal* **7**, 203–210.

Raineri, D. M., Bottino, P., Gordon, M. P. and Nester, E. W. (1990). *Agrobacterium*-mediated transformation of rice. *Bio/Technology* **8**, 33–38.

Rao, K. V., Rathore, K. S., Hodges, T. K., Fu, X., Stoger, E., Sudhakar, D., Williams, S., Christou, P., Bharathi, M., Bown, DP., Powell, K. S., Spence, J., Gatehouse, A. M.R. and Gatehouse, J. A. (1998). Expression of snowdrop lectin (GNA) in transgenic rice plants confers resistance to rice brown planthopper. *Plant Journal* **15**, 469–477.

Rasco-Gaunt, S., Riley, A., Barcelo, P. and Lazzeri, P. A. (1999c). A facile method for screening for phosphinothricin (PPT)-resistant transgenic wheats. *Molecular Breeding* **5**, 255–262.

Rasco-Gaunt, S., Grierson, D., Lazzeri, P. A. and Barcelo, P. (1999a). Genetic manipulation of S-adenosyl methionine decarboxylase (SAMDC) in wheat. *In* "Proceedings of the 9th International Wheat Genetics Symposium", Saskatoon, Canada (A. E. Slinkard, ed.), pp. 262–264.

Rasco-Gaunt, S. and Barcelo, P. (1999). Immature inflorescence culture of cereals: a highly responsive system for regeneration and transformation. *In* "Methods in Molecular Biology: Plant Cell Culture Protocols" (R. D. Hall, ed.), pp. 71–81. Humana Press Inc., Totowa, NJ.

Rasco-Gaunt, S., Riley, A., Barcelo, P. and Lazzeri, P. A. (1999b). Analysis of particle bombardment parameters to optimise DNA delivery into wheat tissues. *Plant Cell Reports* **19**, 118–127.

Rasco-Gaunt, S., Riley, A., Cannell, M., Barcelo, P. and Lazzeri, P. A. (2001). Procedures allowing the transformation of a range of European elite wheat varieties via particle bombardment. *Journal of Experimental Botany* (in press).

Rashid, H., Yokoi, S., Toriyama, K. and Hinata, K. (1996). Transgenic plant production mediated by *Agrobacterium* in indica rice. *Plant Cell Reports* **15**, 727–730.

Rech, E., Vainstein, M. and Davey, M. R. (1991). An electrical particle acceleration gun for gene transfer into cells. *Technique* **3**, 143–149.

Reed, J. N., Chang, Y. F., D. D., McNamara, Beer, S. and Miles, P. J. (1999). High frequency transformation of wheat with the selectable marker mannose-6-phosphate isomerase (PMI). *In Vitro* **35**, 57A.

Register, J. C., Peterson, D. J., Bell, P. J., Bullock, W. P., Evans, I. J., Frame, B., Greenland, A. J., Higgs, N. S., Jepson, I., Jiao, S., Lewnau, C. J., Sillick, J. M. and Wilson, H. M. (1994). Structure and function of selectable and non-selectable transgenes in maize after introduction by particle bombardment. *Plant Molecular Biology* **25**, 951–961.

Rethmeier, N., Kramer, E., van Montagu, M. and Cornelissen, M. (1998). Identification of cat sequences required for intron-dependent gene expression in maize cells. *Plant Journal* **13**, 831–835.

Rhodes, C. A., Pierce, D. A., Mettler, I. J., Mascarenhas, D. and Detmer, J. J. (1988). Genetically transformed maize plants from protoplasts. *Science* **240**, 204–207.

Ritala, A., Apegren, K., Kurten, U., Salmenkallio-Marttila, M., Mannonen, L., Hannus, R., Kauppinene, V., Teeri, T. H. and Enari, T. (1994). Fertile transgenic barley by particle bombardment of immature embryos. *Plant Molecular Biology* **24**, 317–325.

Rothstein, S. J., Dimaio, J., Strand, M. and Rice, D. (1987). Stable and inheritable inhibition of the expression of nopaline synthase in tobacco expressing antisense RNA. *Proceedings of the National Academy of Sciences (USA)* **84**, 8439–8443.

Rouwendal, G. J.A., Mendes, O., Wolbert, E. J.H. and De Boer, A. D. (1997). Enhanced expression in tobacco of the gene encoding green fluorescent protein by modification of its codon usage. *Plant Molecular Biology* **33**, 989–999.

Russell, D. A. and Fromm, M. E. (1997). Tissue-specific expression in transgenic maize of four endosperm promoters from maize and rice. *Transgenic Research* **6**, 157–168.

Salgueiro, S., Gil, J., Savazzini, F., Steele, S., Lazzeri, P. A. and Barcelo, P. (1998). Tagging of regulatory elements in cereals. *In* "Proceedings of the 9th International Wheat Genetics Symposium", Vol. 3, Saskatoon, Canada, 2–7 August, pp. 151–153. University Extension Press, University of Saskatchewan.

Salomon, S. and Puchta, H. (1998). Capture of genomic and T-DNA sequences during double-strand break repair in somatic plant cells. *EMBO Journal* **17**, 6086–6095.

Sanford, J. C., Klein, T. M., Wolf, E. D. and Allen, N. (1987). Delivery of substances into cells and tissues using a particle bombardment process. *Journal of Particle Science and Technology* **5**, 27–37.

Sanford, J. C., Smith, F. D. and Russell, J. A. (1993). Optimising the biolistic process for different biological applications. *Methods Enzymology* **217**, 483–509.

Sautter, C. (1993) Development of a microtargetting device for particle bombardment of plant meristems. *Plant Cell Tissue and Organ Culture* **33**, 251–257.

Schaefer, W., Goerz, A. and Kahl, G. (1987). T-DNA integration and expression in a monocot crop plant after induction of *Agrobacterium*. *Nature* **327**, 529–530.

Schiebel, W., Pélissier, T., Riedel, L., Thalmeir, S., Schiebel, R., Kempe, D., Lottspeich, F., Sänger, H. L. and Wassenegger, M. (1998). Isolation of an RNA-directed RNA polymerase-specific cDNA clone from tomato. *Plant Cell* **10**, 2087–2101.

Scrable, H. and Stambrook, P. J. (1999). A genetic program for deletion of foreign DNA from the mammalian genome. *Mutation Research – Fundamental and Molecular Mechanisms of Mutagenesis* **429**, 225–237.

Serik, O., Ainur, I., Murat, K., Tetsuo, M. and Masaki, I. (1996). Silicon carbide fiber-mediated DNA delivery into cells of wheat (*Triticum aestivum* L.) mature embryos. *Plant Cell Reports* **16**, 133–136.

Shimamoto, K., Terada, R., Izawa, T. and Fujimoto, H. (1989). Fertile transgenic rice plants regenerated from transformed protoplasts. *Nature* **338**, 274–276.

Sivamani, E., Shen, P., Opalka, N., Beachy, R. N. and Fauquet, C. M. (1996). Selection of large quantities of embryonic calli from indica rice seeds for production of fertile transgenic plants using the biolistic method. *Plant Cell Reports* **15**, 322–327.

Smith, R. H. and Hood, E. E. (1995). Review and interpretation: *Agrobacterium tumefaciens* transformation of monocotyledons. *Crop Science* **35**, 301–309.

Somers, D., Rines, H. W., Gu, W., Kaeppler, H. F. and Bushnell, W. R. (1992). Fertile transgenic oat plants. *Bio/Technology* **10**, 1589–1594.

Songstad, D. D., Somers, D. A. and Griesbach, R. J. (1995). Advances in alternative DNA delivery techniques. *Plant Cell Tissue and Organ Culture* **40**, 1–15.

Srivastava, V., Anderson, O. D. and Ow, D. W. (1999). Single-copy transgenic wheat generated through the resolution of complex integration patterns. *Proceedings of the National Academy of Sciences (USA)* **96**, 11117–11121.

Stam, M., de Bruin, R., van Blockland, R., van der Hoorn, A. L., Mol, J. N.M., and Kooter, J. M. (2000). Distinct features of post-transcriptional gene silencing by antisense transgenes in single copy and inverted T-DNA repeat loci. *Plant Journal* **21**, 27–42.

Stam, M., Viterbo, A., Mol, J. N.M. and Kooter, J. M. (1998). Position-dependent methylation and transcriptional silencing of transgenes in inverted T-DNA repeats, Implications for posttranscriptional silencing of homologous host genes in plants. *Molecular and Cell Biology* **18**, 6165–6177.

Stark-Lorenzen, P., Nelke, B., Haenssler, G., Muehlbach, H. P. and Thomzik, J. E. (1997). Transfer of a grapevine stilbene synthase gene to rice (*Oryza sativa* L.). *Plant Cell Reports* **16**, 668–673.

Stoger, E., Williams, S., Keen, D. and Christou, P. (1998). Molecular characterization of transgenic wheat and the effect on transgene expression. *Transgenic Research* **7**, 463–471.

Stoger, E., Williams, S., Christou, P., Down, R. E. and Gatehouse, J. A. (1999a). Expression of the insecticidal lectin from snowdrop (*Galanthus nivalis* agglutinin; GNA) in transgenic wheat plants: effects on predation by the grain aphid *Sitobion avenae*. *Molecular Breeding* **5**, 65–73.

Stoger, E., Williams, S., Keen, D. and Christou, P. (1999b). Constitutive versus seed specific expression in transgenic wheat: temporal and spatial control. *Transgenic Research* **8**, 73–82.

Sugita, K., Matsunaga, E. and Ebinuma, H. (1999). Effective selection system for generating marker-free transgenic plants independent of sexual crossing. *Plant Cell Reports* **18**, 941–947.

Tada, Y., Nakase, M., Adachi, T., Nakamura, R., Shimada, H., Takahashi, M., Fujimura, T. and Matsuda, T. (1996). Reduction of 14–16kDa proteins in transgenic rice plants by antisense gene. *FEBS Letters* **391**, 341–345.

Takano, M., Egawa, H., Ikeda, J. E. and Wakasa, K. (1997). The structures of integration sites in transgenic rice. *Plant Journal* 11, 353–361.

Takeuchi, Y., Dotson, M. and Keen, N. T. (1992). Plant transformation: a simple particle device based on flowing helium. *Plant Molecular Biology* 18, 835–839.

Takumi, S. and Shimada, T. (1996). Production of transgenic wheat through particle bombardment of scutellar tissues: frequency is influenced by culture duration. *Journal of Plant Physiology* **149**, 418–423.

Tanaka, A., Mita, S., Ohta, S., Kyozuka, J., Shimamoto, K. and Nakamura, K. (1990). Enhancement of foreign gene expression by a dicot intron in rice but not in tobacco is correlated with an increased level of mRNA and an efficient splicing of the intron. *Nucleic Acids Research* 18, 6767–6770.

Tang, K. X., Tinjuangjun, P., Xu, Y., Sun, X. F., Gatehouse, J. A., Ronald, P. C., Qi, H. X., Lu, X. G., Christou, P. and Kohli, A. (1999). Particle-bombardment-mediated co-transformation of elite Chinese rice cultivars with genes conferring resistance to bacterial blight and sap-sucking insect pests. *Planta* **208**, 552–563.

Thomas, C. M., Jones, D. A., English, J. J., Carrol, B. J., Bennetzen, J. L., Harrison, K., Burbidge, A., Bishop, G. J. and Jones, J. D.G. (1994). Analysis of the chromosomal distribution of transposon-carrying T-DNAs in tomato using the inverse polymerase chain reaction. *Molecular and General Genetics* **242**, 573–585.

Tingay, S., McElroy, D., Kalla, R., Fieg, S., Wang, M., Thornton, S. and Brettell, R. (1997). *Agrobacterium tumefaciens*-mediated barley transformation. *Plant Journal* **11**, 1369–1376.

Toki, S. (1997). Rapid and efficient *Agrobacterium*-mediated transformation in rice. *Plant Molecular Biology Reporter* **15**, 16–21.

Topping, J. F., Wei, W. and Lindsey, K. (1991). Functional tagging of regulatory elements in the plant genome. *Development* **112**, 1009–1019.

Torbert, K. A., Rines, H. W. and Somers, D. A. (1995). Use of paromomycin as a selective agent for oat transformation. *Plant Cell Reports* **14**, 635–640.

Torbert, K. A., Rines, H. W. and Somers, D. A. (1998). Transformation of oat using mature embryo-derived tissue cultures. *Crop Science* **38**, 226–231.

US Department of Agriculture (1998). Cah marker aids plant transformation. *Agricultural Research* **46**, 12–13.

Uze, M., Potrykus, I. and Sautter, C. (1999). Single-stranded DNA in the genetic transformation of wheat (*Triticum aestivum* L.): transformation frequency and integration pattern. *Theoretical Applied Genetics* **99**, 487–495.

Vaden, V. R. and Melcher, U. (1990). Recombination sites in CaMV DNAs: implications for mechanisms of recombination. *Virology* **177**, 717–726.

Vain, P., Keen, N., Murillo, J., Rathus, C., Nemes, C. and Finer, J. J. (1993a). Development of the particle inflow gun. *Plant Cell Tissue and Organ Culture* **33**, 237–246.

Vain, P., McMullen, M. D. and Finer, J. J. (1993b). Osmotic treatment enhances particle bombardment-mediated transient and stable transformation of maize. *Plant Cell Reports* **12**, 84–88.

Vain, P., Worland, B., Kohli, A., Snape, J. W. and Christou, P. (1998). The green fluorescent protein (GFP) as a vital screenable marker in rice transformation. *Theoretical Applied Genetics* **96**, 164–169.

Vain, P., Worland, B., Kohli, A., Snape, J. W., Christou, P., Allen G. C. and Thompson, W. F. (1999). Matrix attachment regions increase transgene expression levels and stability in transgenic rice plants and their progeny. *Plant Journal* **18**, 233–242.

van Blockland, R., ten Lohuis, M. and Meyer, P. (1997). Condensation of chromatin in transcriptional regions of an inactivated plant transgene, evidence for an active role of transcription in gene silencing. *Molecular and General Genetics* **257**, 1–13.

van Eldik, G. J., Litiere, K., Jacobs, J. J., van Montagu, M. and Cornelissen, M. (1998). Silencing of β-1,3-glucanase genes in tobacco correlates with an increased abundance of RNA degradation intermediates. *Nucleic Acids Research* **26**, 5176–5181.

Vasil, V., Castillo, A. M., Fromm, M. E. and Vasil, I. K. (1992). Herbicide resistant fertile transgenic wheat plants obtained by microprojectile bombardment of regenerable embryonic callus. *Bio/Technology* **10**, 667–674.

Vaucheret, H., Béclin, C., Elmayan, T., Feuerbach, F., Godon, C., Morel, J-B., Mourrain, P., Palauqui, J-C. and Vernhettes, S. (1998). Transgene-induced gene silencing in plants. *Plant Journal* **16**, 651–659.

Waldman, B. C. and Waldman, A. S. (1990). Illegitimate and homologous recombination in mammalian cells: differential sensitivity to an inhibitor of poly(ADP-ribosylation). *Nucleic Acids Research* **18**, 5981–5988.

Wan, Y. and Lemaux, P. G. (1994). Generation of large numbers of independently transformed fertile barley plants. *Plant Physiology* **104**, 37–48.

Wang, J., Lewis, M. E., Whallon, J. H. and Sink, K. C. (1995a). Chromosomal mapping of T-DNA inserts in transgenic *Petunia* by *in situ* hybridization. *Transgenic Research* **4**, 241–246.

Wang, K., Drayton, P., Frame, B., Dunwell, J. and Thompson, J. (1995b). Whisker-mediated plant transformation: an alternative technology. *In Vitro Cell and Developmental Biology* **31**, 101–104.

Wassenegger, M. and Pélissier, T. (1998). A model for RNA-mediated gene silencing in higher plants. *Plant Molecular Biology* **37**, 349–362.

Wassenegger, M. and Pélissier, T. (1999). Signalling in gene silencing. *Trends in Plant Science* **4**, 207–209.

Waterhouse, P. M., Graham, M. W. and Wang, M.-B. (1998). Virus resistance and gene silencing in plants can be induced by simultaneous expression of sense and antisense RNA. *Proceedings of the National Academy of Sciences (USA)* **95**, 13959–13964.

Waterhouse. P. M., Smith, N. A. and Wang, M.-B. (1999). Virus resistance and gene silencing, killing the messenger. *Trends in Plant Science* **4**, 452–457.

Weeks, J. T., Anderson, O. D. and Blechl, A. E. (1993). Rapid production of multiple independent lines of fertile transgenic wheat (*Triticum aestivum*). *Plant Physiology* **102**, 1077–1084.

Xu, D., Lei, M. and Wu, R. (1995). Expression of the rice *Osgrp1* promoter-Gus reporter gene is specifically associated with cell elongation/expansion and differentiation. *Plant Molecular Biology* **28**, 455–471.

Xu, X. and Li, B. (1994). Fertile transgenic Indica rice plants obtained by electroporation of seed embryo cells. *Plant Cell Reports* **13**, 237–242.

Xu, Y., Zhu, Q., Panbangreb, W., Shirasu, K. and Lamb, C. (1996). Regulation, expression and function of a new basic chitinase gene in rice (*Oryza sativa* L.). *Plant Molecular Biology* **30**, 387–401.

Yoder, J. I. and Goldsbrough, A. P. (1994). Transformation systems for generating marker-free transgenic plants. *Bio/Technology* **12**, 263–267.

Yokoi, S., Tsuchiya, T., Toriyama, K. and Hinata, K. (1997). Tapetum-specific expression of the *Osg6B* promoter-beta-glucuronidase gene in transgenic rice. *Plant Cell Reports* **16**, 363–367.

Zhang, L., Rybczynski, J. J., Langenberg, W. G., Mitra, A. and French, R. (2000). An efficient wheat transformation procedure: transformed calli with long-term morphogenic potential for plant regeneration. *Plant Cell Reports* **19**, 241–250.

Zhang, S., Cho, M. J., Koprek, T., Yun, R., Bregitzer, P. and Lemaux, P. G. (1999). Genetic transformation of commercial cultivars of oat (*Avena sativa* L.) and barley (*Hordeum vulgare* L.) using in vitro shoot meristematic cultures derived from germinated seedlings. *Plant Cell Reports* **18**, 959–966.

Zhang, W., McElroy, D. and Wu, R. (1991). Analysis of rice *Act1* 5′ region activity in transgenic rice plants. *Plant Cell* **3**, 1155–1165.

Zheng, Z., Kawagoe, Y., Xiao, S., Li, Z., Okita, T., Hau, T. L., Lin, A. and Murai, N. (1993). 5′ distal and proximal *cis*-acting regulatory elements are required for developmental control of a rice seed storage protein glutelin gene. *Plant Journal* **4**, 357–366.

Zhong, H., Zhang, S., Warkentin, D., Sun, B., Wu, T., Wu, R. and Sticklen, M. B. (1996). Analysis of the functional activity of the 1.4-Kb 5′-region of the rice actin 1 gene in stable transgenic plants of maize (*Zea mays* L.). *Plant Science* **116**, 73–84.

Zhou, H., Arrowsmith, J. W., Fromm, M. E., Hironaka, C. M., Taylor, M. L., Rodriguez, D., Pajeau, M. E., Brown, S. M., Santino, C. G. and Fry, J. E. (1995). Glyphosate-tolerant CP4 and GOX genes as a selectable marker in wheat transformation. *Plant Cell Reports* **15**, 159–163.

Opportunities for the Manipulation of Development of Temperate Cereals

JOHN R. LENTON

IACR–Long Ashton Research Station, Department of Agricultural Sciences, University of Bristol, Long Ashton, Bristol BS41 9AF, UK

I. INTRODUCTION: FEEDING THE WORLD

As the world population reached six billion in 1999, it is worthwhile reflecting that, with the exception of sub-Saharan Africa, these mouths have by and large been fed despite the dire predictions of the Rev. Thomas Malthus just over 200 years ago. In his first anonymous "Essay on the Principle of Population", Malthus warned that food supply would only increase arithmetically whereas population growth would increase geometrically. At that time, the population was less than 1 billion and it is fortunate that Malthus' "dismal theorem" has so far been proved wrong. Will this still be the case as we attempt to double food supply in order to feed the predicted population of 10 billion by mid-21st century? A successful

Advances in Botanical Research Vol. 34
incorporating Advances in Plant Pathology
ISBN 0-12-005934-7

outcome will depend much on the ability to improve world cereal production. Despite the prophets of doom highlighting a decline in world per capita cereal production since the mid-1980s, a more thorough analysis of past and future population growth and cereal production trends paints a relatively optimistic picture, at least to 2025 (Dyson, 1999). However, it is predicted that the mismatch between expansion in world demand for cereals, compared with regional supply capacity, will require an estimated tripling in volume of trade

In his erudite book, Evans (1998) has charted the changes in agriculture that have led to or supported the increase in world population. The most recent advance, resulting in considerable increases in cereal yields world-wide, was the introduction of dwarfing genes into wheat and rice by conventional plant breeding techniques, coupled with the escalating use of nitrogen fertilizer and the development of herbicides, that led to the so-called "green revolution" of the 1960s. As Evans rightly points out, the generation of these new dwarf varieties adapted to subtropical regions was a gradual "evolution" rather than the more evocative "revolution". However, the question arises as to what technologies are going to be employed to maintain a sustainable food supply, in particular to support the predicted population of over 8 billion in currently designated less developed regions? Added to this scenario is the uncertain impact of global climate change on food production, limited water supply, salinization and a demographic shift towards an ageing population in most industrialized countries, all of which exacerbate the problem. These issues highlight the urgency of the challenge ahead (Fedoroff and Cohen, 1999) to produce sufficient food in a sustainable manner in order to eliminate malnutrition, particularly as there is evidence that genetic yield potential of current cereal varieties appears to be reaching an upper limit (Mann, 1999).

II. APPROACHES TO A SOLUTION: COMPARATIVE GENETICS, BIOTECHNOLOGY AND THE PHENOTYPE GAP

Although the vast majority of the required increase in crop yields will probably have to be generated by conventional and advanced marker-selected breeding methods, difficulties arise in predicting "the consequences of a particular gene transfer across different environments and into different genetic backgrounds" (Law, 1995). However, the means to improve agronomic traits of crop plants may be provided by genetic variation present in germplasm banks, coupled with detailed genetic linkage maps allowing the study of quantitative trait loci (QTL) (Tanksley and McCouch, 1997). The identification of genes and QTLs that control a process such as flowering time in cereals has been of practical value to plant breeders for adaptation of different crops to specific environments (Law and Worland, 1997). In future, a more detailed understanding of the flowering process in cereals (Laurie, 1997) should become available based on knowledge gained from the complex genetic pathways and gene interactions that regulate the transition to flowering in *Arabidopsis* (Levy and Dean, 1998). A more targeted approach to manipulate such an important developmental process as flowering time in cereals will be possible with such information, coupled with that on the conservation of genome organization across members of the Poaceae (Devos and Gale, 1997).

Besides the obvious targets of pest and disease resistance, plant biotechnology offers the opportunity to manipulate the pattern of plant development in novel ways once the function of key genes has been established. Whether or not it will be used to aid food production in developing countries is a matter of keen debate (Conway and Toenniessen, 1999; http://www.biotech-info.net/gordon_conway.html; Macilwain, 1999). In terms of plant development, biotechnology has proved more applicable to expanding the diversity of end products by directly changing the composition and functional properties of storage reserves, such as wheat glutenin (Barro *et al.*, 1997). In addition, renewable sources of feedstocks of specific plant constituents for nutritional, pharmaceutical or industrial purposes have been generated by genetic engineering (Mazur *et al.*, 1999). An obvious example of altered end-product quality was silencing the gene for the cell wall degrading enzyme, polygalacturonase, in transgenic tomatoes, resulting in fruits with extended shelf life and from which more viscous tomato paste could be processed. These products arose from the application of established physiological knowledge on the mode of action of the plant hormone, ethylene, in initiating ripening in climacteric fruits. In practice, tomato fruit ripening could be delayed either by blocking production or action of the signalling hormone, or by directly down-regulating the key enzyme involved in cell wall breakdown. Such approaches are ideally suited to manipulation of processes affecting functionality of harvested storage organs. In principle, a similar model might serve as a useful framework for the manipulation of other aspects of crop performance that have also been the targets of plant breeders and the plant growth regulator sector of the agrochemical industry for decades. In this review, attention is focused on selected developmental processes affecting structure and function of temperate cereals that may provide potential targets for future modification.

The developmental genetic basis of changes in plant form relating to the generation of the diversity of plant phenotypes during the course of evolution may provide important lessons for the selection of potential target genes for future manipulation of plant development (Baum, 1998). A consensus appears to be emerging that changes in the expression of regulatory genes at relatively few loci may have had profound effects on the evolution of different phenotypes. However, the determination of plant form is complex and confounded by the non-linearity of developmental pathways, coupled with gene duplication allowing specialization for particular functions or co-option of genes for new roles. Clearly, the selection of suitable target genes affecting different aspects of plant development is no trivial issue and in practice will depend much on empirical approaches. In order to be able to capitalize on the potential benefits of manipulating genes, either in more conventional plant breeding or by transformation technology, a much clearer understanding of the relationship between the plant genotype and how it performs in the field (the phenotype) is required. One of the distinguishing features of higher plants is that they exhibit developmental plasticity by retaining the capacity to alter their final form in response to environmental cues, such as light quality, temperature, water and nutrient status, to say nothing of pathogen and pest attack. The complexity of the signalling networks affecting the pattern of gene expression that result ultimately in a modification of plant form often involves one or more of the different groups of endogenous plant hormones and secondary messengers.

An example of the interaction between developmental and environmental signalling involves one of the groups of plant hormones, the gibberellins (GAs), that are important regulators of several aspects of plant development (Hooley, 1994). The identification of GA as the causal agent of shoot overgrowth or "bakanae" (foolish seedling) disease of rice infected with the fungus, *Gibberella fujikuori*, provided the first evidence that GAs are promoters of shoot growth in higher plants. Other key events confirming a role for GA in determining plant stature included the identification of GAs in non-infected plants and the demonstration that some dwarf mutants are blocked at different steps in the GA biosynthesis pathway. In contrast, other dwarf mutants are unable to respond to GA and may or may not be blocked at some point in the GA perception and signal transduction chain, whereas certain "slender", overgrowth mutants express the GA response pathway constitutively (Ross *et al.*, 1997). Of the various environmental factors, light quality, acting via the photosensitive phytochrome proteins, affects GA production at different stages of plant development (Kamiya and Garcia-Martinez, 1999). For example, a maturity allele in sorghum, ma_3^R, which originally was thought to affect GA biosynthesis leading to overproduction of active hormone (Beall *et al.*, 1991), was shown, subsequently, to be a null *phyB-1* mutant (Childs *et al.*, 1997). In addition to environmental factors affecting GA production, both phytochrome action (Nick and Furuya, 1993; Toyomasu *et al.*, 1994) and temperature (Stoddart and Lloyd, 1986; Pinthus *et al.*, 1989; Pinthus and Abraham, 1996) also affect GA responsiveness in cereals. Equally convincing cases could be made for changes in the production or responsiveness of other groups of plant hormones in different aspects of developmental and environmental signalling.

III. MANIPULATION OF PLANT STATURE BY AFFECTING GIBBERELLIN SIGNALLING: LESSONS FROM THE "GREEN REVOLUTION" GENES

The development of short-strawed varieties of breadwheat is not in fact a recent innovation but started in Japan in the 19th century (Gale and Youssefian, 1985). Two main sources of reduced height (*Rht*) dwarfing genes are present in most current wheat varieties grown world-wide. These have been derived either from Norin 10 (*Rht1* and *Rht2*), a variety that is unable to respond to GA, or, to a much lesser extent, from Akakomugi (*Rht8* and *Ppd1*), a variety that is GA-responsive (Yamada, 1990). The main source of dwarfing in rice can be traced back to a Chinese Indica cultivar, Dee-geo-woo-gen, used by Japanese plant breeders in Taiwan, with different alleles at the *sd1* locus being present in many current commercial rice cultivars (Futsuhara and Kikuchi, 1997). The single recessive *sd1* gene confers a GA deficient phenotype (Suge, 1975). In these instances, Japanese plant breeders, selecting for beneficial traits in wheat and rice, unwittingly chose genes encoding enzymes that were components of the GA biosynthesis or signal transduction pathways. With the benefit of hindsight, this is not too surprising since one of the main actions of GAs is to promote shoot growth by overcoming the action of a growth-inhibiting signalling pathway (Harberd *et al.*, 1998).

Not only were the GA non-responsive semi-dwarf wheat genotypes, derived from Norin 10, able to resist the damaging effects of rain and wind that cause stem collapse (lodging) in taller crops but also they were inherently higher yielding in the absence of lodging (Gale and Youssefian, 1985). The agronomic potential of these semi-dwarf wheat varieties became fully realized following a transition from the use of farmyard manure to inorganic fertilizers and with the introduction of herbicides to combat weed competition (Evans, 1998). The development of near-isogenic lines containing a range of the GA non-responsive *Rht* genes (Fig. 1) in different genetic backgrounds has proved a valuable resource for examining various aspects of *Rht* gene action (Gale and Youssefian, 1985). The semi-dwarfing genes, *Rht1* and *Rht2*, are homologues present on chromosomes 4BS and 4DS, respectively (Table I). (For convenience, the "old" nomenclature is retained throughout this article.) These genes are present in combination (*Rht1+2*) in the variety Norin 10. The potent dwarfing gene, *Rht3*, derived from "Tom Thumb", is an alternative allele at the same locus as *Rht1* and the combination of *Rht2+Rht3* leads to an even more severe dwarf phenotype (Fig. 1). More recently, *Rht10*, derived from the Chinese variety "Ai-bian1", was shown to be a more severe allele at the same locus as *Rht2* (Table I) (Börner *et al.*, 1997). In this allelic series of *Rht* dwarfing genes the severity of the block in responsiveness to applied GA was proportional to the degree of dwarfism (Pinthus *et al.*, 1989).

J. R. LENTON

Interestingly, these tall and semi-dwarf isolines were equally GA-responsive in terms of α-amylase production in mature aleurone, whereas *Rht3* and *Rht2+3* lines remained GA non-responsive (J. E. Flintham, personal communication). In barley, a potent GA insensitivity dwarfing gene, *Dwf2*, mapped to the same position as the multiallelic *Rht* loci of wheat, described above, suggesting synteny of GA-insensitive dwarfing genes within the Triticeae (Ivandic *et al.*, 1999).

In terms of development, the timing of leaf and spikelet primordium initiation at the shoot apex was the same in near-isogenic lines, carrying *rht* (tall), *Rht1*, *Rht2* or *Rht3* alleles in a genetic background of spring and winter wheat varieties grown under UK conditions (Youssefian *et al.*, 1992a). The primary effect of these dwarfing genes was to decrease the rate of elongation and final lengths of leaves and individual stem internodes by blocking the responsiveness of these vegetative tissues to GA. Since the dwarfing genes did not affect the expansion of internodes that form the rachis of the ear, or final ear size, the competition for assimilates between stem and ear growth was reduced in these dwarf genotypes, compared with tall, GA-responsive, lines. As a consequence, a greater proportion of dry mass was partitioned to developing ears of *Rht* genotypes to support the survival of florets (potential grain sites), which also had a greater carpel mass at anthesis (Youssefian *et al.*, 1992b). Typically, 6–10 florets were initiated within each spikelet of young developing wheat ears but death of more distal florets occurred during expansion of last two vegetative stem internodes such that only 3–4 florets per spikelet survived at anthesis to produce viable grains. A greater availability of assimilates in ears of semi-dwarf genotypes was assumed to be

Fig. 1. Near-isogenic lines of wheat seedlings differing at reduced height (dwarfing) gene loci in the background variety, Maris Huntsman. Seed stocks were generously supplied by Professor M. D. Gale, John Innes Centre, Norwich.

TABLE I
Nomenclature and source of some reduced height, gibberellin 'insensitivity' genes of wheat

Chromosome 4BS			Chromosome 4DS		
"Old"	"New"	Source	"Old"	"New"	Source
rht	*Rht-B1a*	Wild type	*rht*	*Rht-D1a*	Wild type
Rht1	*Rht-B1b*	Norin 10	*Rht2*	*Rht-D1b*	Norin 10
Rht3	*Rht-B1c*	Tom Thumb	*Rht10*	*Rht-D1c*	Ai-bian 1

the main factor responsible for decreased floret die-back during ear development, thus resulting in a greater grain set per ear at anthesis. However, information on the vascular pathways and carbohydrate status between florets within a spikelet and between young developing ears and subtending internodes is lacking. It has also been suggested that early-formed florets within a spikelet may exert a hormonally based inhibitory effect on the development of more distal florets (Evans *et al.*, 1972). Although individual grains of semi-dwarf genotypes may have a smaller final mass, compared with tall lines, overall ear mass was often greater, and shoot dry matter less, resulting in a greater harvest index (ratio of grain dry mass to total shoot dry mass) (Youssefian *et al.*, 1992b). A detailed agronomic study of near-isogenic *Rht* genotypes, in four different genetic backgrounds, grown under field conditions in eastern England and central Germany, revealed optimum grain yields at intermediate plant heights with intrinsically taller genotypes requiring more potent *Rht* alleles to achieve maximum potential grain yield (Flintham *et al.*, 1997).

Near-isogenic lines have also proved a valuable tool to determine the physiology of *Rht* gene function at the cellular level. In general, the semi-dwarfing genes, *Rht1* and *Rht2* decreased cell expansion in vegetative tissues in proportion to gene dosage, although there were subtle differences between effects on adaxial versus abaxial leaf surfaces and between laminae and leaf sheaths (Keyes *et al.*, 1989). The semi-dwarfing genes, alone and in combination, also had a greater effect on cell elongation in stem internodes than leaves (Hoogendoorn *et al.*, 1990; Miralles *et al.*, 1998), but maternal grain pericarp cell dimensions were unaffected (Miralles *et al.*, 1998). The longitudinal component of plastic wall extensibility of cells in the elongation zone of the leaf decreased in these *Rht* genotypes, compared with tall lines (Keyes *et al.*, 1990). Consistent with these observations, cell wall extensibility and overall sheath elongation in tall seedlings treated with a GA biosynthesis inhibitor was decreased, whereas GA application increased wall extensibility in control but not the *Rht* genotypes. These responses to semi-dwarf *Rht* gene dosage and plant growth regulator applications were reflected in change in size of the leaf extension zone (Paolillo *et al.*, 1991) and similar

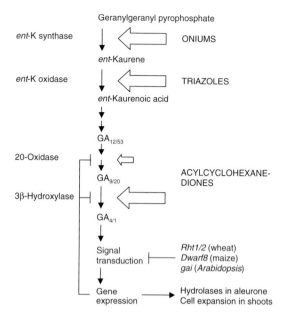

Fig. 2.   Scheme outlining gibberellin (GA) biosynthesis and response pathway in cereals. The sites of action of different classes of plant growth retardants (open arrows), feedback down-regulation of GA biosynthesis genes as a consequence of GA action, and the orthologous semi-dominant mutations in wheat, maize and *Arabidopsis* that block GA responsiveness and relieve feedback regulation of GA pathway are indicated (⊣).

results were obtained with the more potent *Rht3* dwarfing allele (Tonkinson *et al.*, 1995). It appeared from these analyses that processes involved in the cessation of cell elongation were promoted either by blocking GA action, as in *Rht* genotypes, or by reducing GA content with applied growth retardants. Conversely, applied GA, rather than increasing the maximal rate of cell expansion, increased the duration of expansion by delaying cell maturation. A key enzyme activity preventing cell maturation in wheat has yet to be identified but, in barley, abundance of specific xyloglucan endo-transglucosidase transcripts at the leaf base was associated with increased leaf elongation rate (Schünmann *et al.*, 1997). Manipulation of certain members of this gene family may provide a direct means of affecting cell elongation and shoot stature in cereals.

Paradoxically, as well as exhibiting decreased cell expansion and being non-responsive to applied GA, wheat *Rht* genotypes contained elevated levels of biologically active GA_1 in vegetative tissues in proportion to the potency of the dwarfing allele (Lenton *et al.*, 1987; Appleford and Lenton, 1991; Webb *et al.*, 1998). The increased GA_1 content of shoots of *Rht* genotypes was thought to be due to the release of a regulatory step on the biosynthetic pathway to active GA resulting from blocked GA action (Fig. 2) rather than inherently slower growth rate. In contrast, GA_1 did not accumulate in developing ears of *Rht* genotypes, compared with tall lines, implying that developing ears either do not produce GA-responsive cell types or that expression of *Rht* genes is restricted to vegetative tissues. Similar increased contents of biologically active GAs were also found in vegetative tissues of GA non-responsive dwarf mutants of maize (*D8*) (Fujioka *et al.*, 1988) and *Arabidopsis* (*gai*) (Talon *et al.*, 1990a). In contrast, "slender" barley, a constitutive GA response mutant (Chandler, 1988; Lanahan and Ho, 1988), contained low amounts of GA_1, compared with normal segregants (Croker *et al.*, 1990).

At the molecular level, the recent cloning and mapping of the "green revolution" genes showed that several alleles at the wheat *rht* locus were orthologues of the maize *dwarf-8* and *Arabidopsis gibberellin insensitive* (*GAI*) genes (Peng *et al.*, 1999). The genes responsible for the *Rht1* and *Rht2* mutations contained nucleotide substitutions that introduced early stop codons, suggesting that the encoded products were truncated in the N-terminal domain that normally confers GA specificity to a protein involved in the GA signalling pathway (Fig. 2). It will be interesting to learn the molecular basis of more potent, *Rht3* and *Rht10*, dwarfing alleles. The value of these GA non-responding genes for manipulating plant stature was elegantly demonstrated by transformation of Basmati rice with the *Arabidopsis* mutant *gai* gene behind a maize ubiquitin promoter (Peng *et al.*, 1999). Rice seedlings expressing the transgene were relatively non-responsive to applied GA and mature plants were dwarfed. These GA insensitivity genes have obvious potential for manipulating harvest index and increasing yields of other transformable arable crops, such as barley, potatoes and oilseed rape.

In rice, the Daikoku dwarfs were shown to be relatively GA insensitive both in terms of shoot growth and α-amylase production (Mitsunaga *et al.*, 1994). Mutations in the coding region of the α-subunit of a heterotrimeric G protein accounted for five alleles of Daikoku dwarf (*d-1*) and expression of a gene encoding the GTP-binding protein in transgenic *d1* mutant restored the normal phenotype. In addition, antisense suppression of the α-subunit of G protein in transgenic rice produced a dwarf phenotype with small seeds (Ashikari *et al.*, 1999; Fujisawa *et al.*, 1999). Dwarfing of the shoot was associated with decreased cell production in the intercalary meristem of internodes without affecting cell length, implying that G-protein signalling is involved in cell production but not cell elongation. In deepwater rice, the primary site of GA action was considered to be a stimulation of cell production in the intercalary meristem but cell elongation was also promoted in the elongation zone (Kende *et al.*, 1998). Hence, G proteins appear to be involved in GA signalling in relation to progression through the cell cycle in the intercalary meristem but not in processes involved in cell expansion in the elongation zone. In addition, G proteins have also been implicated in GA signalling in relation to α-amylase production in wild oat aleurone, a non-growing tissue (Jones *et al.*, 1998).

IV. MANIPULATION OF PLANT STATURE BY BLOCKING GIBBERELLIN PRODUCTION: AN ALTERNATIVE TO SPRAYING PLANT GROWTH RETARDANTS?

The single gene, recessive, dwarf mutants of maize proved instrumental in defining individual steps on the early 13-hydroxylation GA pathway and established GA_1 as the active compound essential for shoot elongation (Phinney, 1984). Additional GA responsive dwarf mutants have also been characterized in diploid cereals such as rice (Kobayashi *et al.*, 1989) and barley (Chandler and Robertson, 1999). Quantitative analyses of GAs in three alleles of a deficient mutant of pea established a log:linear relationship between GA_1 content and stem internode growth (Ingram *et al.*, 1986) and a similar result was found for leaf growth in *rht* (tall) wheat treated with a GA biosynthesis inhibitor (Lenton *et al.*, 1987). Whilst these observations confirmed the importance of GAs as essential regulators of plant stature, they did not address the issue of whether or not GAs are sensitive regulators of growth in normal, untreated plants.

Many arable, horticultural and ornamental crops are routinely sprayed with plant growth retardants that act by blocking various steps along the biosynthetic pathway to active GA (Fig. 2). Virtually all the UK cereal acreage is treated with plant growth regulators since, in a bad year, lodging costs growers up to £110m in lost yield. It is not immediately clear why this should be the case for wheat varieties already containing semi-dwarfing genes that confer blocked responsiveness to GA. The causes of lodging are complex and involve aspects of root development and stem thickening as well as stem height reduction.

Chlormequat chloride (CCC), one of an onium class of growth retardants (Fig. 2), was the first compound used commercially to control stem lodging in wheat. Even in trials where lodging was prevented, application of CCC to a tall wheat isoline reduced stem height and increased grain number, mimicking the effect of the semi-dwarfing genes (Gale and Youssefian, 1983). Although CCC inhibits GA production in wheat (Graebe *et al.*, 1992; Lenton *et al.*, 1987), it may also have additional modes of action that help to prevent crop lodging and account for its continued use in cereal production.

In contrast, the triazole group of plant growth retardants (Fig. 2) that inhibit *ent*-kaurene oxidase at an early step on the GA biosynthesis pathway (Hedden and Graebe, 1985; Rademacher *et al.*, 1987) are potent inhibitors of GA biosynthesis in wheat (Lenton *et al.*, 1994) and barley (Croker *et al.*, 1990). However, at higher concentrations, these compounds also inhibit

C-14 demethylation on the sterol biosynthesis pathway in plants (Burden *et al.*, 1987). In general, this class of growth retardant appears to be too persistent for commercial use on arable crops.

Members of another group of plant growth retardants, the acylcyclohexanediones, inhibit the soluble 2-oxoglutarate dependent dioxygenases involved in the later stages of GA biosynthesis (Fig. 2). These compounds interact with the 2-oxoglutarate binding site of the GA dioxygenases with GA 3β-hydoxylase being the main target (Adams *et al.*, 1992; Kamiya *et al.*, 1992), although activities of GA 20-oxidase (Rademacher *et al.*, 1992) and GA 2-oxidase (Griggs *et al.*, 1991) are also inhibited. One member of this group of compounds, trinexapac ethyl (Moddus®), has been approved for commercial use to prevent lodging in winter wheat and winter barley, including malting varieties. More specific inhibitors of the GA dioxygenases based on potential competitive GA substrate inhibitors have proved effective growth retardants, particularly for cereals and related grasses (Evans *et al.*, 1994; Junttila *et al.*, 1997).

Alternative strategies for manipulating crop stature have become available following the recent cloning of several genes encoding enzymes involved in the main GA biosynthesis pathway (Hedden and Kamiya, 1997; Lange, 1998; Hedden, 1999; Hedden and Proebsting, 1999). Progress in this area has been built on detailed chemical and biochemical knowledge that defined the GA biosynthetic pathways in the fungus, *Gibberella fujikuroi*, and vegetative and reproductive tissues of several higher plants (MacMillan, 1997). A GA biosynthesis gene, GA 20-oxidase, cloned from pumpkin embryos, encoded a multifunctional enzyme that oxidized carbon 20 from a methyl group through to a carboxylic acid (Lange *et al.*, 1994). However, subsequent cloning of GA 20-oxidases from other species confirmed earlier reports that the aldehyde group was lost predominantly as CO_2, resulting in the formation of a C-19 GA (Fig. 3), with only small amounts of the C-20 carboxylic acid being formed (Hedden and Kamiya, 1997). In *Arabidopsis*, GA 20-oxidase was present as a small gene family with a tissue-specific pattern of expression that was suppressed by the presence of active GA (Fig. 2) (Phillips *et al.*, 1995). A GA 20-oxidase clone mapped tightly to the *GA5* locus, mutation of which caused a semi-dwarf phenotype containing decreased content of active GA (Talon *et al.*, 1990b; Xu *et al.*, 1995). Overexpression of GA 20-oxidase in transgenic *Arabidopsis* produced taller plants that flowered earlier in both long and short days (Coles *et al.*, 1999). Promotion of shoot growth was associated with increased GA_4 content in vegetative rosettes. Conversely, suppression of the activity of a stem-specific GA 20-oxidase decreased GA_4 content in rosettes and shoot tips. These plants had slower rates of stem elongation and delayed flowering. Taken together, the results clearly demonstrate that altering expression of a GA biosynthesis gene is a viable route to manipulating plant stature and flowering time.

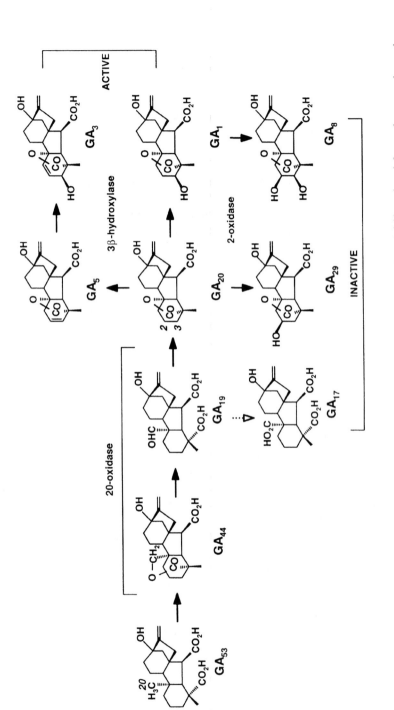

Fig. 3. Later stages of the early 13-hydroxylation gibberellin (GA) pathway in cereals. Multifunctional 2-oxoglutarate-dependent dioxygenases are involved in the biosynthesis (GA 20-oxidase and GA 3β-hydroxylase) and catabolism (GA 2-oxidase) of active GA.

A notable feature of many cereals is that one of the intermediates of the GA 20-oxidase reaction sequence, GA_{19} (Fig. 3), accumulates to relatively high levels in vegetative tissues. In wheat, it was proposed that the conversion of GA_{19} to GA_{20} was a rate-limiting step in the biosynthesis of GA_1 (Lin and Stafford, 1987), suggesting that wheat GA 20-oxidase would be a good target gene for manipulation in transgenic plants. Detailed kinetic studies with recombinant protein from one of three GA 20-oxidase clones isolated from wheat scutella showed that the enzyme was multifunctional, with the K_m for GA_{19} being about 150-fold greater than that for GA_{53} (N. E. J. Appleford, unpublished observations). This result provided a possible explanation for the accumulation of GA_{19} in wheat. Expression of GA 20-oxidase was high in scutella of germinating grain but appeared to be confined to the region of the intercalary meristem in vegetative stem internodes. Comparison of GA contents of various cereal GA mutants suggested that the conversion of GA_{19} to GA_{20} was normally suppressed as a consequence of GA action in GA responsive genotypes (Croker et al., 1990; Appleford and Lenton, 1991; Hedden and Croker, 1992). Characterization of a GA 20-oxidase from rice confirmed that expression was up-regulated in the absence and suppressed in the presence of active GA in shoot tissues (Toyomasu et al., 1997). However, expression of the currently cloned GA 20-oxidases from wheat does not appear to be regulated by GA (N. E. J. Appleford and J. R. Lenton, unpublished observations).

The final step in the formation of active GAs is the 3β-hydroxylation of the immediate precursors, GA_9 and GA_{20}, to GA_4 and GA_1, respectively (Fig. 3). GA 3β-hydroxylase is encoded by the *GA4* gene of *Arabidopsis* (Chiang et al., 1995; Williams et al., 1998) and the *LE* gene of pea (Lester et al., 1997; Martin et al., 1997). The *le* dwarf mutation in pea, one of the characters studied by Mendel when formulating his laws of inheritance, arose as a consequence of a single base substitution that caused an amino-acid substitution close to the enzyme active site. Again, both the *Arabidopsis* and pea GA 3β-hydroxylase genes were feedback regulated by active GA (Fig. 2) (Martin et al., 1996, 1997; Cowling et al., 1998; Ross et al., 1999). Transcripts of the GA 3β-hydroxylase homologue from tobacco were expressed in actively dividing and elongating cells, suggesting that active GA is produced close to the site of action in GA-responsive tissues (Itoh et al., 1999). Manipulation of the activity of GA 3β-hydroxylase represents a further target for perturbing plant stature in transgenic plants.

The initial step in the deactivation of biologically active GAs involves oxidation of the A ring at carbon-2 (Fig. 3). The recent cloning of a GA 2-oxidase from runner bean and heterologous expression in *E. coli* confirmed that it was a multifunctional enzyme with the ability to oxidize certain GA substrates through to a keto group (Thomas et al., 1999). In addition to deactivating active GAs, such as GA_4 and GA_1, their immediate precursors, GA_9 and GA_{20}, were also effective substrates for the enzyme. Interestingly,

Fig. 4.   An extreme dwarf phenotype in a primary transformant of wheat, variety Canon, overexpressing runner bean GA 2-oxidase (right), compared with a non-transformed control (left). Photograph courtesy of Dr Mark Wilkinson, IACR–Rothamsted.

gibberellic acid (GA_3), another active GA that is produced as a side reaction of GA 3β-hydroxylase activity in maize (Fig. 3) (Fujioka *et al.*, 1990), was not an effective substrate for GA 2-oxidase. Application of active GA increased the abundance of GA 2-oxidase transcripts in *Arabidopsis*, suggesting a transcriptionally activated feedforward mechanism decreasing the content of active GA in tissues. GA 2-oxidase has also been cloned from pea seed and the *sln* mutation, which caused an accumulation of GA_{20} in mature seed, contained a single base deletion that resulted in the production of a truncated protein (Lester *et al.*, 1999; Martin *et al.*, 1999). By deactivating both immediate precursors and active GAs, except GA_3, overexpression of GA 2-oxidase in transgenic plants would be expected to provide a powerful means of dwarfing plants. In practice, phenotypic dwarfs in *Arabidopsis* similar to the GA-deficient mutant, *ga1–3*, were produced by overexpression of a bean GA 2-oxidase (S. G. Thomas, personal communication). Similarly, a dwarf phenotype was also obtained by overexpression of bean GA 2-oxidase in primary transformants of wheat (Fig. 4), with the more extreme phenotypes being induced to flower by GA_3 application (M. D. Wilkinson, personal communication).

In the short period of time that the genes have been available, the relative merits of blocking GA production or GA action as the most suitable means of manipulating plant stature have yet to be evaluated. In both cases, it may be important to restrict expression of a transgene to specific cell type(s) in stem internode tissue in order to avoid unwanted pleiotropic effects of GA on other aspects of plant development. In all probability, a range of stable height phenotypes could be generated given the availability of different *gai*-like alleles that restrict the extent of GA action. The situation may be more complex for blocking GA_1 accumulation since the hormone may be synthesized in tissues other than the site of action and the homeostatic mechanism regulating GA_1 content would be expected to counteract the action of an inserted GA biosynthesis transgene. Preliminary data from wheat treated with a growth retardant suggested that GA_1 levels had to be reduced considerably before a decrease in leaf elongation was observed (Lenton *et al.*, 1987; Tonkinson *et al.*, 1997). Certainly in practice, with the use of "strong" promoters, it has proved possible to manipulate the expression of GA biosynthesis genes successfully to generate a range of dwarf phenotypes in several species. One advantage of perturbing GA production in transgenic plants is that, in certain situations, such as restoring male fertility or encouraging bolting, it would be advantageous to be able to restore growth by GA_3 application.

V.   OTHER ROUTES TO MANIPULATION OF PLANT ARCHITECTURE: PHYTOCHROME GENES AND TRANSCRIPTION FACTORS

One of the most notable features of many plant species grown in close proximity is an increase in stem growth, at the expense of leaf and storage organ growth, in response to changes in light quality. This example of developmental plasticity is a consequence of plants having a modular architecture and an indeterminate growth habit. In direct sunlight, the ratio of red:far-red (R:FR) radiation is high but decreases markedly under a canopy of vegetation that absorbs R and transmits FR wavelengths (Smith, 1995). The decreased R:FR ratio is sensed by neighbouring plants well before canopy cover occurs and is part of a "shade avoidance response" that includes rapid increases in stem elongation. In green plants, FR light, acting mainly through phytochrome B, promotes cell extension whereas in dark-grown seedlings, FR light, mediated by phytochrome A, inhibits cell expansion. Normally, phytochrome A is degraded in the light but overexpression of oat *phyA* in transgenic tobacco was sufficient to disable the shade avoidance response resulting in a dwarfed phenotype and a diversion of assimilates from stems to leaves when plants were grown in close proximity (Robson *et al.*, 1996). This modification of crop architecture and assimilate partitioning in tobacco through manipulation of the phytochrome signalling system represented a novel approach to increasing plant productivity and was built on many years of research on phytochrome physiology.

A shade avoidance response was also observed with the wild progenitor of domesticated maize, teosinte, grown under shaded conditions (Doebley *et al.*, 1995). In this case, plants showed strong apical dominance with short lateral branches tipped by ears whereas normally, under more favourable conditions, teosinte is characterized by highly branched shoot architecture with long lateral shoots tipped by tassels. Domestication of maize has involved a marked increase in apical dominance, similar to that shown by teosinte grown under shaded conditions. A major QTL was identified as the main contributor to the difference in phenotype between teosinte grown under non-competitive conditions and domesticated maize. Expression of a gene, *teosinte branched1* (*tb1*), associated with this QTL, determined the fate of axillary meristems repressing growth of lateral organs and enabling the formation of female inflorescences (Doebley *et al.*, 1997). In teosinte, *tb1* was only turned on under competitive conditions as part of a shade avoidance response causing increased apical dominance. In contrast, the *tb1* allele of maize was expressed constitutively at twice the level of the teosinte allele, suggesting that this change in pattern of gene expression accounted for the main difference in apical dominance and sex expression between domesticated maize and its wild progenitor. By selecting favoured phenotypes, early Aztec farmers enriched the harvested crop with alleles at loci controlling traits of interest until they became "fixed" in the population. In the case of *tb1*, the surprising feature is that selection was restricted to sequence changes in the promoter, rather than the coding region of the gene (Wang *et al.*, 1999). These results are a striking example of changes in expression of a single gene associated with a QTL having a profound influence on plant architecture and evolution of a crop species. Since *tb1* showed sequence homology with the *cycloidea* gene of snapdragon, both of which contain putative nuclear localization signals, it was suggested that these genes encode transcription factors (Doebley *et al.*, 1997).

The more general proposition that transcriptional regulators are subject to evolutionary modification resulting in new spatial or temporal patterns of expression of downstream target genes that result in novel phenotypes provides an interesting framework to consider manipulation of plant morphology (Doebley and Lukens, 1998). Such an approach is considered preferable to perturbing expression of genes involved in upstream signalling pathways that tend to generate more pleiotropic mutant phenotypes. Homeodomain proteins are a group of transcription factors involved in the regulation of plant morphogenesis. One distinct class within the group, the *knotted1*-type homeobox (*knox*) genes, is expressed within the shoot apical meristem where it maintains cells in an undifferentiated state (Martin, 1996). A loss of function mutation of a rice *knox* gene, *OSH15*, from a retrotransposon-tagged line, mapped to the same position and had a similar phenotype as three independent deletion alleles of the *d6* dwarf mutant of rice (Sato *et al.*, 1999). These conventional recessive mutants were characterized by short internodes below the peduncle and were complemented by a genomic fragment containing *OSH15*. The cause of the mutant phenotype was related to abnormally shaped epidermal and hypodermal cells and an unusual arrangement of the small vascular bundles in internodes, indicating that members of this family of transcription factors may also function outside the shoot apical meristem (Sentoku *et al.*, 1999). The downstream target genes of *knox* gene expression have not been identified but the altered phenotype involved changes in the relative amounts of plant hormones (Kusaba *et al.*, 1998b). Overexpression of a *knox* gene was associated with the accumulation of cytokinins at a tissue level that directly related to changes in leaf morphology in lettuce (Frugis *et al.*, 1999). On the other hand, insertion of the *knox* genes, *OSH1* (Kusaba *et al.*, 1998a) and *NTH15* (Tanaka-Ueguchi *et al.*, 1998) into transgenic tobacco suppressed GA 20-oxidase gene expression and decreased GA_1 content but the phenotype was not fully restored by treatment with active GA. At present, it is far from certain that these changes in plant hormones represent primary responses to ectopic expression of *knox* genes. The phenotype of such transgenic plants in which the delicate balance of transcriptional activators has been disrupted should be interpreted with some caution.

VI. GRAIN DEVELOPMENT: MANIPULATION OF ENDOSPERM CELL NUMBER?

In principle, increasing the size of a harvested storage organ is a more attractive proposition than affecting earlier stages of development that may become compromised subsequently by other internal or external factors. The potential storage capacity of cereal grain is dependent on the final number and size of cells produced in the endosperm. In both wheat (Brocklehurst, 1977) and barley (Cochrane and Duffus, 1983), there was a close association between final weight of individual grains and the number of endosperm cells produced. In wheat, this relationship applied to ancient versus modern varieties grown in controlled conditions (Dunstone and Evans, 1974) and to positional differences in grain size within an ear (Singh and Jenner, 1982). The effects of shading intact plants and variation in nutrient supply in detached, cultured ears (Singh and Jenner, 1984) also supported the relationship between endosperm cell number and grain size and highlighted the importance of assimilate supply for optimizing grain size.

Whilst the number of endosperm cells provides a physical constraint to sink size in wheat and barley, it is important to realize that developing storage organs also compete for a finite supply of assimilate. In addition, it is not uncommon for developing fruits to exhibit dominance relationships such that, once a certain size has been achieved, the dominant organ survives at the expense of a neighbour. This has been termed the Matthew principle (Evans, 1993), "Unto every one that hath shall be given, and he shall have abundance: but from him that hath not shall be taken away even that which he hath" (Matt. 25:29). Leaving aside this unlikely biblical message, it is uncertain whether the physiological basis of dominance between developing fruits is assimilate supply, hormonal status or a combination of both (Bangerth, 1989). In general, final fruit and seed size are influenced by controls acting on morphogenesis around the time of fertilization with pollination, providing the stimulus for expansion of maternal tissues surrounding the embryo sac. That these controlling factors involve the action of plant hormones is evident from the long-established ability of applied GAs, auxin and cytokinin to induce parthenocarpic fruit formation in several species.

Despite detailed anatomical and morphological descriptions of endosperm development in cereals (Evers, 1970; Huber and Grabe, 1987a,b; Olsen *et al.*, 1992), only relatively few genes affecting different stages of cellular differentiation have been characterized (Olsen *et al.*, 1999). In higher plants, there is a double fertilization event with a sperm nucleus fusing with two polar nuclei of the central cell, to form the primary endosperm nucleus, and another sperm nucleus fusing with the egg cell that becomes the diploid embryo. In temperate cereals, the triploid endosperm nucleus divides synchronously, without wall formation, and daughter nuclei migrate to the peripheral cytoplasm of a rapidly expanding central cell of the embryo sac (Bennett *et al.*, 1973, 1975). A group of cells, called the antipodals, which are formed in the embryo-sac 3 days before anthesis, become highly endopolyploid before completing degeneration about three days after anthesis (Nutman, 1939; Bennett *et al.*, 1973; Engell, 1994). These extremely large cells are thought to provide the protein synthesizing machinery to support the free nuclear division stage with waves of nuclear divisions being initiated from the region of the antipodal cells. The formation of cell walls commences in the periphery of the endosperm 3 days after fertilization (Morrison and O'Brien, 1976; Morrison *et al.*, 1978; Olsen *et al.*, 1995) and approximately 5000 cells are present in endosperm at day five, compared with less than 100 in the embryo. In turn, the endosperm transfers nutrients to the developing embryo via vesicles that coalesce above the base of the embryo sac (Huber and Grabe, 1987b).

Rapid longitudinal growth of the carpel also occurs during the first 6 days after anthesis and is accompanied by degeneration of maternal nucellar tissue, except in the region of the ventral groove (Nutman, 1939). The remaining nucellar "pillar" grows by cell elongation in the region of the degenerating antipodals and contains an activated intercalary meristem associated with the tip of the advancing vascular bundle that is the route of nutrient movement to the endosperm. The degenerating antipodal cells are thought to provide a hormonal stimulus both for nuclear division in endosperm and for promoting extension growth in the nucellar "pillar". In his seminal work on grain development in rye, Nutman (1939) recognized that cell death within the carpel was an integral part of embryo and endosperm development. It should also be appreciated that the genes and signalling molecules involved in regulating the complex pattern of development of maternal tissues surrounding the embryo sac may impose the ultimate constraint on final grain size. In other words, attempts to directly manipulate endosperm cell number may not be realized without co-ordinating processes involved in development of surrounding pericarp and testa as well as the enclosing lemma and palea.

Following cellularization of wheat endosperm, cell divisions continue at a much slower rate until 16–20 days after anthesis giving a final population of 120–150 000 cells (Briarty et al., 1979; Gao et al., 1992; Singh and Jenner, 1982). In wheat and barley, aleurone and sub-aleurone layers differentiate around the periphery of the endosperm 10–14 days after anthesis (Morrison et al., 1978; Olsen et al., 1995) and storage reserves begin to accumulate in embryo and endosperm. In maize, the cessation of cell division in endosperm is associated with a period when cells undergo endoreduplication and become polyploid (Grafi and Larkins, 1995). The mechanism by which this is achieved involves induction of S phase kinases and the inactivation of a cdc2/cyclin B kinase required for entry into mitosis. Expression of a protein kinase, ZmWee1, that inactivated both M and S phase cyclin-dependent kinase activity, coincided with the cessation in endosperm mitosis (Sun et al., 1999). Interestingly, one proposed action of cytokinin in tobacco pith parenchyma and suspension cultures is the control of the G2/M transition of the cell cycle by stimulation of a cdc25 phosphatase that dephosphorylates cdc2/cyclin B kinase (Zhang et al., 1996). Additional evidence from tobacco BY2 cell suspensions showed that both cytokinin biosynthesis and cell mitosis were inhibited by blocking 3-hydroxy-3-methylglutaryl-coenzyme A (HMGCoA) reductase activity, a response that was overcome by applying the cytokinin, zeatin (Laureys et al., 1998). If cytokinin acted in the same manner in maize endosperm it would be expected that cdc25 phosphatase would compete with ZmWee1 protein kinase to control the G2/M transition.

An increase in phosphorylation of a retinoblastoma (Rb)-like protein was also associated with endoreduplication in maize endosperm (Grafi et al., 1996). In maize leaves, interaction of a Rb protein homologue with cyclin D inactivated the transcription factor E2F (Huntley *et al.*, 1998). Phosphorylation of Rb protein by cyclin D kinases released E2F to initiate the commitment to cell division. In *Arabidopsis* , another action of cytokinin involved activation of cyclin D3 gene expression and stimulation of cell division (Riou-Khamlichi *et al.*, 1999), by driving the cell cycle beyond the restriction point into the S phase (Frank and Schmülling, 1999; Mironov *et al.*, 1999). In addition to the importance of cytokinin, sucrose has also been implicated in regulation of other cyclin D genes (Huntley and Murray, 1999). These proposed actions of cytokinin support the close relationship observed between endogenous cytokinin content and mitotic activity during maize endosperm development (Dietrich *et al.*, 1995). Although cessation in mitotic activity in maize endosperm was associated with a decrease in cytokinin content, other antagonistic plant hormones may also be involved. For example, abscisic acid (ABA) application decreased endosperm cell numbers without affecting cell size and, at high concentrations, inhibited the transition to endoreduplication (Mambelli and Setter, 1998). Whether ABA action in maize endosperm involves induction of a cyclin-dependent kinase inhibitor that interacts with cdc2a and cyclin D3 inhibiting activity of the complex, as occurred in *Arabidopsis* (Wang *et al.*, 1998), remains to be determined. During the period of cell endoreduplication in maize, ethylene-induced programmed cell death is initiated in the upper central endosperm and expands outwards and downwards (Young *et al.*, 1997). A similar response also occurred in starchy endosperm of wheat but in this case continued in a more random fashion with both embryo and aleurone remaining viable (Young and Gallie, 1999).

A genetic basis for the relationship between endosperm and embryo development is becoming evident from experiments on parental gene dosage in *Arabidopsis* (Scott *et al.*, 1998a,b). Increasing paternal genome dosage stimulated mitosis and delayed cellularization resulting in a large endosperm whereas the converse was observed with an increased maternal genome contribution. In addition, the *MEDEA* (*MEA*) gene product that regulates fertilization-independent endosperm development (Grossniklaus *et al.*, 1998), suppressed both replication of the central cell prior to fertilization and late endosperm nuclear division (Kiyosue *et al* ., 1999). In endosperm, only the maternal *MEA* allele was expressed and the paternal allele was silenced whereas transcripts for both alleles were detected in embryos (Kinoshita *et al.*, 1999). The question arises, would manipulation of a corresponding gene in cereals provide a means of increasing endosperm cell number and deferring cellularization? Such an approach may be as valid as attempting to manipulate the cell cycle machinery directly or via the action of plant hormones. Whatever the case, even a slight delay in the initiation of cell wall formation may have a profound influence on the number of cells produced in cereal endosperm.

VII.   PRE-HARVEST SPROUTING: A DELICATE BALANCE BETWEEN EMBRYO DORMANCY AND GERMINATION

A characteristic of many wild species is an extended period of seed maturation and dispersal coupled with an effective dormancy mechanism to ensure maximum survival. Wild cereals that originated in the Near East are characterized by a dormancy pattern that is adapted to the short growing season before the shed seed passes the long, hot summer in the dormant state (MacKey, 1976). During domestication, the gradual northerly movement of cereals to more marginal maritime climates would have highlighted the requirement for an effective dormancy mechanism. However, besides selecting for non-shattering and larger seeds that ripened over a shorter period of time, early farmers may have also selected material with low dormancy in order to increase germination vigour of subsequent crops. The question arises, has the potential for pre-harvest sprouting (PHS) increased during domestication and the more recent breeding of cereal varieties, compared with the dormancy characteristics of related wild grass species?

The occurrence of PHS in commercially grown cereals is due to insufficient embryo or coat-imposed dormancy, resulting in pre-germination in the ear (Fig. 5) and is prevalent in seasons with cool damp autumns or heavy rainfall around harvest time (Gale, 1989; King, 1989). Because of a strong interaction between genotype and environmental conditions, PHS has proved a difficult character to select for in conventional breeding programmes. In general, wheat and barley plants grown under cool conditions throughout grain development produce more dormant embryos than their counterparts grown under warm conditions (Walker-Simmons and Sesing, 1990; Schuurink et al., 1992b; Garello and Le Page-Degivry, 1999). The main reason that PHS is detrimental to grain quality is the production of hydrolytic enzymes that accelerate the degradation of endosperm storage reserves, as occurs during normal germination of fully after-ripened grain. In barley, PHS renders grain unfit for malting whereas in wheat the presence of α-amylase in flour causes dextrin production during baking resulting in loaves with sticky crumb structure. The background relating to PHS has focused mainly on the genetic interactions involved in the maturation and germination programmes during grain development (McCarty, 1995) as well as the production and counteractive effects of GA and ABA on embryo germination and α-amylase formation in barley (Jacobsen et al., 1995) and wheat (Appleford and Lenton, 1997).

Classically, primary dormancy is considered to be either coat-imposed, involving a source of inhibitors and/or a physical barrier to water and gas exchange, or true embryo dormancy where excised embryos fail to germinate under otherwise favourable environmental conditions. Clearly, an ability to regulate PHS in cereals will depend on understanding how and when dormancy is induced, what genes are involved and how they interact with environment, as well as the process by which dormancy is broken and how and when the germination programme begins. There is a vast body of literature covering the physiology of seed dormancy (Bewley, 1997) and several genetic loci involved in seed maturation and dormancy induction have been characterized recently (Holdsworth *et al.*, 1999; Wobus and Weber, 1999), including dormancy traits in grasses (Li and Foley, 1997). Compelling evidence for the role of ABA in preventing precocious germination has come from developmental mutants of tomato and *Arabidopsis* that are blocked either in the production of, or responsiveness to, the hormone (Hilhorst, 1995; Karssen, 1995). This conclusion was

Fig. 5.   Severe pre-harvest sprouting in the ear of wheat, variety Boxer, grown under cool, moist conditions during the later stages of grain growth.

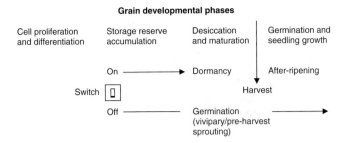

Fig. 6.   Scheme of cereal grain developmental phases associated with the switch between the pathway to embryo dormancy or pre-harvest sprouting.

strengthened by the manipulation of seed dormancy in transgenic tobacco plants following the expression of a single chain Fv antibody directed against ABA (Phillips *et al.*, 1997) or by perturbing expression of an ABA biosynthesis gene (Frey *et al.*, 1999). Based on the results from the developmental mutants, a clear distinction was made between the pivotal role of ABA in primary dormancy, compared with that of GA in germination, with changes in hormone sensitivity being affected by environmental factors (Hilhorst and Karssen, 1992). It is also important to appreciate that the study of seed dormancy is complicated by variation in sensitivity thresholds to water availability and hormonal responses between individual seeds within a population (Ni and Bradford, 1993; Trewavas, 1988). The idea that individuals have different basal response thresholds has provided a useful conceptual framework for analysing how plant hormones and other factors may modify developmental time scales (Bradford and Trewavas, 1994).

In terms of developmental physiology, the later stages of cereal grain growth include cessation of embryo differentiation, acquisition of desiccation tolerance and induction of dormancy in both embryo and aleurone (Fig. 6). Desiccation tolerance of wheat embryos was induced by a relatively small loss in water content but without an increase in ABA or the oligosaccharide, raffinose, both of which had been thought previously to be involved in this process (Black *et al.*, 1999). Instead, the accumulation of a dehydrin protein was positively related to the initiation of desiccation tolerance of developing embryos of intact wheat grains. Besides arresting embryo development, desiccation is an important signal inducing embryo dormancy of wheat, thereby preventing initiation of the germination programme (King, 1993). In this case, the ABA content of embryos declined during the imposition of desiccation-induced dormancy, compared with those maintained in moist conditions, suggesting that the hormone is not directly involved in the process of dormancy induction in wheat. Other evidence supports the view that a change in ABA responsiveness, rather than content, is more directly related to the maintenance of embryo dormancy. Developing wheat embryos isolated from varieties showing strong dormancy were more ABA responsive than those susceptible to PHS (Walker-Simmons, 1987). The same was also true for varieties of both wheat and barley grown under relatively cool conditions that induce stronger dormancy than grains from plants grown under warmer conditions (Walker-Simmons and Sesing, 1990; van Beckum et al., 1993; Garello and Le Page-Degrivry, 1999). Expression of a gene, *Per1*, that encoded an antioxidant protein thought to protect cells from reactive oxygen species during desiccation, correlated with the degree of dormancy in both barley embryo and aleurone tissues (Stacy et al., 1996). Transcripts were up-regulated by ABA and osmotica and suppressed by GA. In barley and wild oat, the degree of embryo dormancy was also related to the GA responsiveness of aleurone with less dormant aleurone being more GA responsive (Hooley, 1992; Schuurink et al., 1992a). An induced mutant of a malting barley cultivar showed reduced embryo dormancy and ABA responsiveness as well as increased α-amylase activity after micromalting (Molina-Cano et al., 1999). Decreased embryo sensitivity to ABA during the later stages of grain development was also associated with a low dormancy phenotype in selected wheat mutants (Kawakami et al., 1997).

The capacity of mature embryos of wheat (Ried and Walker-Simmons, 1990) and barley (Wang et al., 1995) to reinitiate ABA synthesis following imbibition suggested a potential overlap of ABA from dormancy into the germination programme. High-temperature-induced dormancy of imbibed mature cereal grains may also reflect ABA production that can be blocked by the carotenoid biosynthesis inhibitor, fluridone (Garello and Le Page-Degivry, 1999). Conversely, rapid germination of an induced barley mutant was related to a faster metabolism of ABA with no difference in ABA responsiveness of group 2 *LEA* gene expression, compared with the parent line (Visser et al., 1996). The potential for positive involvement of ABA in preventing germination of mature cereal grain begs the question of a possible role for GA during the later stages of grain development. In *Arabidopsis*, overexpression of GA 20-oxidase decreased seed dormancy, measured as the period of after-ripening required for near complete germination (Huang et al., 1998). Recent evidence from mutants of barley selected on the basis of grain shrivelling showed a positive GA dependency for pre-maturity α-amylase formation during grain development (Green et al., 1997). Similarly, the constitutive GA-response mutant, "slender", in the background variety Himalaya barley, showed precocious germination in the ear (P. M. Chandler, personal communication). Overall, it would appear that, at least in barley and wheat, the antagonistic functions of ABA and GA have the potential to overlap during grain maturation and germination.

In many regions world-wide, certain wheat genotypes have a predisposition to produce high α-amylase activity in developing grains, even in the absence of visible sprouting (Gale, 1989). Genetic analysis indicated a major component of pre-maturity α-amylase (PMAA) production was controlled by a single recessive allele in triploid endosperm, highlighting the difficulty in selecting homozygous low α-amylase lines (Mrva and Mares, 1996a). Although the presence of the GA-insensitivity gene, *Rht3*, reduced PMAA activity in developing wheat grains (Flintham and Gale, 1982; Mrva and Mares, 1996b), no direct evidence was obtained for GA involvement in PMAA production in a GA-responsive line growing under field conditions (Gale and Lenton, 1987). At present, the precise tissue origin and the regulatory mechanisms involved in PMAA formation require clarification before effective strategies for gene silencing, in a manner analogous to that for polygalacturonase during tomato fruit ripening, could be developed. Claims that the groove aleurone of large-sized grains was the main source of PMAA (Cornford *et al.*, 1987; Evers *et al.*, 1995) was not supported by other evidence that favoured a germination model with initial production of enzyme activity in the embryo and subsequent synthesis in aleurone (Gale and Lenton, 1987; Marchylo and Kruger, 1987). The extent to which hormonal and/or sugar signals (Perata *et al.*, 1997; Umemura *et al.*, 1998) are involved in PMAA synthesis from the scutellar epithelium and aleurone remains to be fully evaluated.

As with the related ABA mutants of dicots, the viviparous mutants of maize that are blocked in ABA production or responsiveness do not undergo normal seed maturation and enter a default germination programme (Fig. 6) (McCarty, 1995). Precocious germination of embryos on the cob of the ABA-insensitive *vp1* mutant is phenotypically similar to PHS in temperate cereals, suggesting possible disruption of VP1 function, which in wheat and barley becomes evident particularly after unfavourable weather conditions (Holdsworth *et al.*, 1999). In maize, the transcription factor, VP1, up-regulates genes involved in the maturation programme and confers responsiveness of tissues to ABA (McCarty, 1995). In addition, maize VP1 represses germination-associated genes, such as α-amylase, in aleurone (Hoecker *et al.*, 1995), a property that may have important implications for the regulation of PMAA production as well as embryo dormancy. More recent genetic evidence indicated that wild-type embryos were also a source of inhibitory signals that acted independently of VP1 action in maize aleurone (Hoecker *et al.*, 1999). Embryo sensitivity to ABA was associated with one of two genes regulating embryo dormancy in the Australian white wheat, AUS 1408 (Mares, 1996). These two independent recessive genes mapped to chromosome 3, the same chromosomal location as that found subsequently for the wheat *Vp1* homologue (*TaVp1*) (Bailey *et al.*, 1999). Recent evidence suggested that the relatively weak embryo dormancy of modern wheat varieties may be related to incorrect splicing of *TaVp1* transcripts that would result in translation of non-functional VP1 protein lacking a DNA binding domain (McKibbin *et al.*, 1999). In addition, the abundance of *TaVp1* transcripts declined, whereas α-amylase gene expression increased, in non-sprouted embryos from grains of plants grown under sprout-inducing conditions. In marked contrast, expression of the homologous *Vp1* gene in inbred lines of the arable weed, wild oat, was closely associated with the depth of embryo dormancy, declining during after-ripening and increasing again during the induction of secondary dormancy (Jones *et al.*, 1997). Such a close association between the pattern of expression of the wild oat *Vp1* gene and embryo dormancy suggested that it may be a useful gene for generation of prototype transgenic wheat lines with altered embryo dormancy characteristics, a proposal currently being explored within the Institute for Arable Crops Research.

VIII. PROSPECTS

There can be no doubt that there have been remarkable increases in crop yields during the latter half of the 20th century but more has to be achieved in order to feed the expanding world population adequately. Whilst much of the increased yields has come about by mechanization and the introduction of artificial fertilizers and crop protection chemicals, a considerable proportion has also been due to the introduction of higher yielding genotypes with increased pest and disease resistance characters. Conventional and novel breeding technologies, including plant transformation, rather than improvements in agronomic practices, are the most likely means by which the increased demand for cereals will have to be met in future. Given the rapid pace of advances in genome sequencing and gene technology, there are good reasons to be cautiously optimistic about the potential to produce novel crops tailored to specific environments. As the function of more of the key genes regulating specific developmental processes become known, there remains a considerable gap in exploiting this information for the production of improved phenotypes. In other words, knowledge of gene function also requires a similar degree of understanding of processes at higher levels of organization, culminating in the interaction with environmental factors that affect the performance of a population of individuals in the field. Perhaps it is in these currently unfashionable areas of science that the greatest challenges lie ahead.

ACKNOWLEDGEMENTS

The Institute of Arable Crops Research receives grant-aided support from the Biotechnology and Biological Sciences Research Council. The author thanks too many colleagues to mention individually for enlightenment over the years but accepts sole responsibility for the viewpoints expressed in this review.

REFERENCES

Adams, R., Kerber, E., Pfister, K. and Weiler, E. W. (1992). Studies on the action of CGA 163′935 (Cimectacarb). *In* "Progress in Plant Growth Regulation" (C. M. Karssen, L. C. van Loon and D. Vreugdenhil, eds), pp. 818–827. Kluwer Academic Publishers, Dordrecht.

Appleford, N. E. J. and Lenton, J. R. (1991). Gibberellins and leaf expansion in near-isogenic wheat lines containing *Rht1* and *Rht3* dwarfing alleles. *Planta* **183**, 229–236.

Appleford, N. E. J. and Lenton, J. R. (1997). Hormonal regulation of α-amylase gene expression in germinating wheat (*Triticum aestivum*) grains. *Physiologia Plantarum* **100**, 534–542.

Ashikari, M., Wu, J., Yano, M., Sasaki, T. and Yoshimura, A. (1999). Rice gibberellin-insensitive dwarf mutant gene *Dwarf 1* encodes the α-subunit of GTP-binding protein. *Proceedings of the National Academy of Sciences (USA)* **96**, 10284–10289.

Bailey, P. C., McKibbin, R. S., Lenton, J. R., Holdsworth, M. J., Flintham, J. E. and Gale, M. D. (1999). Genetic map locations for orthologous *Vp1* genes in wheat and rice. *Theoretical and Applied Genetics* **98**, 281–284.

Bangerth, F. (1989). Dominance among fruits/sinks and the search for a correlative signal. *Physiologia Plantarum* **76**, 608–614.

Barro, F., Rooke, L., Békés, F., Gras, P., Tatham, A. S., Fido, R., Lazzeri, P. A., Shewry, P. R. and Barceló, P. (1997). Transformation of wheat with high molecular weight subunit genes results in improved functional properties. *Nature Biotechnology* **15**, 1295–1299.

Baum, D. A. (1998). The evolution of plant development. *Current Opinion in Plant Biology* **1**, 79–86.

Beall, F. D., Morgan, P. W., Mander, L. N. and Babb, K. H. (1991). Genetic regulation of development in *Sorghum bicolor*. V. The ma_3^R allele results in gibberellin enrichment. *Plant Physiology* **95**, 116–125.

Bennett, M. D., Rao, M. K., Smith, J. B. and Bayliss, M. W. (1973). Cell development in the anther, ovule and young seed of *Triticum aestivum* L., var. Chinese Spring. *Philosophical Transactions of the Royal Society of London* **266**, 39–81.

Bennett, M. D., Smith, J. B. and Barclay, I. (1975). Early seed development in the triticeae. *Philosophical Transactions of the Royal Society of London* **272**, 199–227.

Bewley, J. D. (1997). Seed germination and dormancy. *Plant Cell* **9**, 1055–1066.

Black, M., Corbineau, F., Gee, H. and Côme, D. (1999). Water content, raffinose, and dehydrins in the induction of desiccation tolerance in immature wheat embryos. *Plant Physiology* **120**, 463–471.

Börner, A., Röder, M. and Korzun, V. (1997). Comparative molecular mapping of GA insensitive *Rht* loci on chromosomes 4B and 4D of common wheat (*Triticum aestivum* L.). *Theoretical and Applied Genetics* **95**, 1133–1137.

Bradford, K. J. and Trewavas, A. J. (1994). Sensitivity thresholds and variable time scales in plant hormone action. *Plant Physiology* **105**, 1029–1036.

Briarty, L. G., Hughes, C. E. and Evers, A. D. (1979). The developing endosperm of wheat – a stereological analysis. *Annals of Botany* **44**, 641–658.

Brocklehurst, P. A. (1977). Factors controlling grain weight in wheat. *Nature* **266**, 348–349.

Burden, R. S., Carter, G. A., Clark, T., Cooke, D. T., Deas, A. H. B., Hedden, P., James, C. S. and Lenton, J. R. (1987). Comparative activity of the enantiomers of triadimenol and paclobutrazol as inhibitors of fungal growth and plant sterol and gibberellin biosynthesis. *Pesticide Science* **21**, 253–267.

Chandler, P. M. (1988). Hormonal regulation of gene expression in the "slender" mutant of barley (*Hordeum vulgare* L.). *Planta* **175**, 115–120.

Chandler, P. M. and Robertson, M. (1999). Gibberellin dose-response curves and the characterisation of dwarf mutants of barley. *Plant Physiology* **120**, 623–632.

Chiang, H-H., Hwang, I. and Goodman, H. M. (1995). Isolation of the *Arabidopsis GA4* locus. *Plant Cell* **7**, 195–201.

Childs, K. L., Miller, F. R., Cordonnier-Pratt, M. M., Pratt, L. H., Morgan, P. W. and Mullet, J. E. (1997). The *Sorgum bicolor* photoperiod sensitive gene, *Ma₃*, encodes a phytochrome B. *Plant Physiology* **113**, 611–619.

Cochrane, M. P. and Duffus, C. M. (1983). Endosperm cell number in cultivars of barley differing in grain weight. *Annals of Applied Biology* **102**, 177–181.

Coles, J. P., Phillips, A. L., Croker, S. J., Garcia-Lepe, R., Lewis, M. J. and Hedden, P. (1999). Modification of gibberellin production and plant development in *Arabidopsis* by sense and antisense expression of gibberellin 20-oxidase genes. *Plant Journal* **17**, 547–556.

Conway, G. and Toenniessen, G. (1999). Feeding the world in the twenty-first century. *Nature* **402**, C55-C58.

Cornford, C. A., Black, M. and Chapman, J. (1987). Sensitivity of developing wheat grains to gibberellin and production of alpha-amylase during grain development and maturation. *In* "Fourth International Symposium on Pre-harvest Sprouting in Cereals" (D. J. Mares, ed.), pp. 283–292. Westview Press, Boulder, CO.

Cowling, R. J., Kamiya, Y., Seto, H. and Harberd, N. P. (1998). Gibberellin dose–response regulation of *GA4* gene transcript levels in *Arabidopsis*. *Plant Physiology* **117**, 1195–1203.

Croker, S. J., Hedden, P., Lenton, J. R. and Stoddart, J. l. (1990). Comparison of gibberellins in normal and slender barley seedlings. *Plant Physiology* **94**, 194–200.

Devos, K. M. and Gale, M. D. (1997). Comparative genetics of grasses. *Plant Molecular Biology* **35**, 3–15.

Dietrich, J. T., Kaminek, M., Blevins, D. G., Reinbott, T. M. and Morris, R. O. (1995). Changes in cytokinins and cytokinin oxidase activity in developing maize kernels and the effect of exogenous cytokinin on kernel development. *Plant Physiology and Biochemistry* **33**, 327–336.

Doebley, J. and Lukens, L. (1998). Transcriptional regulators and the evolution of plant form. *Plant Cell* **10**, 1075–1082.

Doebley, J., Stec, A. and Gustus, C. (1995). *teosinte branched1* and the origin of maize: evidence for epistasis and the evolution of dominance. *Genetics* **141**, 333–346.

Doebley, J., Stec, A. and Hubbard, L. (1997). The evolution of apical dominance in maize. *Nature* **386**, 485–488.

Dunstone, R. L. and Evans, L. T. (1974). Rôle of changes in cell size in the evolution of wheat. *Australian Journal of Plant Physiology* **1**, 157–165.

Dyson, T. (1999). World food trends and prospects to 2025. *Proceedings of the National Academy of Sciences (USA)* **96**, 5929–5936.

Engell, K. (1994). Embryology of barley. IV. Ultrastructure of the antipodal cells of *Hordeum vulgare* L. cv. Bomi before and after fertilisation of the egg cell. *Sexual Plant Reproduction* **7**, 333–346.

Evans, L. T. (1993). "Crop Evolution, Adaption and Yield". Cambridge University Press, Cambridge.

Evans, L. T. (1998). "Feeding the Ten Billion. Plants and Population Growth". Cambridge University Press, Cambridge.

Evans, L. T., Bingham, J. and Roskams, M. A. (1972). The pattern of grain set within ears of wheat. *Australian Journal of Biological Sciences* **25**, 1–8.

Evans, L. T., King, R. W., Mander, L. N., Pharis, R. P. and Duncan, K. A. (1994). The differential effects of C-16,17-dihydro gibberellins and related compounds on stem elongation and flowering in *Lolium temulentum*. *Planta* **193**, 107–114.

Evers, A. D. (1970). Development of the endosperm of wheat. *Annals of Botany* **34**, 547–555.

Evers, A. D., Flintham, J. and Kotecha, K. (1995). Alpha-amylase and grain size in wheat. *Journal of Cereal Science* **21**, 1–3.

Fedoroff, N. V.and Cohen, J. E. (1999). Plants and population: Is there time? *Proceedings of the National Academy of Sciences (USA)* **96**, 5903–5907.

Flintham, J. E. and Gale, M. D. (1982). The Tom Thumb dwarfing gene, *Rht3* in wheat, 1. Reduced pre-harvest damage to breadmaking quality. *Theoretical and Applied Genetics* **62**, 121–126.

Flintham, J. E., Börner, A., Worland, A. J. and Gale, M. D. (1997). Optimising wheat grain yield: effects of *Rht* (gibberellin-insensitive) dwarfing genes. *Journal of Agricultural Science, Cambridge* **128**, 11–25.

Frank, M. and Schmülling, T. (1999). Cytokinin cycles cells. *Trends in Plant Science* **4**, 243–244.

Frey, A., Audran, C., Marin, E., Sotta, B. and Marion-Poll, A. (1999). Engineering seed dormancy by the modification of zeaxanthin epoxidase gene expression. *Plant Molecular Biology* **39**, 1267–1274.

Frugis, G., Giannino, D., Mele, G., Nicolodi, C., Innocenti, M., Chiappetta, A., Bitonti, M. B., Dewitte, W., van Onckelen, H. and Mariotti, D. (1999). Are homeobox *knotted*-like genes and cytokinins the leaf architects? *Plant Physiology* **119**, 371–373.

Fujioka, S., Yamane, H., Spray, C. R., Katsumi, M., Phinney, B. O., Gaskin, P., MacMillan, J. and Takahashi, N. (1988). The dominant non-gibberellin-responding dwarf mutant (*D8*) of maize accumulates native gibberellins. *Proceedings of the National Academy of Sciences (USA)* **85**, 9031–9035.

Fujioka, S., Yamane, H., Spray, C. R., Phinney, B. O., Gaskin, P., MacMillan, J. and Takahashi, N. (1990). Gibberellin A_3 is biosynthesised from gibberellin A_{20} via gibberellin A_5 in shoots of *Zea mays* L. *Plant Physiology* **94**, 127–131.

Fujisawa, Y., Kato, T., Ohki, S., Ishikawa, A., Kitano, H., Sasaki, T., Asahi, T. and Iwasaki, Y. (1999). Suppression of the heterotrimeric G protein causes abnormal morphology, including dwarfism, in rice. *Proceedings of the National Academy of Sciences (USA)* **96**, 7575–7580.

Futsuhara, Y. and Kikuchi, F. (1997). Dwarf characters. *In* "Science of the Rice Plant. Vol. 3. Genetics" (T. Matsuo, Y. Futsuhara, F. Kikuchi and H. Yamaguchi, eds), pp. 300–317. Food and Agriculture Policy Research Center, Tokyo.

Gale, M. D. (1989). The genetics of preharvest sprouting in cereals, particularly in wheat. *In* "Preharvest Field Sprouting in Cereals" (N. F. Derera, ed.), pp. 85–110. CRC Press, Boca Raton, FL.

Gale, M. D. and Lenton, J. R. (1987). Preharvest sprouting in wheat – a complex genetic and physiological problem affecting breadmaking quality of UK wheats. *Aspects of Applied Biology* **15**, 115–124.

Gale, M. D. and Youssefian, S. (1983). Pleiotropic effects of the Norin 10 dwarfing genes, *Rht1* and *Rht2*, and interactions in response to chlormequat. *In* "Proceedings of the Sixth International Wheat Genetics Symposium," pp. 271–277. Kyoto, Japan.

Gale, M. D. and Youssefian, S. (1985). Dwarfing genes in wheat. *In* "Progress in Plant Breeding" (G. E. Russell, ed.), pp. 1–35. Butterworths, London.

Gao, X., Francis, D., Ormrod, J. C. and Bennett, M. D. (1992). Changes in cell number and cell division activity during endosperm development in allohexaploid wheat, *Triticum aestivum* L. *Journal of Experimental Botany* **43**, 1603–1609.

Garello, G. and Le Page-Degivry, M. T. (1999). Evidence for the role of abscisic acid in the genetic and environmental control of dormancy in wheat (*Triticum aestivum* L.). *Seed Science Research* **9**, 219–226.

Graebe, J. E., Böse, G., Grosselindemann, E., Hedden, P., Aach, H., Schweimer, A., Sydow, S. and Lange, T. (1992). The biosynthesis of *ent*-kaurene in germinating seeds and the function of 2-oxoglutarate in gibberellin biosynthesis. *In* "Progress in Plant Growth Regulation" (C. M. Karssen, L. C. van Loon and D. Vreugdenhil, eds), pp. 545–554. Kluwer Academic Publishers, Dordrecht.

Grafi, G. and Larkins, B. A. (1995). Endoreduplication in maize endosperm: involvement of M-phase promoting factor inhibition and induction of S phase-related kinases. *Science* **269**, 1262–1264.

Grafi, G., Burnett, R. J., Helentjaris, T., Larkins, B. A., DeCaprio, J. A., Sellers, W. R. and Kaelin, W. G. (1996). A maize cDNA encoding a member of the retinoblastoma protein family: involvement in endoreduplication. *Proceedings of the National Academy of Sciences (USA)* **93**, 8962–8967.

Green, L. S., Færgestad, E. M., Poole, A. and Chandler, P. M. (1997). Grain development mutants in barley. α-Amylase production during grain maturation and its relation to endogenous gibberellic acid content. *Plant Physiology* **114**, 203–212.

Griggs, D. L., Hedden, P., Temple-Smith, K. E. and Rademacher, W. (1991). Inhibition of gibberellin 2β-hydroxylases by acylcyclohexanedione derivatives. *Phytochemistry* **30**, 2513–2517.

Grossniklaus, U., Vielle-Cazada, J-P., Hoeppner, M. A. and Gagliano, W. B. (1998). Maternal control of embryogenesis by *MEDEA*, a polcomb-group gene in *Arabidopsis*. *Science* **280**, 446–450.

Harberd, N. P., King, K. E., Carol, P., Cowling, R. J., Peng, J. and Richards, D. E. (1998). Gibberellin: inhibitor of an inhibitor of...? *BioEssays* **20**, 1001–1008.

Hedden, P. (1999). Recent advances in gibberellin biosynthesis. *Journal of Experimental Botany* **50**, 553–563.

Hedden, P and Croker, S. J. (1992). Regulation of gibberellin biosynthesis in maize seedlings. *In* "Progress in Plant Growth Regulation" (C. M. Karssen, L. C. van Loon and D. Vreugdenhil, eds), pp. 534–544. Kluwer Academic Publishers, Dordrecht.

Hedden, P. and Graebe, J. E. (1985). Inhibition of gibberellin biosynthesis by cell-free homogenates of *Cucurbita maxima* endosperm and *Malus pumila* embryos. *Journal of Plant Growth Regulation* **4**, 111–122.

Hedden, P. and Kamiya, Y. (1997). Gibberellin biosynthesis: enzymes, genes and their regulation. *Annual Review of Plant Physiology and Plant Molecular Biology* **48**, 431–460.

Hedden, P. and Proebsting, W. M. (1999). Genetic analysis of gibberellin biosynthesis. *Plant Physiology* **119**, 365–370.

Hilhorst, H. W. M. (1995). A critical update on seed dormancy. 1. Primary dormancy. *Seed Science Research* **5**, 61–73.

Hilhorst, H. W. M. and Karssen, C. M. (1992). Seed dormancy and germination: the role of abscisic acid and gibberellins and the importance of hormone mutants. *Plant Growth Regulation* **11**, 225–238.

Hoecker, U., Vasil, I. K. and McCarty, D. R. (1995). Integrated control of seed maturation and germination programs by activator and repressor functions of Viviparous-1 of maize. *Genes and Development* **9**, 2459–2469.

Hoecker, U., Vasil, I. K. and McCarty, D. R. (1999). Signaling from the embryo conditions *Vp1*-mediated repression of α-amylase genes in the aleurone of developing maize seeds. *Plant Journal* **19**, 371–377.

Holdsworth, M., Kurup, S. and McKibbin, R. (1999). Molecular and genetic mechanisms regulating the transition from embryo development to germination. *Trends in Plant Science* **4**, 275–280.

Hoogendoorn, J., Rickson, J. M. and Gale, M. D. (1990). Differences in leaf and stem anatomy related to plant height of tall and dwarf wheat (*Triticum aestivum* L.). *Journal of Plant Physiology* **136**, 72–77.

Hooley, R. (1992). The responsiveness of *Avena fatua* aleurone protoplasts to gibberellic acid. *Plant Growth Regulation* **11**, 85–89.

Hooley, R. (1994). Gibberellins: perception, transduction and responses. *Plant Molecular Biology* **26**, 1529–1555.

Huang, S., Raman, A. S., Ream, J. E., Fujiwara, H., Cerny, R. E. and Brown, S. M. (1998). Overexpression of 20-oxidase confers a gibberellin-overproduction phenotype in *Arabidopsis*. *Plant Physiology* **118**, 773–781.

Huber, A. G. and Grabe, D. F. (1987a). Endosperm morphogenesis in wheat: transfer of nutrients from antipodals to the lower endosperm. *Crop Science* **27**, 1248–1252.

Huber, A. G. and Grabe, D. F. (1987b). Endosperm morphogenesis in wheat: termination of nuclear division. *Crop Science* **27**, 1252–1256.

Huntley, R. P. and Murray, J. A. H. (1999). The plant cell cycle. *Current Opinion in Plant Biology* **2**, 440–446.

Huntley, R., Healy, S., Freeman, D., Lavender, P., de Jager, S., Greenwood, J., Makker, J., Walker, E., Jackman, M., Xie, Q., Bannister, A. J., Kouarides, T., Gutiérrez, C., Doonan, J. H. and Murray, J. A. H. (1998). The maize retinoblastoma protein homologue ZmRb-1 is regulated during leaf development and displays conserved interactions with G1/S regulators and plant cyclin D (CycD) proteins. *Plant Molecular Biology* **37**, 155–169.

Ingram, T. J., Reid, J. B. and MacMillan, J. (1986). The quantitative relationship between GA$_1$ and internode growth in *Pisum sativum* L. *Planta* **168**, 414–420.

Itoh, H., Tanaka-Ueguchi, M., Kawaide, H., Chen, X., Kamiya, Y. and Matsuoka, M. (1999). The gene encoding tobacco gibberellin 3β-hydroxylase is expressed at the site of GA action during stem elongation and flower organ development. *Plant Journal* **20**, 15–24.

Ivandic, V., Malyshev, S., Korzun, V., Graner, A. and Börner, A. (1999). Comparative mapping of a gibberellic acid-insensitive dwarfing gene (*Dwf2*) on chromosome 4HS in barley. *Theoretical and Applied Genetics* **98**, 728–731.

Jacobsen, J. V., Gubler, F. and Chandler, P. M. (1995). Gibberellin action in germinated cereal grains. *In* "Plant Hormones. Physiology, Biochemistry and Molecular Biology" (P. J. Davies, ed.), pp. 246–271. Kluwer Academic Publishers, Dordrecht.

Jones, H. D., Peters, N. C. B. and Holdsworth, M. J. (1997). Genotype and environment interact to control dormancy and differential expression of the *VIVIPAROUS1* homologue in embryos of *Avena fatua*. *Plant Journal* **12**, 911–920.

Jones, H. D., Smith, S. J., Desikan, R., Plakidou-Dymock, S., Lovegrove, A. and Hooley, R. (1998). Heterotrimeric G proteins are implicated in gibberellin induction of α-amylase gene expression in wild oat aleurone. *Plant Cell* **10**, 245–253.

Junttila, O., King, R. W., Poole, A., Kretschmer, G., Pharis, R. P. and Evans, L. T. (1997). Regulation in *Lolium temulentum* of the metabolism of gibberellin A_{20} and gibberellin A_1 by 16,17-dihydro GA_5 and by the growth retardant, LAB 198 999. *Australian Journal of Plant Physiology* **24**, 359–369.

Kamiya, Y. and Garcia-Martinez, J. L. (1999). Regulation of gibberellin biosynthesis by light. *Current Opinion in Plant Biology* **2**, 398–403.

Kamiya, Y., Nakayama, I. and Kobayashi, M. (1992). Useful probes to study the biosynthesis of gibberellins. *In* "Progress in Plant Growth Regulation" (C. M. Karssen, L. C. van Loon, and D. Vreugdenhil, eds), pp. 555–565. Kluwer Academic Publishers, Dordrecht.

Karssen, C. M. (1995). Hormonal regulation of seed development, dormancy and germination studied by genetic control. *In* "Seed Development and Germination" (J. Kigel and G. Galili, eds), pp. 333–350. Marcel Dekker, New York.

Kawakami, N., Miyake, Y. and Noda, K. (1997). ABA sensitivity and low ABA levels during seed development of non-dormant wheat mutants. *Journal of Experimental Botany* **48**, 1415–1421.

Kende, H., van der Knaap, E. and Cho, H-T. (1998). Deepwater rice: a model to study stem elongation. *Plant Physiology* **118**, 1105–1110.

Keyes, G. J., Paolillo, D. J. and Sorrells, M. E. (1989). The effects of dwarfing genes *Rht1* and *Rht2* on cellular dimensions and rate of leaf elongation in wheat. *Annals of Botany* **64**, 683–690.

Keyes, G. J., Sorrells, M. E. and Setter, T. L. (1990). Gibberellic acid regulates cell extensibility in wheat (*Triticum aestivum* L.). *Plant Physiology* **92**, 242–245.

King, R. W. (1989). Physiology of sprouting resistance. *In* "Preharvest Field Sprouting in Cereals" (N. F. Derera, ed.), pp. 27–60. CRC Press, Boca Raton, FL.

King, R. W. (1993). Manipulation of grain dormancy in wheat. *Journal of Experimental Botany* **44**, 1059–1066.

Kinoshita, T., Yadegari, R., Harada, J. J., Goldberg, R. B. and Fischer, R. L. (1999). Imprinting of the *MEDEA* polcomb gene in the *Arabidopsis* endosperm. *Plant Cell* **11**, 1945–1952.

Kiyosue, T., Ohad, N., Yadegari, R., Hannon, M., Dinneny, J., Wells, D., Katz, A., Margossian, L., Harada, J. J., Goldberg, R. B. and Fischer, R. L. (1999). Control of fertilisation-independent endosperm development by *MEDEA* polcomb gene in *Arabidopsis*. *Proceedings of the National Academy of Sciences (USA)* **96**, 4186–4191.

Kobayashi, M., Sakuri, A., Saka, H. and Takahashi, N. (1989). Quantitative analysis of endogenous gibberellins in normal and dwarf cultivars of rice. *Plant and Cell Physiology* **30**, 963–969.

Kusaba, S., Fukumoto, M., Honda, C., Yamaguchi, I., Sakamoto, T. and Kano-Murakami, Y. (1998a). Decreased GA_1 content caused by overexpression of *OSH1* is accompanied by suppression of GA 20-oxidase gene expression. *Plant Physiology* **117**, 1179–1184.

Kusaba, S., Kano-Murakami, Y., Matsuoka, M., Tamaoki, M., Sakamoto, T., Yamaguchi, I. and Fukumoto, M. (1998b). Alteration of hormone levels in transgenic tobacco plants overexpressing the rice homeobox gene *OSH1*. *Plant Physiology* **116**, 471–476.

Lanahan, M. B. and Ho, T.-H. D. (1988). Slender barley: a constitutive gibberellin response mutant. *Planta* **175**, 107–114.

Lange, T. (1998). Molecular biology of gibberellin synthesis. *Planta* **204**, 409–419.

Lange, T., Hedden, P. and Graebe, J. E. (1994). Expression cloning of a gibberellin 20-oxidase, a multifunctional enzyme involved in gibberellin biosynthesis. *Proceedings of the National Academy of Sciences (USA)* **91**, 8552–8556.

Laureys, F., Dewitte, W., Witters, E., van Montague, M., Inzé, D. and van Onckelen, H. (1998). Zeatin is indispensable for the G_2-M transition in tobacco BY-2 cells. *Federation of European Biochemical Societies* **426**, 29–32.

Laurie, D. A. (1997). Comparative genetics of flowering time. *Plant Molecular Biology* **35**, 167–177.

Law, C. N. (1995). Genetic manipulation in plant breeding – prospects and limitations. *Euphytica* **85**, 1–12.

Law, C. N. and Worland, A. J. (1997). Genetic analysis of some flowering time and adaptive traits in wheat. *New Phytologist* **137**, 19–28.

Lenton, J. R., Hedden, P. and Gale, M. D. (1987). Gibberellin insensitivity and depletion in wheat – consequences for development. In "Hormone Action in Plant Development – A Critical Appraisal" (G. V. Hoad, J. R. Lenton, M. B. Jackson and R. K. Atkin, eds), pp. 145–160. Butterworths, London.

Lenton, J. R., Appleford, N. E. J. and Temple-Smith, K. E. (1994). Growth retardant activity of paclobutrazol enantiomers in wheat seedlings. *Plant Growth Regulation* **15**, 281–291.

Lester, D. R., Ross, J. J., Davies, P. J. and Reid, J. B. (1997). Mendel's stem length gene (*Le*) encodes a gibberellin 3β-hydroxlase. *Plant Cell* **9**, 1435–1443.

Lester, D. R., Ross, J. J., Smith, J. J., Elliott, R. C. and Reid, J. B. (1999). Gibberellin 2-oxidation and the *SLN* gene of *Pisum sativum*. *Plant Journal* **19**, 65–73.

Levy, Y. Y. and Dean, C. (1998). The transition to flowering. *Plant Cell* **10**, 1973–1989.

Li, B. and Foley, M. E. (1997). Genetic and molecular control of seed dormancy. *Trends in Plant Science* **2**, 384–389.

Lin, J.-T. and Stafford, A. E. (1987). Comparison of endogenous gibberellins in the shoots and roots of vernalised and non-vernalised Chinese Spring wheat seedlings. *Phytochemistry* **26**, 2485–2488.

Macilwain, C. (1999). Access issues may determine whether agri-biotech will help the world's poor. *Nature* **402**, 341–345.

MacKey, J. (1976). Seed dormancy in nature and agriculture. *Cereal Research Communications* **4**, 83–91.

MacMillan, J. (1997). Biosynthesis of gibberellin plant hormones. *Natural Products Reports* **14**, 221–243.

Mambelli, S. and Setter, T. L. (1998). Inhibition of maize endosperm cell division and endoreduplication by exogenously applied abscisic acid. *Physiologia Plantarum* **104**, 266–272.

Mann, C. C. (1999). Crop scientists seek a new revolution. *Science* **283**, 310–314.

Marchylo, B. A. and Kruger, J. E. (1987). Degradation of starch granules in maturing wheat and its relationship to alpha-amylase production by the embryo. *In* "Fourth International Symposium on Pre-harvest Sprouting in Cereals" (D. J. Mares, ed.), pp. 483–493. Westview Press, Boulder, CO.

Mares, D. J. (1996). Dormancy in white wheat: mechanism and location of the genes. *In* "PreHarvest Sprouting in Cereals 1995" (K. Noda and D. J. Mares, eds), pp. 179–184. Center for Academic Societies Japan, Osaka.

Martin, C. (1996). Transcription factors and the manipulation of plant traits. *Current Opinion in Biotechnology* **7**, 130–138.

Martin, D. N., Proebsting, W. M., Parks, T. D., Dougherty, W. G., Lange, T., Lewis, M. J., Gaskin, P. and Hedden, P. (1996). Feed-back regulation of gibberellin biosynthesis and gene expression in *Pisum sativum*. *Planta* **200**, 159–166.

Martin, D. N., Proebsting, W. M. and Hedden, P. (1997). Mendel's dwarfing gene: cDNAs from the *Le* alleles and function of the expressed proteins. *Proceedings of the National Academy of Sciences (USA)* **94**, 8907–8911.

Martin, D. N., Proebsting, W. M. and Hedden, P. (1999). The *SLENDER* gene of pea encodes a gibberellin 2-oxidase. *Plant Physiology* **121**, 775–781.

Mazur, B., Krebbers, E. and Tingey, S. (1999). Gene discovery and product development for grain quality traits. *Science* **285**, 372–375.

McCarty, D. R. (1995). Genetic control and integration of maturation and germination pathways in seed development. *Annual Review of Plant Physiology and Plant Molecular Biology* **46**, 71–93.

McKibbin, R., Bailey, P., Flintham, J., Gale, M., Lenton, J. and Holdsworth, M. (1999). Molecular analysis of the wheat *VIVIPAROUS1* (*VP1*) orthologue. *In* "Eighth International Symposium on Pre-Harvest Sprouting in Cereals" (D. Weipert, ed.), pp. 113–118. Association of Cereal Research, Detmold, Germany.

Miralles, D. J., Calderini, D. F., Pomar, K. P. and D'Ambrogio, A. (1998). Dwarfing genes and cell dimensions in different organs of wheat. *Journal of Experimental Botany* **49**, 1119–1127.

Mironov, V., de Veylder, L., van Montagu, M. and Inzé, D. (1999). Cyclin-dependent kinases and cell division in plants – the nexus. *Plant Cell* **11**, 509–521.

Mitsunaga, S., Tashiro, T. and Yamaguchi, J. (1994). Identification and characterisation of gibberellin-insensitive mutants selected from among dwarf mutants of rice. *Theoretical and Applied Genetics* **87**, 705–712.

Molina-Cano, J. L., Sopena, A., Swanston, J. S., Casas, A. M., Moralejo, M. A., Ubieto, A., Lara, I., Pérez-Vendrell, A. M. and Romagosa, I. (1999). A mutant induced in malting barley cv Triumph with reduced dormancy and ABA response. *Theoretical and Applied Genetics* **98**, 347–355.

Morrison, I. N. and O'Brien, T. P. (1976). Cytokinesis in the developing wheat grain: division with and without a phragmoplast. *Planta* **130**, 57–67.

Morrison, I. N., O'Brien, T. P. and Kuo, J. (1978). Initial cellularisation and differentiation of the aleurone cells in the ventral region of the developing wheat grain. *Planta* **140**, 19–30.

Mrva, K. and Mares, D. J. (1996a). Inheritance of late maturity α-amylase in wheat. *Euphytica* **88**, 61–67.

Mrva, K. and Mares, D. J. (1996b). Expression of late maturity α-amylase in wheat containing gibberellic acid insensitivity genes. *Euphytica* **88**, 69–76.

Ni, B-R. and Bradford, K. J. (1993). Germination and dormancy in abscisic acid- and gibberellin-deficient mutant tomato (*Lycopersicon esculentum*) seeds. *Plant Physiology* **101**, 607–617.

Nick, P. and Furuya, M. (1993). Phytochrome dependent decrease in gibberellin-sensitivity. *Plant Growth Regulation* **12**, 195–206.

Nutman, P. S. (1939). Studies in vernalisation in cereals. VI. The anatomical and cytological evidence for the formation of growth-promoting substances in the developing grains of rye. *Annals of Botany* **3**, 731–758.

Olsen, O.-A., Potter, R. H. and Kalla, R. (1992). Histo-differentiation and molecular biology of developing cereal endosperm. *Seed Science Research* **2**, 117–131.

Olsen, O.-A., Brown, R. C. and Lemmon, B. E. (1995). Pattern and process of wall formation in developing endosperm. *BioEssays* **17**, 803–812.

Olsen, O.-A., Linnestad, C. and Nichols, S. E. (1999). Developmental biology of the cereal endosperm. *Trends in Plant Science* **4**, 253–257.

Paolillo, D. J., Sorrells, M. E. and Keyes, G. J. (1991). Gibberellic acid sensitivity determines the length of the extension zone in wheat leaves. *Annals of Botany* **67**, 479–485.

Peng, J., Richards, D. E., Hartley, N. M., Murphy, G. P., Devos, K. M., Flintham, J. E., Beales, J., Fish, L. J., Worland, A. J., Pelica, F., Sudhakar, D., Christou, P., Snape, J. W., Gale, M. D. and Harberd, N. P. (1999). "Green revolution" genes encode mutant gibberellin response modulators. *Nature* **400**, 256–261.

Perata, P., Matsukura, C., Vernieri, P. and Yamaguchi, J. (1997). Sugar repression of a gibberellin-dependent signaling pathway in barley embryos. *Plant Cell* **9**, 2197–2208.

Phillips, A. L., Ward, D. A., Uknes, S., Appleford, N. E. J., Lange, T., Huttly, A., Gaskin, P., Graebe, J. E. and Hedden, P. (1995). Isolation and expression of three gibberellin 20-oxidase cDNA clones from *Arabidopsis*. *Plant Physiology* **108**, 1049–1057.

Phillips, J., Artsaenko, O., Fiedler, U., Horstmann, C., Mock, H-P., Müntz,K. and Conrad, U. (1997). Seed-specific immunomodulation of abscisic acid activity induces a developmental switch. *European Molecular Biology Organisation Journal* **16**, 4489–4496.

Phinney, B. O. (1984). Gibberellin A$_1$, dwarfism and the control of shoot elongation in higher plants. *In* "The Biosynthesis and Metabolism of Plant Hormones" (A. Crozier and J. R. Hillman, eds), pp. 17–41. Cambridge University Press, Cambridge.

Pinthus, M. J. and Abraham, M. (1996). Effects of light, temperature, gibberellin (GA$_3$) and their interaction on coleoptile and leaf elongation of tall, semi-dwarf and dwarf wheat. *Plant Growth Regulation* **18**, 239–247.

Pinthus, M. J., Gale, M. D., Appleford, N. E. J. and Lenton, J. R. (1989). Effect of temperature on gibberellin (GA) responsiveness and on endogenous GA$_1$ content in tall and dwarf wheat genotypes. *Plant Physiology* **90**, 854–859.

Rademacher, W., Fritsch, H., Graebe, J. E., Sauter, A. and Jung, J. (1987). Tetcyclacis and triazole plant growth retardants: their influence on the biosynthesis of gibberellins and other metabolic processes. *Pesticide Science* **21**, 241–252.

Rademacher, W., Temple-Smith, K. E., Griggs, D. L. and Hedden, P. (1992). The mode of action of acylcyclohexanediones – a new type of growth retardant. *In* "Progress in Plant Growth Regulation" (C. M. Karssen, L. C. van Loon, and D. Vreugdenhil, eds), pp. 571–577. Kluwer Academic Publishers, Dordrecht.

Ried, J. L. and Walker-Simmons, M. K. (1990). Synthesis of abscisic acid-responsive, heat-stable proteins in embryonic axes of dormant wheat grain. *Plant Physiology* **93**, 662–667.

Riou-Khamlichi, C., Huntley, R., Jacqmard, A. and Murray, J. A. H. (1999). Cytokinin activation of *Arabidopsis* cell division through a D-type cyclin. *Science* **283**, 1541–1544.

Robson, P. R. H., McCormac, A. C., Irvine, A. S. and Smith, H. (1996). Genetic engineering of harvest index in tobacco through overexpression of a phytochrome gene. *Nature Biotechnology* **14**, 995–998.

Ross, J. J., Murfet, I. C. and Reid, J. B. (1997). Gibberellin mutants. *Physiologia Plantarum* **100**, 550–560.

Ross, J. J., MacKenzie-Hose, A. K., Davies, P. J., Lester, D. R., Twitchen, B. and Reid, J. B. (1999). Further evidence for feedback regulation of gibberellin biosynthesis in pea. *Physiologia Plantarum* **105**, 532–538.

Sato, Y., Sentoku, N., Miura, Y., Hirochika, H., Kitano, H. and Matsuoka, M. (1999). Loss-of-function mutations in the rice homeobox gene *OSH15* affect the architecture of internodes resulting in dwarf plants. *European Molecular Biology Organisation Journal* **18**, 992–1002.

Schünmann, P. H. D., Smith, R. C., Lång, V., Matthews, P. R. and Chandler, P. M. (1997). Expression of XET-related genes and its relation to elongation in leaves of barley (*Hordeum vulgare* L.). *Plant, Cell and Environment* **20**, 1439–1450.

Schuurink, R. C., Sedee, N. J. A. and Wang, M. (1992a). Dormancy of the barley grain is correlated with gibberellic acid responsiveness of the isolated aleurone layer. *Plant Physiology* **100**, 1834–1839.

Schuurink, R. C., van Beckum, J. M. M. and Heidekamp, F. (1992b). Modulation of grain dormancy in barley by variation in plant growth conditions. *Hereditas* **117**, 137–143.

Scott, R. J., Spielman, M., Bailey, J. and Dickinson, H. G. (1998a). Parent-of-origin effects on seed development in *Arabidopsis thaliana*. *Development* **125**, 3329–3341.

Scott, R. J., Vinkenoog, R., Spielman, M. and Dickinson, H. G. (1998b). Medea: murder or mistrial? *Trends in Plant Science* **3**, 460–461.

Sentoku, N., Sato, Y., Kurata, N., Ito, Y., Kitano, H. and Matsuoka, M. (1999). Regional expression of the rice *KN1*-type homeobox gene family during embryo, shoot and flower development. *Plant Cell* **11**, 1651–1663.

Singh, B. K. and Jenner, C. F. (1982). Association between concentrations of organic nutrients in the grain, endosperm cell number and grain dry weight within the ear of wheat. *Australian Journal of Plant Physiology* **9**, 83–95.

Singh, B. K. and Jenner, C. F. (1984). Factors controlling cell number and grain dry weight in wheat: effects of shading on intact plants and of variation in nutritional supply to detached, cultured ears. *Australian Journal of Plant Physiology* **11**, 151–163.

Smith, H. (1995). Physiological and ecological function within the phytochrome family. *Annual Review of Plant Physiology and Plant Molecular Biology* **46**, 289–315.

Stacy, R. A. P., Munthe, E., Steinum, T., Sharma, B. and Aalen, R. B. (1996). A peroxiredoxin antioxidant is encoded by a dormancy-related gene, *Per1*, expressed during late development in the aleurone and embryo of barley grains. *Plant Molecular Biology* **31**, 1205–1216.

Stoddart, J. L. and Lloyd, E. J. (1986). Modification by gibberellin of the growth-temperature relationship in mutant and normal genotypes of several cereals. *Planta* **167**, 364–368.

Suge, H. (1975). Complementary genes for height inheritance in relation to gibberellin production in rice plants. *Japan Journal of Genetics* **50**, 121–131.

Sun, Y., Dilkes, B. P., Zhang, C., Dante, R. A., Carneiro, N. P., Lowe, K. S., Jung, R., Gordon-Kamm, W. J. and Larkins, B. A. (1999). Characterisation of maize (*Zea mays* L.) Wee1 and its activity in developing endosperm. *Proceedings of the National Academy of Sciences (USA)* **96**, 4180–4185.

Talon, M., Koornneef, M. and Zeevaart, J. A. D. (1990a). Accumulation of C_{19}-gibberellins in the gibberellin-insensitive dwarf mutant *gai* of *Arabidopsis thaliana* (L.) Heynh. *Planta* **182**, 501–505.

Talon, M., Koornneef, M. and Zeevaart, J. A. D. (1990b). Endogenous gibberellins in *Arabidopsis thaliana* and possible steps blocked in the biosynthetic pathways of the semi-dwarf *ga4* and *ga5* mutants. *Proceedings of the National Academy of Sciences (USA)* **87**, 7983–7987.

Tanaka-Ueguchi, M., Itoh, H., Oyama, N., Koshioka, M. and Matsuoka, M. (1998). Over-expression of a tobacco homeobox gene, *NTH15*, decreases the expression of a gibberellin biosynthetic gene encoding GA 20-oxidase. *Plant Journal* **15**, 391–400.

Tanksley, S. D. and McCouch, S. R. (1997). Seed banks and molecular maps: unlocking genetic potential from the wild. *Science* **277**, 1063–1066.

Thomas, S. G., Phillips, A. L. and Hedden, P. (1999). Molecular cloning and functional expression of gibberellin 2-oxidases, multifunctional enzymes involved in gibberellin deactivation. *Proceedings of the National Academy of Sciences (USA)* **96**, 4698–4703.

Tonkinson, C. L., Lyndon, R. F., Arnold, G. M. and Lenton, J. R. (1995). Effects of the *Rht3* dwarfing gene on dynamics of cell extension in wheat leaves, and its modification by gibberellic acid and paclobutrazol. *Journal of Experimental Botany* **46**, 1085–1092.

Tonkinson, C. L., Lyndon, R. F., Arnold, G. M. and Lenton, J. R. (1997). The effects of temperature and the *Rht3* dwarfing gene on growth, cell extension, and gibberellin content and responsiveness in the wheat leaf. *Journal of Experimental Botany* **48**, 963–970.

Toyomasu, T., Yamane, H., Murofushi, N. and Nick, P. (1994). Phytochrome inhibits the effectiveness of gibberellins to induce cell elongation in rice. *Planta* **194**, 256–263.

Toyomasu, T., Kawaide, H., Sekimoto, H., von Numers, C., Phillips, A. L., Hedden, P. and Kamiya, Y. (1997). Cloning and characterisation of a cDNA encoding gibberellin 20-oxidase from rice (*Oryza sativa*) seedlings. *Physiologia Plantarum* **99**, 111–118.

Trewavas, A. J. (1988). Timing and memory processes in seed embryo dormancy – a conceptual paradigm for plant development questions. *BioEssays* **6**, 87–92.

Umemura, T-a., Perata, P., Futsuhara, Y. and Yamaguchi, J. (1998). Sugar sensing and α-amylase gene repression in rice embryos. *Planta* **204**, 420–428.

van Beckum, J. M. M., Libbenga, K. R. and Wang, M. (1993). Abscisic acid and gibberellic acid-regulated responses of embryos and aleurone layers isolated from dormant and non-dormant barley grains. *Physiologia Plantarum* **89**, 483–489.

Visser, K., Vissers, A. P. A., Çağirgan, M. I., Kijne, J. W and Wang, M. (1996). Rapid germination of a barley mutant is correlated with a rapid turnover of abscisic acid outside the embryo. *Plant Physiology* **111**, 1127–1133.

Walker-Simmons, M. (1987). ABA levels and sensitivity in developing wheat embryos of sprouting resistant and susceptible cultivars. *Plant Physiology* **84**, 61–66.

Walker-Simmons, M. and Sesing, J. (1990). Temperature effects on embryonic abscisic acid levels during development of wheat grain dormancy. *Journal of Plant Growth Regulation* **9**, 51–56.

Wang, H., Qi, Q., Schorr, P., Cutler, A. J., Crosby, W. L. and Fowke, L. C. (1998). ICK1, a cyclin-dependent protein kinase inhibitor from *Arabidopsis thaliana* interacts with both Cdc2a and CycD3, and its expression is induced by abscisic acid. *Plant Journal* **15**, 501–510.

Wang, M., Heimovaara-Dijkstra, S. and van Duijn, B. (1995). Modulation of germination of embryos isolated from dormant and non-dormant barley grains by manipulation of endogenous abscisic acid. *Planta* **195**, 586–592.

Wang, R.-L., Stec, A., Hey, J., Lukens, L. and Doebley, J. (1999). The limits of selection during maize domestication. *Nature* **398**, 236–239.

Webb, S. E., Appleford, N. E. J., Gaskin, P. and Lenton, J. R. (1998). Gibberellins in internodes and ears of wheat containing different dwarfing alleles. *Phytochemistry* **47**, 671–677.

Williams, J., Phillips, A. L., Gaskin, P. and Hedden, P. (1998). Function and substrate specificity of the gibberellin 3β-hydroxylase encoded by the *Arabidopsis GA4* gene. *Plant Physiology* **117**, 559–563.

Wobus, U. and Weber, H. (1999). Seed maturation: genetic programmes and control signals. *Current Opinion in Plant Biology* **2**, 33–38.

Xu, Y. L., Li, L., Wu, K., Peeters, A. J. M., Gage, D. and Zeevaart, J. A. D. (1995). The *GA5* locus of *Arabidopsis thaliana* encodes a multifunctional gibberellin 20-oxidase: molecular cloning and functional expression. *Proceedings of the National Academy of Sciences (USA)* **92**, 6640–6644.

Yamada, T. (1990). Classification of GA response, *Rht* genes and culm length in Japanese varieties and landraces of wheat. *Euphytica* **50**, 221–239.

Young, T. E. and Gallie, D. R. (1999). Analysis of programmed cell death in wheat endosperm reveals differences in endosperm development between cereals. *Plant Molecular Biology* **39**, 915–926.

Young, T. E., Gallie, D. R. and deMasson, D. A. (1997). Ethylene-mediated programmed cell death during maize endosperm development of wild-type and *shrunken2* genotypes. *Plant Physiology* **115**, 737–751.

Youssefian, S., Kirby, E. J. M. and Gale, M. D. (1992a). Pleiotropic effects of the GA-insensitive *Rht* dwarfing genes of wheat. 1. Effects on development of the ear, stem and leaves. *Field Crops Research* **28**, 179–190.

Youssefian, S., Kirby, E. J. M. and Gale, M. D. (1992b). Pleiotropic effects of the GA-insensitive *Rht* dwarfing genes in wheat. 2. Effects on leaf, stem and floret growth. *Field Crops Research* **28**, 191–210.

Zhang, K., Letham, D. S. and John, P. C. L. (1996). Cytokinin controls the cell cycle at mitosis by stimulating the tyrosine dephosphorylation and activation of p34[cdc2]-like H1 histone kinase. *Planta* **200**, 2–12.

Manipulating Cereal Endosperm Structure, Development and Composition to Improve End-use Properties

PETER R. SHEWRY[1] and MATTHEW MORELL[2]

[1] *IACR–Long Ashton Research Station, Department of Agricultural Sciences, University of Bristol, Long Ashton, Bristol BS41 9AF, UK*
[2] *CSIRO Division of Plant Industry, Institute of Plant Production and Processing, PO Box 1600, Canberra, ACT 2601, Australia*

Advances in Botanical Research Vol. 34
incorporating Advances in Plant Pathology
ISBN 0-12-005934-7

I. INTRODUCTION

The cereal grain plays crucial biological roles in the reproduction, dispersal and survival of the species. In order to fulfil these, it accumulates reserves of storage compounds, chiefly starch and protein but also lipid, which are mobilized during germination to support the growth of the seedling. These reserves have been exploited by mankind for many millennia, with empirical selection and scientific plant breeding over some 10 000 years, giving rise to a range of varieties that are high yielding and adapted to a wide range of environmental conditions. Consequently, cereal grains are currently the most important sources of starch and protein for human consumption and livestock feed, with three species (wheat, rice and maize) accounting for some 85% of the total production.

II. THE CEREAL GRAIN

The cereal grain is, in botanical terms, a single-seeded fruit called a caryopsis. It results from fusion of the fruit coat (pericarp) and seed coat (testa) to form a protective outer layer to the mature seed (Fig. 1). This layer encloses the embryo and endosperm, which arise from separate fertilization events. The embryo is diploid and arises from the fusion of one male nucleus from the pollen with the oosphere (ovum). A second male nucleus fuses with two polar nuclei from the female gametophyte, giving rise to the endosperm, which is consequently triploid. At maturity the embryo consists of the embryonic axis (including the root and shoot) and a single cotyledon (the scutellum). The endosperm also differentiates during development, giving rise to an outer layer of aleurone cells (usually three cells thick in barley but one in wheat, rice, maize and other cereals) surrounding the starchy endosperm. The embryo accounts for only about 2.5–4% of the total grain weight in barley and wheat (Tallberg, 1977; Pomeranz, 1988) and about 10–12% in maize (Watson, 1987).

In wheat the major botanical tissues are crudely separated by milling, which gives fractions corresponding broadly to the starchy endosperm cells (white flour), embryo (germ) and the aleurone and outer layers (bran). However, the bran and germ fractions can be highly contaminated with other tissues and compositions reported must be treated with caution.

Analyses of carefully isolated organs show large differences in composition; see Watson (1987) and Pomeranz (1988) for maize and wheat,

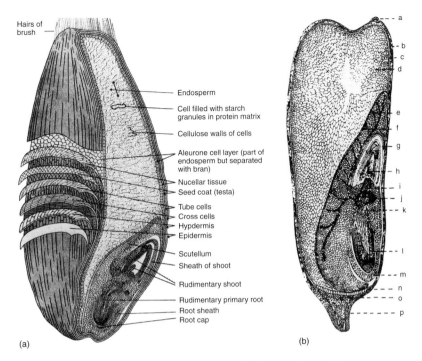

Hairs of brush

Endosperm

Cell filled with starch granules in protein matrix

Cellulose walls of cells

Aleurone cell layer (part of endosperm but separated with bran)

Nucellar tissue
Seed coat (testa)

Tube cells
Cross cells
Hypdermis
Epidermis

Scutellum
Sheath of shoot

Rudimentary shoot

Rudimentary primary root
Root sheath
Root cap

(a)

(b)

Fig. 1.   Structures of mature cereal grains. Vertical sections of the mature grain of wheat (a) and maize (b). a, silk scar; b, pericarp; c, aleurone; d, endosperm; e, scutellum; f, glandular layer of scutellum; g, coleoptile; h, plumule with stem and leaves; i, first internode; j, lateral seminal root; k, scutellar node; l, primary root; m, coleorhiza; n, basal conducting cells; o, brown abscission layer; p, pedicel or flower stalk. (Part (a) reproduced, with permission, from Fennema (1985) by courtesy of Marcel Dekker Inc.; and part (b) reproduced, with permission, from Rédei (1982), Macmillan Publishing Co.)

respectively. The embryos are rich in protein (about 25% dry weight in wheat, 18% dry weight in maize), which also has high nutritional quality, containing over 5 mol% lysine. They are also rich in oil, about 33% in maize and 25–30% in wheat, but contain less than 10% starch. The high oil content and large size of the maize embryos allows their commercial exploitation for edible oil production. The embryos of other cereals are not exploited on a large scale but wheat germ is widely used as a nutritional supplement.

The aleurone cells can be seen by microscopy to be rich in oil and protein and to lack starch granules, indicating that they are more similar to the scutellar cells in composition than the starchy endosperm cells. They also have thick cell walls and contain high levels of phosphate associated with phytin (the mixed potassium and magnesium salt of *myo*-inositol hexaphosphoric acid) present as globoid deposits within the protein bodies (called aleurone grains). Aleurone cells are difficult to isolate in substantial quantities and few detailed quantitative studies have been carried out. However, Briggs and co-workers have made detailed analyses of barley aleurone preparations, showing about 40–45% cell walls, 20% triacylglycerols (representing 67% of the lipid in the whole grain) and 17–20% protein (see Briggs, 1987).

The starchy endosperm forms the major storage tissue of the cereal grain, accounting for some 80–90% of the whole grain. The cells have thin walls and are rich in starch (~80–90%) and protein (~6–12%) but generally contain only about 1–2% lipid. However, the starchy endosperm of oats contains about 7% lipid, accounting for about 53% of the total present in the grain (Welch, 1995). There are, however, major differences between the cells in different parts of the endosperm, with the subaleurone cells in particular being smaller with higher contents of protein and less starch; see Watson (1987) and Evers and Bechtel (1988) for more detailed discussions of maize and wheat, respectively. The starchy endosperm cells are disrupted and dead in the mature grain, in contrast with the aleurone layer cells, which remain alive and secrete enzymes into the starchy endosperm to digest cell walls and mobilize storage compounds during germination.

Because the starchy endosperm is the major tissue in the cereal grain, the number and size of its component cells and their contents of storage products are major determinants of the yield and end-use quality of the whole grain. Consequently, the patterns of endosperm development, including the regulation of cell division and of the expression of genes responsible for the synthesis of structural components and storage products, are of considerable interest in relation to manipulating the cereal grain. This article will therefore focus on the starchy endosperm, and in particular on events that determine the structure, composition and end-use properties of the mature grain. Similarly, most of the work discussed is on wheat and barley, but work on other cereal species will also be discussed where relevant.

A. ENDOSPERM DEVELOPMENT IN CEREALS

Endosperm development follows a similar pattern in all cereals, although the timing and duration of the phases vary with the species, genotype and environmental conditions, particularly temperature. Six phases can be recognized, as summarized for maize in Fig. 2.

The primary endosperm nucleus divides mitotically, with the products of the first two divisions defining the left/right halves of the endosperm and the dorsal/proximal poles, respectively (see DeMason, 1997). Further nuclear divisions then occur, which are initially synchronous to give a syncytium. These divisions continue in the Triticeae (wheat, barley and rye) for about 72 hours, resulting in the presence of about 1–2000 nuclei (Bennett et al., 1975).

Plate 1. A structural model for barley β-glucan. An instantaneous view of a barley (1→3),(1→4)-β-glucan chain conformation was chosen as typical from among the countless conformations the chain will adopt in solution. The model consists of 50 β-glucosyl residues linked through (1→3) and (1→4) linkages, which are arranged in proportions and sequences defined by Woodward *et al.* (1983). The extended, worm-like conformation of the polysaccharide is evident, and this molecular asymmetry accounts for the propensity of the molecule to form solutions of high viscosity. Chain flexibility arises principally from the isolated (1→3)-β-linkages. The image was modified from Buliga *et al.* (1986) and kindly provided by Professor G. Fincher (Adelaide, Australia).

is easy to draw the simplistic conclusion that prolonging the phase of cell division within the endosperm should lead to larger grain and higher yields. However, a range of studies have shown that seed number per unit area is more consistently correlated with yield than seed size (as discussed by Egli, 1998). Also, a more recent study of wheat suggests that the final cell number in the wheat endosperm is determined by the balance between the number of cells produced during the cell division phase and the number lost during late endosperm development (Gao et al., 1992). These two lines of study indicate that increasing the yield of cereal grains by manipulation of endosperm cell number may not be readily achieved.

III.   ENDOSPERM CELL WALLS

The composition and properties of the cell walls of the cereal starchy endosperm are of importance in relation to the nutritional quality and processing properties of the grain. They comprise two major classes of polysaccharide, $(1\rightarrow3),(1\rightarrow4)$-β-D-glucans (β-glucans) and arabino-$(1\rightarrow4)$-β-D-xylans (arabinoxylans), with smaller amounts of cellulose and other polysaccharides (Stone, 1996). The starchy endosperms of maize, barley and oats contain high proportions of β-glucan and low proportions of arabinoxylans, while the converse occurs in wheat and rye (Table I). These differences have major impacts on the utilization of the different species.

A.   ARABINOXYLANS

1.   Structure and Composition
Total arabinoxylans (AX) account for about 6–12% of the dry weight of the whole rye grain (Saastamoinen et al., 1989) and about 4–9% in wheat (Saulnier et al., 1995; Hashimoto et al., 1987; Hong et al., 1989). However, in both species they are concentrated in the pericarp, testa and aleurone layer with lower levels in the starchy endosperm and in white flour: about 4% in rye (Henry, 1987; Glitsø and Bach Knudsen, 1999) and 2–3% in wheat (Henry, 1987; Lineback and Rasper, 1988).

Arabinoxylans consist of a chain of $(1\rightarrow4)$-linked β-D-xylopyranose residues, with some of the xylose residues carrying a terminal O-3-linked L-arabinofuranose residue. A small proportion of the xylose may be disubstituted at the O-2 and O-3 positions or with other sugar residues, while the arabinose may be esterified with ferouloyl or p-coumaroyl residues. The degree of polymerization of the endosperm AX ranges from about 100 to 1200 corresponding to a molecular mass of 13 000–160 000 (Stone, 1996; Lineback and Rasper, 1988). The ratio of xylose:arabinose also varies between the grain tissues, with a higher proportion of arabinose in the endosperm (Henry, 1987).

TABLE I

Approximate proportions of cellulose, β-glucan and arabinoxylan in the cell walls of the starchy endosperm of cereals

	Cellulose	β-Glucan	Arabinoxylan	Reference
Barley	~2	75	20	Fincher (1975)
Wheat	~2	29	65	Bacic and Stone (1980)
Maize	10	65	15	Newton *et al.* (1994)
Rye[a]	8	12	65	Glitsø and Bach Knudsen (1999)
Oats	?	72[b]	28[b]	Henry (1987)

[a] Expressed as % non-starch polysaccharides.
[b] Approximate values calculated from analyses of endosperm showing 0.7% dry weight pentosan and 1.8% dry weight β-glucan.

Part of the AX present in endosperm cell walls is extractable in water at temperatures of 40°C or above but either alkaline solvents or enzyme treatments are required to dissolve the remaining AX. The proportions of water-soluble:water-insoluble AX vary between tissues, being higher in starchy endosperm than in bran fractions of rye (Vinkx and Delcour, 1996). The water-soluble and water-insoluble AX also vary in their ratios of arabinose:xylose, with the water-soluble arabinoxylans of the starchy endosperm having A:X ratios of about 0.75–0.8 in rye and 0.6–0.7 in wheat compared with A:X ratios of about 0.8–1.1 and 0.6–1.1 for alkali- or enzyme-extractable fractions from rye and wheat, respectively (Delcour *et al.*, 1989; Lineback and Rasper, 1988).

2. Properties and Impact on Processing

Arabinoxylans have high water-binding capacities – reported as 0.47 g/g dry matter for the water-extractable fraction of rye (Girhammar and Nair, 1992) – and form highly viscous solutions. Together with β-glucans they form part of the dietary fibre of food and feed, and may have positive or negative nutritional effects, as discussed below.

The high water-binding capacity of AX means that they play a role in determining the water adsorption of flour during dough mixing. In addition, oxidative gelation of the AX via cross-linking of ferulic acid residues leads to increased viscosity and water-binding capacity and may result in changes in dough rheology, notably a decrease in extensibility. Although it is generally accepted that soluble AX have a positive impact on breadmaking quality, the precise basis for this is not clear. Thus, although viscosity, water-holding and oxidative gelation have all been proposed to contribute, this has recently been questioned by Courtin and Delcour (1998).

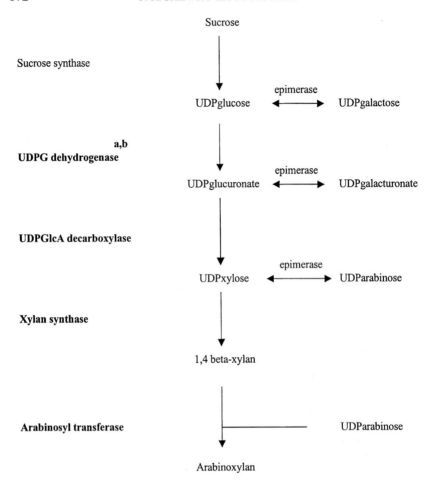

Fig. 3.   The biosynthetic steps to arabinoxylans. The enzymes involved in the synthesis of the xylan backbone and arabinosyl transferase are likely to be targets for genetic manipulation and are shown in bold. There are two forms of uridine diphosphate glucose (UDGP) dehydrogenase: (a) a high specificity form with homology to mammalian UDPG dehydrogenase; and (b) a low specificity vascular-localized form with homologies to alcohol dehydrogenase. There are also three epimerases that interconvert UDP sugars between those used in pectin synthesis and those used in hemicellulose biosynthesis. All the enzymes providing the precursor UDP sugars for the backbone are cytosolic, while the xylan synthase and arabinosyl transferase are Golgi located. UDPGlcA, UDP glucuronate. (Modified, with permission, from Gregory *et al.*, 1998.)

In addition to playing a role in breadmaking, AX may also be exploited to develop novel ingredients, such as viscosity control agents in sauces, and to enhance foam stability by forming cross-links with proteins (Sarker *et al.*, 1998).

3.   Synthesis and Manipulation

Analyses of non-cereal systems show that arabinoxylans are synthesized from sucrose via a simple pathway, as shown in Fig. 3 (Gregory *et al.*, 1998). UDP glucose is initially synthesized by sucrose synthase and then converted into 1,4-β-xylan by three enzymic steps. Arabinose residues are then added to this xylan backbone from UDP-arabinose catalysed by arabinosyl transferase. All of the enzymes providing the UDP-sugars appear to be cytosolic while the xylan synthase and arabinosyl transferase are located in the Golgi. All of these enzymes have been purified from *Phaseolus* (bean) by Bolwell and co-workers (see Gregory *et al.*, 1998), while the UDP-glucose dehydrogenase has been cloned from soybean (Tenhaken and Thulke, 1996). We, therefore, have the tools to isolate corresponding cloned cDNAs and genes from wheat, rye and other cereals in order to study their expression patterns and to up- or down-regulate their activities in transgenic plants. However, little or nothing is currently known about the enzymes that introduce other substituents, such as D-glucuronic acid, 4–O-methyl D-glucuronic acid, *p*-coumaric acid and ferulic acid, which may have major effects on the nutritional and functional properties of AX.

B.   B-GLUCAN

1.   Structure

(1→3),(1→4)-β-D-Glucans are linear polymers comprising β-D-glucopyranosyl monomers joined by both (1→3) and (1→4) linkages, usually in a ratio of about 2:1 (Stone, 1996). They range in molecular weight, with averages of about 1.13×10^6, 2.14×10^6 and 3.00×10^6 reported for rye, barley and oats, respectively (Wood *et al.*, 1994). They have extended "snake-like" conformations in solution (Plate 1), resulting from the presence of interspersed (1→3) links within the chains of (1→4) links (Stone, 1996).

Extraction of the starchy endosperm cell walls of barley, oats and rye with water at 40°C results in extraction of part of the β-glucans as a viscous solution, but dilute alkali is required to extract β-glucans from endosperm cell walls of wheat and from aleurone cell walls of wheat and barley (Stone, 1996). The reader is referred to excellent reviews by MacGregor and Fincher (1993), Fincher (1992), Fincher and Stone (1986) and Stone (1996) for more detailed discussions of β-glucan structure and properties.

2.   Impact on Processing

The major impact of β-glucans is on the malting and brewing performance of barley grain. High levels of β-glucan in the grain may limit modification during malting, resulting in malt with incompletely degraded cell walls and reduced mobilization of starch and proteins. Consequently, the extract values are low with insufficient nutrients available for the growth of yeast during brewing (MacGregor and Fincher, 1993).

Furthermore, the presence of β-glucans in wort leads to increased viscosity, resulting in slow filtration and production delays. When present in beer, β-glucans can be involved in the formation of haze and precipitates, but also contribute positively to "body" (MacGregor and Fincher, 1993).

It is not surprising, therefore, that β-glucans have been targeted in plant breeding programmes for malting barley, with selection either for low levels of β-glucan in the mature grain or high levels of β-glucanase production during malting (see below). At present, we know little about the glycan synthases that catalyse β-glucan synthesis and it is therefore not possible to manipulate β-glucan synthesis by genetic engineering.

C. NUTRITIONAL PROPERTIES OF CELL WALL COMPONENTS

The cell wall carbohydrates are not digested in the stomach or small intestine of humans or other monogastic animals but may be digested by bacteria in the colon. They form a major part of the dietary fibre intake of humans, with the soluble components in particular having important physiological effects. At a gross level they contribute to faecal bulk, regulating intestinal function (see Betschart, 1988). They also appear to slow down the absorption of monosaccharides and oligosaccharides, possibly by raising the viscosity of the gut contents, decreasing post-prandial levels of blood glucose and insulin (Wood, 1992; Asp *et al.*, 1993). Furthermore, β-glucans have well-established and well-publicized effects on lowering blood cholesterol, when present in the diet at high levels such as those in barley and oats (Newman and Newman, 1992; Kahlon and Chow, 1997).

However, cell wall carbohydrates may also have anti-nutritional effects. Both AX and β-glucans may decrease the absorption of nutrients in diets for chickens, probably due to their high viscosity and indigestibility (Antoniou *et al.*, 1981; Choct and Annison, 1992; Newman and Newman, 1992). Furthermore, high levels of β-glucan may lead to "sticky droppings" in chickens, which is probably due to high caecal microbial populations as the β-glucan itself is almost completely degraded (Fincher and Stone, 1986; Stone, 1996).

A full discussion of the nutritional and anti-nutritional impacts of β-glucans and AX is clearly outside the scope of this chapter and the reader is referred to Betschart (1988), Newman and Newman (1992), Stone (1996) and Vinkx and Delcour (1996) for more detailed accounts.

A major aim of biotechnology will clearly be to manipulate the relative amounts, structures and properties of AX and β-glucans in the starchy endosperms of cereals, in order to optimize nutritional and functional properties for different end-uses.

IV. STARCH

A. INTRODUCTION

In 1941, Harris and Sibbitt wrote that

the starch in wheat flour has not received the consideration that it deserves from the standpoint of flour quality. This constituent is normally present in flour in a concentration of at least 70% by weight and it is only reasonable to expect that some effect must be exerted upon baking strength by variations in starch properties due to wheat variety or environmental condition... .

This situation essentially remained until the 1990s when the tools and background became available to revisit the synthesis of wheat and barley starch and to begin to understand the linkages between the presence and role of specific genes and specific end-uses. In contrast, these types of studies have been actively conducted for the seed storage proteins for several decades.

The analysis of natural variation for starch properties reveals important sources of natural diversity in hexaploid wheat but does not reveal the wide range of mutations that are available in diploid species such as maize. This is probably because the presence of three genomes masks mutations in individual genomes. The probability of finding a spontaneous or induced triple null line is vanishingly small. A number of excellent reviews of aspects of wheat starch synthesis and end use have been written (Lineback and Rasper, 1988; Morrison, 1989; Jenner *et al.*, 1991; Evers *et al.*, 1999; Rahman *et al.*, 2000). However, the purpose of this review is to provide an update of important aspects of wheat starch structure and function research, and the rapidly changing picture that is emerging of the biosynthesis of starch in wheat and barley.

B. STRUCTURE AND PHYSICAL PROPERTIES

The basic elements of the chemical structure of wheat and barley starch are similar to those of other cereal starches and it is the assembly of the starch polymers into the granule that defines the unique properties of the starches from these species. A major challenge for future research is to define precisely how differences in gene expression and properties within and between cereal species translate into differences in starch fine structure, starch granules assembly and structure and, ultimately, starch properties.

1. Amylose to Amylopectin Ratio

Extensive surveys of natural diversity in amylose to amylopectin ratios have been carried out in wheat (Mohammadkhani *et al.*, 1998; Raeker *et al.*, 1998). While there is natural variation within a range from 15% to 35% amylose content (as measured by iodine blue colour), naturally occurring waxy or high amylose types have not been found. In barley, the classical waxy mutation has been widely used and a mutant with elevated amylose content (up to 45–50%), Glacier AC38, has been characterized (Walker and Merritt, 1968; Schondelmaier *et al.*, 1992). The fine structure of wheat amylopectin has been investigated in greater detail as improved methods for chain length analysis have been developed. Hizukuri and Maehara (1990) have used enzymatic digestion to examine the chain length distribution of wheat starch and the manner in which external unsubstituted chains of amylopectin (the "A" chains) are linked to the internal substituted chains (the "B" chains"). Nagamine and Komae (1996) have used high-performance liquid chromatography (HPLC)-based anion-exchange separation at high pH to investigate diversity in wheat starch amylopectin and, more recently, Morell *et al.* (1998) have used fluorophore-assisted carbohydrate electrophoresis to investigate the fine structure of cereal amylopectin. The determination of the molecular weight profile of amylopectin molecules is difficult because of their large size and solubility properties. However, You *et al.* (1999) have recently reported the separation and analysis of amylopectin polymers of wheat using size exclusion chromatography and multiangle laser light scattering detection and analysis. The weight average molecular weights of amylopectins ranged from 29×10^6 to 349×10^6. Hizukuri (1988) has reported that, for a limited range of wheat varieties, the molecular size and intrinsic viscosity of wheat amylopectin is lower than for corn starch and considerably lower than for potato or tapioca starch. However, there is limited information available concerning the diversity of structure of wheat amyloses from different sources.

2. Presence and Identity of Minor Non-glucan Components – Lipids and Phosphate

Cereal starch granules contain two types of non-glucan components, proteins (0.5–1%) and lipids (0.77–1.17%; Morrison, 1988b) and, unlike some tuber starches, contain essentially no starch-phosphate esters. The proteins present within the matrix of the starch granule are starch biosynthetic enzymes, and their nature and roles will be discussed in later sections on starch biosynthesis.

Wheat and barley starch granules contain lipids, predominantly in the form of lysophospholipids, which are associated with the amylose fraction as an amylose/lipid clathrate (Gidley and Bociek, 1988; Morrison *et al.*, 1993). The lysophospholipids comprise approximately 70% lysophosphatidylcholine, 20% lysophosphatidylethanolamine and 10% lysophosphatidylglycerol, and the fatty acid moieties, ranked from most to least abundant, include linoleic, palmitic, oleic, linolenic and stearic acids (Morrison, 1988b). In contrast, maize contains predominantly free fatty acids in the starch granule (Morrison, 1988b). The lipid fraction is thought to be important in controlling starch granule swelling, as complexed amylose interferes with the swelling of the starch granules (Shi and Seib, 1989; Morrison *et al.*, 1993). For further detail, the reader is referred to the reviews of Morrison (1988a,b, 1989).

C.   GRANULE SIZE, SHAPE AND STRUCTURE

A major distinguishing feature of wheat, barley and rye starch is the bimodal starch granule distribution, generally described as "A" starch granules (>10 μm) and "B" granules (<10 μm). A third group of "C" granules, of smaller size and initiated after the "B" granule fraction, has also been reported (Bechtel *et al.*, 1990). The structure of the starch granules in wheat and barley was described using the light microscope by Sandstedt (1946) and by Buttrose (1960, 1963) and these observations have been confirmed and extended through the use of higher resolution tools such as the scanning electron microscope (Evers, 1971). Buttrose (1960, 1963) described the presence of alternating concentric "rings" of 0.5–1 μm diameter in the wheat starch granule, which are now proposed to be composed of alternating layers of starch differing in crystallinity (Gallant *et al.*, 1997; Buleon *et al.*, 1998). The crystallinity of wheat and barley starch shows the typical "A" X-ray diffraction pattern of the cereal starches (Buleon *et al.*, 1998).

The measurement of starch granule size distributions in wheat and barley has been carried out using a range of techniques including microscopy (Sanstedt, 1946), sieving (Evers, 1973), the Coulter counter (Brocklehurst and Evers, 1977; Evers and Lindley, 1977; Morrison and Scott, 1986), image analysis (Evers and Lindley, 1977; Bechtel *et al.*, 1990; Oliveira *et al.*, 1994) and laser diffraction technology (Blumenthal *et al.*, 1994; Stoddard, 1999a). The changes in starch granule size distribution during endosperm development of wheats and barleys grown in a variety of environments have been described using a range of techniques (Karlsson *et al.*, 1983; Baruch *et al.*, 1979, 1982; Bechtel *et al.*, 1990). Stoddard has recently investigated developmental differences in A to B granule ratio within a single wheat head (Stoddard, 1999a) and differences between accessions of wheat using laser light scattering (Stoddard, 1999b). The separation of barley (Takeda *et al.*, 1999) and wheat granules (Peng *et al.*, 1999) of different sizes has been carried out, and the structure of the granule and the starch in differing granule size classes analysed, showing that amylose content is 1–4% higher in A granules than B granules.

D.   STARCH CONTENT

The starch content of wheat and barley is determined by an interplay between environment, development and genetics. In the starchy endosperm, starch is laid down from the beginning of endosperm development whereas storage proteins are synthesized from mid-endosperm development through to the end of grain filling. As a consequence, protein concentration is highest, and starch content and A-granule size lowest (Briarty *et al.*, 1979), in the youngest endosperm cells, which are those closest to the aleurone layer (Gaines *et al.*, 1985). As much of the world's wheat is sold on the basis of protein content, genetic selection for protein content is a feature of many breeding programmes. However, as protein content and starch content are generally found to be inversely related (Hucl and Chibbar, 1996), selection for protein content will also select for factors that depress starch content. As heat stress reduces starch biosynthesis more than protein synthesis, high temperatures lead to higher protein content (low starch content) wheats. However, the higher protein content is associated with decreased quality of the protein fraction (Blumenthal *et al.*, 1994).

In barley, increased starch content in the grain is associated with increased hot water extract potential. Factors, such as heat stress, that decrease starch content are therefore deleterious to barley malting quality (Wallwork *et al.*, 1998).

E. FUNCTIONAL PROPERTIES

It is not possible in this review to discuss how the properties of wheat and barley starches influence the full range of end uses of these cereals. Instead, the broad physicochemical characteristics of starch are described, with particular reference to wheat and barley starch and to selected end uses.

In contrast to the protein components of wheat flours, comparatively little is known about the potential for optimization of wheat starch for particular processes. However, it is worth noting that a number of studies have shown that wheat starch can be replaced by barley, rye and triticale starches in baking with satisfactory results, but not by starches from other sources (D'Appolonia and Gilles, 1971; Ciacco and D'Appolonia, 1977).

We can expect a dramatic extension of our knowledge in this area over the next few decades as diverse starches created by the identification of mutations in different genes or by genetic engineering become available on a research and pilot scale.

1. Starch Damage and its Consequences

The milling of wheat or barley inevitably leads to the damage of starch granules, which is an important aspect of starch functionality. Damaged starch granules liberate granule fragments that, in contrast to intact starch granules, are rapidly hydrated and swollen, and can be attacked by degradative enzymes (Ziegler and Greer, 1978). As such, damaged starch affects water absorption and provides a major source of carbohydrate to support yeast activity in leavened products (Ziegler and Greer, 1978). The major genetic factor determining the degree of starch damage in milling is not the starch granule itself but the degree of hardness of the endosperm. In the hard endosperm, fracturing occurs across the endosperm matrix with fracture lines passing through the starch granules, releasing granule fragments. In contrast, in a soft endosperm, fracture lines pass around the granule leading to a lower degree of starch damage (Barlow *et al.*, 1973). The impact on starch damage during milling will be an important practical aspect of the future manipulation of granule size and shape.

2. Gelatinization and Melting

Gelatinization involves the irreversible disruption of ordered structures within the granule on heating starch in the presence of excess water. It is most often observed and quantified using hot-stage microscopy to monitor the loss of birefringence or dye uptake or, more commonly, over the last two decades, differential scanning calorimetry (DSC) (Wang *et al.*, 1998). In contrast, melting also involves the loss of order within the granule, but occurs at lower water content (below 30%) where starch swelling does not occur (Wang *et al.*, 1998). The gelatinization onset temperature in the DSC provides a useful point of comparison of gelatinization temperature, which, unlike gelatinization peak temperature, is relatively insensitive to starch concentration (Wang *et al.*, 1998). Gelatinization temperature is influenced by amylopectin branching and amylose content (Jane *et al.*, 1999) and granule structure and is also strongly influenced by environment and the temperature regime under which the grain is stored or extracted (Morrison, 1988a). Although surveys of diversity in wheat starch gelatinization have been carried out across germplasm collections (Zeng *et al.*, 1997; Wootton *et al.*, 1998), it is unclear just how much of the diversity found has a purely genetic basis.

3. Pasting and Viscosity, and the Impact of α-Amylases

The viscosity of a solution formed by a starch/water slurry during heating and stirring is an important indicator of the performance of a starch in a range of end uses and instruments such as the amylograph and the rapid visco-analyser (RVA) have been used to quantify the development of viscosity under different experimental protocols (Batey *et al.*, 1989). Comparisons of different starches in the amylograph or RVA define a fundamental drawback for the utilization of wheat and barley starches compared with maize starches; their hot paste viscosity (measured as the peak viscosity in the RVA) is considerably lower than the viscosity of maize starch (Shi and Seib, 1989). The molecular reasons for this difference have been suggested to reflect the lower swelling power of wheat starch granules and the lower intrinsic viscosity of the amylose released into the liquid phase during the pasting of wheat starch relative to maize starch (Shi and Seib, 1989). There is a considerable range of pasting and viscosity properties within commercial hexaploid wheats (McCormick *et al.*, 1995) with further diversity within landraces (Bhattacharya *et al.*, 1997). However, the range of variability does not approach the levels of pasting properties achieved by maize starch. The development of waxy and partially waxy wheats has introduced a new source of variation in viscosity parameters; however, peak viscosity is not increased and final viscosity is reduced (Reddy and Seib, 1999).

While there is intrinsic diversity in starch pasting and viscosity in wheats and barleys, an overriding factor can be the presence of α-amylase in the grain due to premature sprouting or the development of late-maturity α-amylase in apparently sound grain. The traditional "falling number" measurement of viscosity was developed as a measure of rain damage leading to production of α-amylase and the RVA was essentially developed as a tool for the more rapid assessment of rain damage (Ross *et al.*, 1987). More recently, test kits based on immunological detection of α-amylase have been developed that provide an even more rapid prediction of the impact of weather damage on starch viscosity (Verity *et al.*, 1999).

4. Swelling
The distinct granule morphology of wheat, barley and rye "A" starch granules leads to a characteristic pattern of swelling behaviour. Swelling of the equatorial "plates" surrounding the granule nucleus leads to more rapid and extensive swelling in the equatorial plane (Bowler *et al.*, 1980; Hermansson and Svegmark, 1996). This in turn leads to a characteristic buckling of the granule and may account for the particular properties of wheat, barley and rye starches in baking (Morrison, 1988a). Similar buckling on swelling is not observed for "B" granules or for other cereal starches. Sasaki and Matsuki (1998) examined structural features of wheat starches that relate to granule swelling and found that amylose content was negatively correlated with swelling, whereas higher proportions of chains of degree of polymerization (DP) 35 in amylopectin, starch gelatinization temperature and gelatinization enthalpy were positively correlated with swelling.

The swelling of a starch in water is a central aspect of the development of viscosity. However, measurements of viscosity using the amylograph and the RVA involve swelling in the presence of shear forces. Tests of starch (or flour) swelling behaviour have been developed that allow for the full uptake of water by gelatinizing starch granules with minimal exposure to shear forces; see Crosbie and Lambe (1993) and references therein. In wheat, Crosbie has developed tests of starch swelling behaviour, which give excellent prediction of suitability of wheats for Udon noodle quality (Crosbie, 1991; Crosbie *et al.*, 1992; Crosbie and Lambe, 1993). These tests were also essential in defining the key role of null mutations in granule bound starch synthase (GBSS) in determining the suitability of wheats for Udon noodle quality (Zhao *et al.*, 1998).

5. Gel-forming Ability

The ability of starches to form gels is controlled to a significant degree by their amylose contents. While the amylose content varies from approximately 18% to 35% among commercial hexaploid wheat cultivars, this variation is narrower than in competing starches such as maize mutants. At below 6% starch content, wheat starch gives lower strength gels than corn starch (Shi and Seib, 1989).

6. Retrogradation and Staling

The behaviour of wheat starches following cooling or baking is important for the texture and appearance of a product. The retrogradation of starches occurs first through the association of amylose chains to form tightly hydrogen-bonded complexes through the alignment of long linear regions of the molecule, and secondly through the slower association of the linear regions of the external chains of amylopectin (Biliaderis, 1991). While retrogradation is an important parameter determining staling and firmness of bread, the specific nature of the retrogradation and other interactions underpinning these process remain to be fully defined; for recent views see Fredriksson et al. (1998) and Every et al. (1998).

7. Chemical Modification

While natural variants of crop starches exist, the major source of variation in starch properties used in the food industry is generated through different chemical modifications of a handful of basic starch types, including normal or waxy maize starch, potato starch, and wheat starch (Rutenberg and Solarek, 1984). The introduction of different functionalities occurs principally through the addition of cross-linking (using bifunctional reagents such as phosphorus oxychloride, epichlorohydrin), charged groups (phosphorylation, acetylation, succinylation, carboxymethylation), or bulky side chains that inhibit retrogradation (hydroxypropylation). The examination of the properties of chemically modified starches is beyond the scope of this review; however, it is important to note that chemical modification is a very important aspect of the use of cereal starches (see Rutenberg and Solarek, 1984; Craig, 1989; De Boer, 1991).

8. Foods and Nutritional Properties

The lack of variation in wheat and barley starches has limited the extent to which particular starches have been targeted for specific uses. The best known example is the use of lines lacking expression of one or more of the GBSS genes for noodle quality (Zhao et al., 1998) and it is expected that fully waxy wheats will also find niche applications in the food industry. Two types of experimental system are likely to be used in the future to examine starch functionality for specific wheat- or barley-based foods. Firstly, the development of further genetic variation through targeted plant breeding or genetic engineering and, secondly, the use of reconstitution techniques to examine the performance of existing starches from diverse sources in foods in order to guide germplasm development and product formulation (see MacRitchie, 1989; Toufeili *et al.*, 1999).

There is particular interest in the use of starches containing high levels of amylose as sources of resistant starch, which can act as dietary fibre. In particular, the survival of high-amylose starch granules of maize during passage through the monogastric digestive system to the large intestine has been found to stimulate the production of long-chain fatty acids, which are associated with bowel health and suppression of bowel cancers. For literature on this subject, see Brown (1996), Topping *et al.* (1999) and Jenkins *et al.* (1998).

F.   BREAKDOWN DURING MALTING

The breakdown of starch during malting is catalysed by a suite of enzymes including α-amylase, β-amylase, limit-dextrinase (a pullulanase-type debranching enzyme) and α-glucosidase. Various types of analyses have been used to examine which of these enzymes is critical with respect to starch degradation in commercial barleys under malting conditions (Sissons and MacGregor, 1994; MacGregor *et al.*, 1999). Further definition of the roles of specific enzymes will be generated by the genetic manipulation of their levels and specificities. Relatively little information is available about how starch granule structure and starch fine structure influence the malting process or the profile of oligosaccharides carried through to the fermented beverage. Further genetic manipulation of barley starch structure and composition will be instructive in this respect.

G.   EFFECTS OF ENVIRONMENT ON STARCH SYNTHESIS AND PHYSICOCHEMICAL PROPERTIES

The synthesis of starch in wheat and barley grain is strongly influenced by the environment, particularly the temperature, which affects the growth and development of the plant and grain (Wardlaw, 1994), and the starch biosynthesis process itself (Rijven, 1986; Jenner and Hawker, 1993; Jenner, 1994; Keeling *et al.*, 1994). The effects of heat stress on the developing grain and starch biosynthesis are clearly complex and pleiotrophic and, while particular enzymes, such as the starch synthases (Keeling *et al.*, 1993; Jenner *et al.*, 1995), appear to be sensitive to heat stress, the demonstration of a causal relationship between the inactivation of specific enzymes and heat stress-induced modification of starch biosynthesis and starch properties requires further research. In maize, heat lability of ADP glucose pyrophosphorylase has been suggested to be the primary cause of reduced starch biosynthesis under elevated temperatures (Greene and Hannah, 1998). The effects of heat stress on starch biosynthesis do not only affect starch content but also alter starch structure and properties. In addition to causing shrivelled grain, elevated temperature reduces the size of A starch granules, reduces the mass of B granules, and can result in deformed starch granules, especially in hard and durum wheats (Shi et al., 1994; Blumenthal et al., 1994; Panozzo and Eagles, 1998). With increasing temperature during grain-filling, starch lipid levels and amylose contents increase, leading to decreased starch swelling, and gelatinization temperatures increase (Shi et al., 1994; Tester et al., 1995). Other environmental effects shown to have an effect on starch content and composition are water stress (Brooks et al., 1982) and elevated CO_2 levels (Rogers et al., 1998; Blumenthal et al., 1996; Tester et al., 1995).

H.   SYNTHESIS AND MANIPULATION

1.   Sucrose Delivery to the Endosperm

In recent years, the steps involved in the movement of sucrose from the phloem into the developing grain have been investigated using a range of techniques and the presence of transporters in subaleurone transfer cells in endosperm adjacent to the endosperm cavity described. For further details, the reader is referred to the series of papers by Wang and literature cited therein (Wang et al., 1993, 1994a,b, 1995a,b). The identification and properties of sucrose transporters in the grain that facilitate the movement of sucrose from the endosperm cavity to the sites of sucrose hydrolysis is an active field of current research.

2.   Transformation of Sucrose into Direct Precursors of Starch Biosynthesis

The synthesis of starch in the cereal endosperm is ultimately dependent on the supply of photosynthate, largely in the form of sucrose, to the grain. Early efforts to study sucrose–starch transformations involved the incubation of endosperm tissues or subfractions with various substrates and the analysis of substrate pools and starch synthesis rates (Rijven and Gifford, 1983; Rijven, 1984; Jenner et al., 1993; Niemietz and Jenner, 1993; Riffkin et al., 1995). The developing cereal grain contains both acid and neutral invertases, and sucrose synthase, enzymes that can potentially hydrolyse sucrose. Sucrose synthase is thought to be a key enzyme in the process and is present in the developing wheat (Duffus and Binnie, 1990; Marana et al., 1990) and barley grain (Martinez de Iladuya et al., 1993).

3.   Localization of the Starch Biosynthetic Enzymes

The preparation of intact amyloplasts from endosperm tissue was an essential tool for studying the localization of starch biosynthetic enzymes.

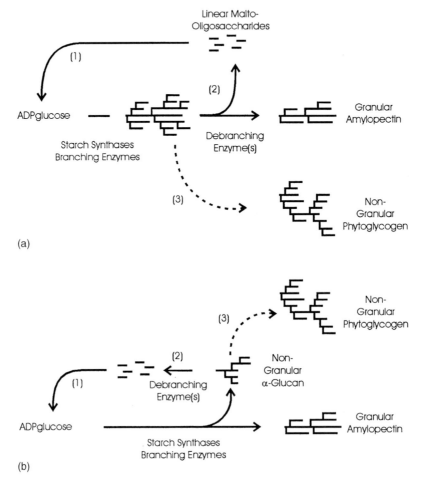

Fig. 4. Alternative models of starch biosynthesis. (a) Model based on the hypothesis proposed by Myers *et al.* (2000). (b) Model based on the hypothesis proposed by Zeeman *et al.* (1998). (1) Pathway of recovery of glucose from malto-oligosaccharides involving hydrolysis to glucose, phosphorylation by hexokinase, conversion of glucose-6-phosphate to glucose-1-phosphate by phosphoglucomutase, and conversion to ADP glucose by ADP glucose pyrophosphorylase. (2) Debranching enzyme reaction blocked by a lesion in the *sugary-1* isoamylase-type debranching enzyme gene. (3) Alternative pathway of α-glucan synthesis invoked by a lesion in the *sugary-1* isoamylase-type debranching enzyme gene.

Methods for preparing intact amyloplasts have been described for wheat (Entwhistle *et al.*, 1988; Mahajan *et al.*, 1990; Tetlow *et al.*, 1993) and used to provide evidence for starch synthesis from glucose-1-phosphate (Tetlow *et al.*, 1994; Tyson and ap Rees, 1988; Tetlow *et al.*, 1998) and not triose phosphates (Keeling *et al.*, 1988).

Transporters involved in the translocation of hexose phosphates into the amyloplast of wheat have been studied by Tetlow *et al.* (1996) and the levels of phosphoglucomutase defined (Esposito *et al.*, 1999).

The synthesis of starch in the endosperm of higher plants occurs within the amyloplast and involves the action of at least four types of enzyme activities, ADP glucose pyrophosphorylase, starch synthases, branching enzymes and debranching enzymes (see Fig. 4). The roles of each of these enzymes will be considered in the following sections.

4.   ADP Glucose Pyrophosphorylase

The first report of ADP glucose pyrophosphorylase in small grain cereals came from the work of Turner (1969). For many years, it was assumed that ADP glucose pyrophosphorylase was entirely located within the amyloplast in cereal grains; however, the work of Denyer and colleagues (1996) demonstrated that this was not the case in maize or barley (Thorbjornsen *et al.*, 1996). The situation in wheat has yet to be resolved. Cereals, like other plants (Morell *et al.*, 1987), contain ADP glucose pyrophosphorylases composed of a large and small subunit. In barley, four types of transcript for ADP glucose pyrophosphorylase are found, for endosperm-specific large and small subunits and for large and small transcripts, which are present in the leaf and at low levels throughout the plant (Doan *et al.*, 1999). cDNAs for large and small subunits of ADP glucose pyrophosphorylase have been cloned from wheat (Olive *et al.*, 1989; Ainsworth *et al.*, 1993b, 1995) and barley (Villand *et al.*, 1992).

5.   Granule Proteins

The proteins associated with the starch granule in wheat have attracted considerable interest. These proteins can be classified into two groups, those associated with the surface of extracted starch granules, and those bound within the interior of the starch granule, also known as the integral proteins. The surface proteins are discussed in an later section with respect to grain hardness. The integral proteins have been shown to be involved in starch biosynthesis. Prominent proteins of M_r 60 000, 75 000, 90 000, 100 000, 108 000 and 115 000 are present (Denyer *et al.*, 1995; Rahman *et al.*, 1995; Yamamori and Endo, 1996; Takaoka *et al.*, 1997). The M_r 60 000 band is the granule-bound starch synthase and is composed of the products of genes on chromosomes 7A, 7D and 4A (Yamamori and Endo, 1996). The M_r 75 000 band is encoded by the starch synthase I gene (Li *et al.*, 1999a) with homologous genes on the short arm of chromosome 7 (Yamamori and Endo, 1996; Li *et al.*, 1999a). The M_r 90 000 band is associated with branching enzyme activity (Denyer *et al.*, 1995) and was shown to cross-react with antibodies to branching enzyme II, but not branching enzyme I (Rahman *et al.*, 1995). The group of three proteins of M_r 100 000, 108 000 and 115 000 are encoded by a homologous set of genes on the short arm of chromosome 7

(Denyer *et al.*, 1995; Yamamori and Endo, 1996) and are associated with starch synthase activity (Denyer *et al.*, 1995). A gene encoding these polypeptides has been cloned and shown to be a class II starch synthase (Li *et al.*, 1999b).

6. Granule-bound Starch Synthase

The manipulation of GBSS in wheat has attracted considerable attention over the past decade, leading to the development of waxy wheats (Nakamura *et al.*, 1995) and to analysis of the properties of wheats containing various gene dosages of the GBSS gene (Zhao *et al.*, 1998). cDNAs and genes encoding GBSS have been isolated (Ainsworth *et al.*, 1993a; Murai *et al.*, 1999). The separation of the GBSS proteins encoded by the homologous genes was originally carried out by two-dimensional gel electrophoresis (Yamamori and Endo, 1996) but more recently one-dimensional sodium dodecylsulphate–polyacrylamide gel electrophoresis (SDS–PAGE) methods have been developed that greatly increase the throughput of analysis (Zhao and Sharp, 1996). These methods have been used to survey the GBSS status of wheats from a variety of sources (Fujita *et al.*, 1996; Demeke *et al.*, 1997; Zhao *et al.*, 1998) and have led to the independent development of waxy wheats by a number of groups (Nakamura *et al.*, 1995; Yasui *et al.*, 1997; Kiribuchi-Otobe *et al.*, 1997; Zhao and Sharp, 1998). The specific contributions of each of the GBSS genes to the synthesis of amylose and starch properties have been investigated (Miura *et al.*, 1994; Muira and Sugawara, 1996; Zhao and Sharp, 1998; Demeke *et al.*, 1999). Confirmation of the importance of GBSS in controlling amylose content has come through QTL analysis, which has shown that the GBSS null allele on chromosome 4A is associated with variation in amylose content (Araki *et al.*, 1999). The mutations leading to one set of GBSS null alleles have been characterized (Vrinten *et al.*, 1999).

The observation that the lack of a single GBSS allele, the null 4A allele, is an important determinant of noodle quality is very interesting. As might be expected, the loss of one of the three homologous GBSS genes has only a minor impact (1–2%) on amylose content because of the buffering effect of the remaining alleles. However, studies of starch swelling behaviour and noodle texture demonstrate that there is a clear phenotype associated with this mutation that is not explained by the minor change in amylose content. For example, wheats containing all three GBSS genes can be found that have amylose contents in the same range as typically found in null 4A wheats; however, those low amylose wheats with normal GBSS do not share the swelling behaviour of the null 4A wheats (Zhao *et al.*, 1998). The impact of the null 4A mutation on functional properties has not yet been explained by examination of the structure of the starch or starch granules.

While the M_r 60 000 GBSS proteins encoded by genes on chromosomes 7A, 7D and 4A, respectively, are required for amylose biosynthesis in the endosperm of wheat and barley, there are two lines of evidence that suggest that a second GBSS is present in wheat and barley but expressed in other tissues. Fujita and Taira (1998) have identified an M_r 56 000 GBSS present in the pericarp, aleurone and embryos of *Triticum monococcum*. Recently, waxy barley, lacking the predominant endosperm GBSS gene, has been shown to contain higher amylose contents than expected for a waxy mutant (about 8%) (Andersson *et al.*, 1999) with starch granules that stain strongly blue with iodine being present in the aleurone and adjacent cells (Andersson *et al.*, 1999). This may represent expression of GBSS from genes other than the major endosperm genes and the recent cloning of a second GBSS gene, located on chromosomes 2AL, 2B and 2D, and expressed in the pericarp and other tissues but not in the endosperm (Vrinten and Nakamura, 2000), provides strong support for this view.

7. Other Starch Synthases (SS)

The characterization of the full complement of starch synthases of wheat and maize has progressed rapidly in recent years, although there is less information available concerning these enzymes in barley. Analysis at the protein, cDNA and genomic DNA levels demonstrates that there are four classes of starch synthases in the wheat and maize endosperm. The non-GBSS starch synthases have been described as "soluble starch synthases"; however, the distribution of SSI and SSII enzymes between the granule and the soluble phrase makes this term misleading (Knight *et al.*, 1998). While a confusing nomenclature for the non-GBSS starch synthases originally evolved in maize, a straightforward system has recently been adopted in which the four classes are defined as GBSS, SSI, SSII and SSIII (Cao *et al.*, 1999).

The SSI gene is located on chromosome 7S of wheat and encodes an M_r 75 000 protein that is distributed between the starch granule and the soluble phase (Li *et al.*, 1999a). No mutations abolishing SSI activity are known in diploid plants and the suppression of this gene has not been reported through genetic engineering technologies, so the contribution of this gene to starch synthesis in the cereal endosperm remains unclear. In wheat, Yamamori and Endo (1996) showed that the forms of SSI encoded on chromosomes 7A, 7B and 7D could be separated by two-dimensional gel electrophoresis and two sources of putative null mutations were identified; however, a triple null line has not been reported.

The SSII class of starch synthases shows interesting differences between wheat and maize. In maize, two members of the SSII class have been isolated and are termed SSIIa and SSIIb (Harn *et al.*, 1998). In wheat, proteins encoded by SSII genes in wheat are present in the starch granule with apparent M_r of 100 000, 108 000 and 115 000 (Denyer *et al.*, 1995; Li *et al.*, 1999b). Comparison of the sequences of starch synthases indicates that the wheat SSII gene identified in the latter study is most similar to the maize

SSIIa gene and no homologue of the maize SSIIb gene has yet been characterized in wheat. Analysis of cDNAs for the wheat SSII proteins indicates that the expressed proteins have essentially identical M_r and that the differences observed by SDS–PAGE analyses of starch granule extracts represent anomalous mobility in SDS–PAGE, a feature noted for starch synthases from a range of sources (Li et al.,1999b). In maize and barley, the corresponding product of the SSII gene is not prominent in the starch granule (Rahman et al., 1995) and is apparently only expressed at low levels in the soluble fraction (Cao et al., 1999). The maize SSIIa gene maps to a location that is consistent with mutations in this gene being the lesion responsible for the *sugary-2* phenotype in maize (Harn et al., 1998), a phenotype characterized by altered chain length distribution of amylopectin and alterations in starch structure. In wheat, Yamamori has recently described the generation of a triple null line lacking the M_r 100 000, 108 000 and 115 000 proteins in the starch granule and preliminary characterization shows that starch granule morphology and starch structure are altered in the triple null line (Yamamori, 1998).

The SSIII gene from maize has been recently shown to be disrupted in the well-known *du1* mutation in maize (Gao et al., 1998; Cao et al., 1999). The product of the SSIII gene has been partially purified from maize endosperm (Pollock and Preiss, 1980; Cao et al., 1999) and was originally described as maize SSII. The gene encodes an approximately 6 kb mRNA that is translated into an enzyme of M_r about 180 000. In wheat, the homologue of the SSIII gene has been been cloned and characterized and the gene assigned to chromosome 1 (Li et al., 2000).

8. Branching Enzymes (BE)

In plants, two classes of genes encode starch branching enzymes, known as BEI and BEII. In cereals, there is strong evidence that the BEII class contains two independent genes, known in maize as BEIIa and BEIIb (Fisher et al., 1993, 1995, 1996b; Gao et al., 1996, 1997; Kim et al., 1998). Morell et al. (1997) partially purified the branching enzymes from wheat endosperm and showed the presence of two classes of BEII polypeptide. Comprehensive analyses of genetic stocks showed that the genes encoding BEI were located on the long arm of chromosome 7 (Morell et al., 1997) and this localization was confirmed by DNA hybridization studies (Rahman et al., 1997, 1999). Analysis of genes from *T. tauschii*, the donor of the D-genome of wheat, shows that multiple copies of the BEI gene are present and that two forms are expressed, one of which corresponds to the endosperm protein (Rahman et al., 1997, 1999). The second cDNA is truncated and may be an expressed pseudogene (Rahman et al., 1997). A cDNA for the endosperm-expressed BEI gene was also described by Repellin et al. (1997) and the expression of the BEI gene studied in further detail by Baga et al. (1999), demonstrating that the transcribed products undergo alternate splicing (Baga *et al.* 1999). In wheat, the BEII gene family has yet to be defined in detail but a cDNA for BEII has been reported (Nair *et al.*, 1997) that has the same N-terminal sequence as the endosperm-expressed BEII prepared from the soluble fraction (Morell *et al.*, 1997). In barley, the partial purification of BEI, BEIIa and BEIIb forms from the soluble fraction of the endosperm has been reported (Sun *et al.*, 1997) and cDNAs and partial genomic sequences for BEIIa and BEIIb described (Sun *et al.*, 1998). A report of an M_r 50 000 branching enzyme in barley has yet to be confirmed (Sun *et al.*, 1996).

In dicotyledonous plants, loss of BEII activity through either mutation (Bhattacharyya *et al.*, 1990) or gene suppression technologies gives rise to starches containing high amylose levels (Safford *et al.*, 1998; Jobling *et al.*, 1999). There is no evidence to date showing that the BEII genes in dicotyledonous plants fall into BEIIa and BEIIb subclasses as in monocots. In *Arabidopsis*, two cDNAs for BEII have been identified (Fisher *et al.*, 1996a), however, sequence comparisons with monocot BEII genes do not allow clear assignment of these genes to subclasses. Further information concerning the full complement of BEII genes in dicotyledonous plants is required.

In monocotyledonous plants, mutations giving rise to high amylose contents are known in maize, rice and barley. In neither rice (Mizuno *et al.*, 1993) nor barley (Schondelmaier *et al.*, 1992) have the high amylose phenotypes been specifically associated with mutations in BEIIa or BEIIb, although in rice there is evidence of a reduction in BEII activity (Mizuno *et al.*, 1993). However, in maize, it is firmly established that the high-amylose phenotype is associated with mutation of the BEIIb gene (Boyer *et al.*, 1980; Boyer and Preiss, 1981; Fisher *et al.*, 1996b). The impact of down-regulation of BEI has been investigated through antisense inhibition in potato tuber and found to alter the properties of the starch, but not gross structural features such as the amylose content (Filpse *et al.*, 1996). In wheat, antisense down-regulation of BEI activity has small but significant effects on starch structure (Baga *et al.*, 1999).

No mutations or gene suppression experiments have resulted in specific reduction in BEIIa activity in plants, although the *du*1 mutation in maize is known to reduce the expression of both BEIIa and starch synthase III. As noted above, the *du*1 mutation is now known to result from mutation of the structural gene for starch synthase III (Gao *et al.*, 1998; Cao *et al.*, 1999).

9. Debranching Enzymes

In maize, the classic *sugary-1* mutation that gives rise to sweet corn has been demonstrated to be caused by a lesion in an isoamylase-type debranching enzyme gene *Su1* (Pan and Nelson, 1984; James *et al.*, 1995). A similar mutation is known in rice where the mutation maps to a location consistent with a lesion affecting the expression of an isoamylase-type debranching enzyme. Interestingly, the expression of both the isoamylase-type and pullulanase-type debranching enzyme genes is suppressed (Nakamura *et al.*, 1996, 1997; Kubo *et al.*, 1999). There is evidence that the rice isoamylase is present in a complex containing several subunits and that the pleiotrophic effects of mutation in the *sugary-1* gene may be mediated through disruption of a complex (Fujita *et al.*, 1999). In barley, both isoamylase cDNA sequences (Sun *et al.*, 1999) and a pullulanase gene (Burton *et al.*, 1999; Kristensen *et al.*, 1999) have been described.

The requirement for debranching enzyme in starch biosynthesis is clearly established by the analysis of mutations in maize (Pan and Nelson, 1984; James *et al.*, 1995), rice (Nakamura *et al.*, 1996, 1997; Kubo *et al.*, 1999), *Chlamydomonas* (Mouille *et al.*, 1996) and *Arabidopsis* (Zeeman *et al.*, 1998). However, the specific roles of debranching enzymes in starch biosynthesis remain unclear. Alternative hypotheses have been proposed by Myers *et al.* (2000) and Zeeman *et al.* (1998). The hypothesis proposed by Myers *et al.* (2000) builds on an earlier hypothesis summarized in Ball *et al.* (1996). This hypothesis, summarized in Fig. 4a, proposes that the primary role of debranching enzyme is to trim or edit the developing amylopectin to form a structure that is capable of crystallizing with other amylopectin molecules to form the blocklets and lamellae of the granule (Buleon *et al.*, 1998; Gallant *et al.*, 1997). The hypothesis of Zeeman *et al.* (1998) (Fig. 4b) proposes that debranching enzymes are not directly involved in amylopectin biosynthesis but rather act to degrade non-granular α-glucan that otherwise might complete with granular starch biosynthesis. In both hypotheses, the result of a lesion in debranching enzyme activity is the accumulation of soluble non-granular phytoglycogen and the suppression or elimination of granular starch deposition. Further evidence is required to clarify the role(s) of debranching enzymes in starch biosynthesis.

10. Roles of Other Enzyme Activities in Starch Biosynthesis

A number of other enzymes and proteins have been suggested to have roles in starch biosynthesis in plants, but their specific roles in starch biosynthesis in the cereal endosperm remain unclear. Starch phosphorylase was originally thought to be responsible for starch biosynthesis. However, the discovery of the sugar nucleotides and ADP glucose pyrophosphorylase, and of mutations in ADP glucose pyrophosphorylase genes that suppress starch synthesis dramatically, led to the current view that the elongation of starch chains is essentially entirely carried out by starch synthases and not starch phosphorylase. However, it remains possible that phosphorylases play a role in shaping starch biosynthesis through the reversibility of the phosphorylase reaction in alternating rounds of synthesis and degradation of starch. Disproportionating enzyme has long been thought to be involved only in starch degradation; however, recent genetic evidence from *Chlamydomonas* suggests that this enzyme is also required for amylopectin biosynthesis

(Colleoni *et al.*, 1999a,b). Further evidence from a variety of higher plant tissues is required to further define the role of disproportionating enzymes in starch biosynthesis in higher plants.

In potatoes, a starch granule associated protein known as the R protein has been shown to be involved in starch biosynthesis (Lorberth *et al.*, 1998) and is particularly associated with the phosphorylation status of potato starch. The biochemical function of the R protein remains unknown and any role of this enzyme in starch biosynthesis in cereals, where phosphorylation is essentially undetectable, is unclear.

11.   Starch Granule Initiation and Development

One of the key characteristics of wheat and barley starch noted above is the bimodal distribution of starch granule size that underlies the definition of "A" and "B" granule populations. Differences between species in starch granule size distribution, granule morphology, and granule packing are critical to functionality of starch but little is really known about the genetic or biochemical basis for these differences. One of the early uses of microscopy was to investigate the variation in starch granule size and shape between different plant species and between different tissues (see references in Wang *et al.*, 1998). The marked differences in granule size and shape occur despite the apparent conservation between species of the primary complement of starch biosynthesis enzymes. Several possible reasons for these species differences may be proposed. First, there are subtle and as yet unrecognized differences between the properties of the primary biosynthetic enzymes between species that influence granule shape. Second, additional genes are involved in aspects of granule initiation and development that remain to be identified and characterized. Third, the interaction between the growing starch granule, the physical environment provided by the amyloplast, and the levels of substrate supply are all important factors in shaping the growth of the starch granule in a species or organ-specific manner.

The propositions that there are further genes involved in granule initiation and development, and that starch granule formation is influenced by the microenvironment of the amyloplast and the flow of substrate supply, can be considered together against the available evidence. The possibility of a proteinaceous initiator being responsible for starch granule initiation has led to the suggestion that an analogue of the glycogen-initiating protein glycogenin, referred to as "amylogenin", might be present in plants. While homologues can be identified through sequence comparison, the assignment of a function in starch biosynthesis to these genes has yet to be made and it remains possible that this family of glycogenin-like genes in plants play roles in the initiation of a range of poylsaccharides in the plant cell. The possible requirement for "amylogenin" for granule initiation in higher plants is therefore an open question with no direct genetic or biochemical evidence.

The appearance of starch granules during endosperm development of wheat has been documented through the use of microscopy (Briarty *et al.*, 1979) and particle size analysis (Bechtel *et al.*, 1990). The A granules are initiated in the first 5–7 days after fertilization as the endosperm passes through its initial coenocytic phase and becomes cellularized (Briarty *et al.*, 1979). There then appears to be a clear reduction, if not cessation, in starch granule initiation events until a second very active burst of granule initiation occurs from approximately 12–15 days after fertilization, giving rise to B granules (Briarty *et al.*, 1979). The precise timing is highly dependent on the environmental conditions, particularly temperature. The A granules in wheat and barley appear to develop through a defined and very curious pathway, observed by Evers and Bechtel (1988). The first granules to be seen are spherical and approximately 1–2 μm in diameter. These granules expand radially until they reach 3–5 μm in diameter, when they typically develop a protuberance on the surface that then grows around the "equator" of the granule as a radial plate. Ultimately the plate encircles the granule and the starch granule has a diameter of 10–15 μm, of which 3–5 μm remain as the initiating sphere. Further expansion of the granule then occurs at the edges of the plate but the principal pattern of deposition shifts to the surface of the plate, extending the granule in a lateral direction. A feature of the equatorial plate is the presence at the rim of the plate of a distinctive "equatorial groove". This groove represents a line of weakness in the granule that is readily attacked by degradative enzymes and may be a built in a zone of weakness allowing rapid degradation of the granule on germination. Briarty *et al.* (1979) noted that invaginations of the amyloplast membrane, the "tubuli", are orientated in the same plane as the equatorial dimension of growth of the starch granule and the great increase in surface area of the tubuli may focus incoming carbon to particular regions of the amyloplast, enhancing local rates of starch biosynthesis such that spatially directed growth of the granule occurs.

The available evidence suggests that the initiation and development of the "B" granules is markedly different. The microcopy work of Parker (1985) showed that B granules are first observed in tubular projections of amyloplasts that already contain A granules. Several nascent B granules are generally observed in each protuberance. The B granules develop spherically but do not progress through the equatorial plate formation stages described for B granules.

Several lines of evidence indicate that there is genetic control of starch granule initiation and development. Firstly, differences in starch granule number per amyloplast are evident between species such as rice and oats. Starch granule initiation is apparently much more active early in development in rice and oats than in wheat or barley, leading to the production of large numbers of small granules. Secondly, the apparently precise regulation of granule initiation over the course of endosperm development in wheat and barley noted above. Thirdly, the existence of wide variation in A to B granule ratios in wheat (Stoddard *et al.*, 1999a) and barley (Oliveira *et al.*, 1994).

12. Manipulation of Starch Properties

The manipulation of wheat flour composition and processing properties has largely focused on the manipulation of protein alleles because the protein products of each genome can be readily separated by electrophoretic techniques and because distinct relationships between glutenin composition and processing qualities can be established (Payne, 1987, and section below). While starch represents 65–70% of the weight of the flour, it is only in the past decade that concerted attempts have been made to manipulate wheat starch composition, largely because the background knowledge of the starch biosynthesis process in plants was still emerging and because there was a need to develop tools to identify natural diversity in starch structure and functionality. The development of efficient wheat and barley transformation systems provides a new opportunity to generate informative and useful diversity.

13. Identification of Natural Mutants

The identification of natural and induced mutants in wheat is more problematic than in a diploid species because of the need to identify the target mutation in each of the homologous genes and combine the mutations through crossing. The major focus of such work in wheat has been the identification of mutations in the GBSS genes in order to develop waxy wheats (Nakamura *et al.*, 1995; Zhao and Sharp, 1998) and lines lacking one or two of the GBSS alleles (Zhao *et al.*, 1998; Demeke *et al.*, 1999), using two-dimensional and later one-dimensional gel electrophoresis techniques (Zhao and Sharp, 1996).

Natural mutations affecting the expression of two other starch biosynthetic genes have been identified. Yamamori (1998) recently reported the generation and preliminary analysis of a line lacking the M_r 100 000, 108 000 and 115 000 starch granule proteins encoded by the SSII genes (Li *et al.*, 1999b). The starch granules of the SSII triple null line were deformed and the amylose content of the starch was elevated to approximately 37%. Mutations affecting the expression of the A and B genome forms of SSI have also been identified through two-dimensional gel electrophoresis (Yamamori and Endo, 1996) but the generation of a SSII triple null line has not yet been reported.

In barley, three mutations affecting starch biosynthesis have been described. The classic GBSS mutation has been identified and widely used to develop waxy barley types. A mutant with elevated amylose content of 45–50%, Glacier AC38, has been identified (Walker and Merritt, 1968). Although the mutation mapped to chromosome 1H (Schondlemeiar *et al.*,

1992), the molecular basis has not yet been identified. A shrunken mutant, *shx*, has been identified (Schulman and Ahokas, 1990) and its starch characteristics defined (Schulman *et al.*, 1995), but neither the map location nor the causal gene have been identified.

A range of mutations affecting the starch biosynthesis pathway have been identified and characterised in maize, including *waxy, amylose extender, sugary-1, sugary-2, brittle-1, brittle-2, shrunken-2* and *dull-1* (see Nelson and Pan, 1995). The generation of similar mutations in wheat and barley would be informative, firstly, because the process of starch granule development in wheat and barley differs from in maize and thus the consequences of the mutations may differ in informative ways, and, secondly, because the end-use properties of wheat and barley starches are very different to those of maize starch.

14. Genetic Engineering Approaches

Advances in transformation technology now allow the possibility of generating sufficient numbers of transgenic wheat and barley plants to thoroughly test the impact of genetic manipulation strategies. Obvious targets for genetic engineering are the suppression of the activities of starch biosynthetic enzymes through antisense, co-suppression or other gene suppression technologies. One example of this work is the suppression of branching enzyme I activity through antisense technology (Baga *et al.*, 1999). The overexpression of starch biosynthetic or starch-modifying genes from heterologous sources in wheat and barley is a second strategy that is likely to generate novel diversity.

The major challenge for the future is to further extend our knowledge of the biochemistry and genetics of starch biosynthesis to include the details of the processes that lead to the development of specific starch granule structures. In turn, we will need to integrate these insights with knowledge of the functional performance of starches in a variety of end uses. Many of the basic tools necessary for these tasks are to hand and the next decade will no doubt be a very interesting and exciting period for starch research.

V.   PROTEINS

A.   TYPES AND PROPERTIES

Although proteins account for less than 15% of the mature cereal grain, they have been studied in much greater detail than other grain components. In fact, wheat gluten was one of the very first proteins to be studied, being first isolated and described by Jacopo Beccari, Professor of Chemistry in the University of Bologna, in 1745. The interest in cereal seed proteins has been largely stimulated by their importance in determining end-use quality, whether the nutritional quality as food for humankind or feed for monogastric livestock (notably pigs and poultry) or the functional properties for processing into foods and beverages. In particular, the gluten proteins of wheat have received the most attention because of their role in determining the viscoelastic properties that enable dough to be processed into bread, pasta and noodles, and other foods. This is discussed in more detail below.

Although cereal proteins were investigated by a number of workers in the late 18th and 19th centuries, their study was put on a systematic basis by the work of T. B. Osborne. Osborne worked at the Connecticut Agricultural Experiment Station from 1886 until 1928 and published studies of seed proteins from 32 plant species. Much of his work is summarized in his monograph "The Vegetable Proteins" published in 1924 (Osborne, 1924).

Osborne classified plant proteins into four groups, subsequently called "Osborne fractions", on the basis of their extraction in a series of solvents. The first two fractions were soluble in water and dilute salt solutions and were called albumins and globulins, respectively. The third group was extracted in alcohol/water mixtures, classically 60/70% (v/v) ethanol, and was termed prolamins to reflect its high content of proline and amide nitrogen (the latter now known to be derived from glutamine residues). Finally, the glutelins are not soluble in any of the above solvents and were initially extracted in dilute acid or alkali. It is more usual today to use a detergent such as sodium dodecyl sulphate or a chaotropic agent such as urea.

The Osborne fractions have provided a framework for modern cereal protein chemistry but the scheme has inevitably been modified in view of our greatly improved knowledge of cereal protein structure. In addition, it is now more usual to classify cereal proteins initially into broad groups based on their functions rather than their solubility. Three such groups can be recognized.

1. Structural and Metabolic Proteins

These proteins contribute to the structure and day-to-day operation of the cell. They therefore include enzymes, structural proteins present in membranes and cell walls, transporters, receptors etc. In Osborne's classification they include albumin, globulin and glutelin proteins.

2. Defensive Proteins

The rich reserves of starch, lipid and protein present in seeds makes them an attractive target for pests and pathogens. Not surprisingly, plants have evolved to provide resistance to such invaders, including a battery of protective proteins. The range of such proteins is wide, even within cereal seeds, and a detailed discussion is beyond the scope of this review. They include inhibitors of hydrolytic enzymes (proteinases and α-amylases), enzymes (β-glucanase, endochitinase), chitin-binding proteins, ribosome-inactivating proteins and surface active proteins, which may act to destabilize membranes. For example, barley grain has been shown to contain at least 11 different types of proteins that are active against pests or pathogens *in vitro*, including three different types of inhibitors of proteinases and/or α-amylases. These proteins may together provide a low level of protection against a broad range of organisms. It should also be noted that none of these proteins appear to have anti-nutritional effects, either in barley or in other cereals. Several of these defensive proteins are discussed below as targets for manipulation while the reader is referred to recent review articles for more detailed accounts (Shewry and Lucas, 1997; Carbonero and García-Olmedo, 1999; García-Olmedo, 1999; Osborn and Broekaert, 1999; Peumans and Van Damme, 1999; Shewry, 1999b).

3. Storage Proteins

These can be defined as proteins whose major function is to act as a store of nitrogen, carbon and sulphur to be mobilized during germination. The embryos and aleurone cells of cereals contain storage proteins with sedimentation coefficients (S_{20w}) of about 7, at least some of which are related to the major 7S storage globulins present in seeds of many other plants including monocots (palms) and dicots (notably legumes but also cottonseed and many other species) (see Kriz, 1999; Casey, 1999). However, these 7S globulins are minor components in the context of the whole grain, the major components being the starchy endosperm-located prolamins and/or 11S globulins.

Storage proteins related to the 11S globulins of dicotyledonous plants form the major storage protein fractions in the starchy endosperms of oats and rice, although the low solubility of the rice proteins means that they are traditionally called "glutelins". Oats and rice also contain prolamin storage proteins but they account for about 10% or less of the total grain nitrogen. In contrast, prolamins are the major storage proteins in the Triticeae (barley, wheat, rye) and in the Panicoideae (maize, sorghum, millets), where they account for up to about 60% of the grain nitrogen.

B. PROLAMINS

Prolamins are unique among the Osborne fractions in that they are restricted to the seeds of one family of plants: the grasses that include cultivated cereals. They include both "classical" prolamins, which are readily soluble in aqueous alcohols, and "glutelins", which are only extracted in the presence of a reducing agent. The insolubility of the "glutelins" in aqueous alcohols results from the formation of high M_r polymers stabilized by interchain disulphide bonds. Although these polymers may be insoluble in aqueous alcohols, the reduced monomeric subunits show typical prolamin solubility properties.

In wheat the monomeric and polymeric prolamins are called gliadins and glutenins, respectively, the two groups together forming the gluten proteins. In other species the monomeric and polymeric forms share the same trivial name, which is usually based on the Latin name for the genus of origin: secalins in rye (*Secale cereale*), hordeins in barley (*Hordeum vulgare*) and zeins in maize (*Zea mays*).

Prolamins vary widely in their M_r (from about 10 000 to 90 000), in their amino-acid compositions and sequences, and in other properties (see Coleman and Larkins, 1999; Leite *et al.*, 1999; Muench *et al.*, 1999; Shewry, 1999a; Shewry *et al.*, 1999). Consequently, it has proved difficult to define properties which apply to all the different types. However, a recent reappraisal (Shewry and Tatham, 1999) has suggested that they share most or all of the following properties:

1. All prolamins are insoluble in water or dilute salt solutions when in the native state. Similarly, all are soluble in alcohol:water mixtures, either in the native state or as reduced subunits. Only one group, the γ-zeins of maize and related prolamins from other panicoid cereals, is soluble in water as reduced subunits.
2. Most prolamins contain high proportions of proline, glutamine and one or more other amino acids (e.g. glycine, phenylalanine, histidine, cysteine). In most cases this unusual composition results from the presence of structural domains, one or more of which comprises repeated amino-acid sequences or is enriched in specific amino-acid residues.
3. All prolamins are deposited in protein bodies, although these may originate from the vacuoles or endoplasmic reticulum and may fuse in the dry mature grain to give a continuous "matrix" (e.g. in wheat).
4. Prolamins have no function apart from storage and their synthesis is, therefore, modulated by the availability of nutrients.

Because of their unusual properties, the prolamins were long considered to form a unique group of proteins present only in cereal grains with no related proteins in other species or tissues. This has now been disproved by the demonstration of sequence homology between the major prolamins of the Triticeae, some prolamins of maize and a range of low M_r cysteine-rich proteins, many of which accumulate in seeds (Kreis *et al.*, 1985a,b; Shewry and Tatham, 1999). The latter include 2S storage albumins of dicotyledonous seeds, cereal inhibitors of α-amylase and/or trypsin, non-specific lipid binding proteins and puroindolines (see below). These proteins are together called the "prolamin superfamily".

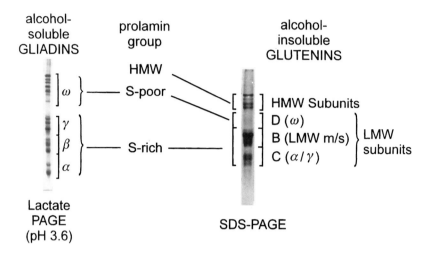

Fig. 5. The classification and nomenclature of wheat gluten proteins separated by sodium dodecylsulphate–polyacrylamide gel electrophoresis (SDS–PAGE) and electrophoresis at low pH. The D group of low-molecular-weight (LMW) subunits are only minor components and are not clearly resolved in the separation shown. HMW, high-molecular-weight. (Reproduced, with permission, from Shewry *et al.*, 1999.)

C. GLUTEN PROTEINS AND WHEAT UTILIZATION

1. Gluten Proteins

Wheat gluten can be readily prepared from dough made with white flour, by gently washing with water. This gives a cohesive mass which comprises (on a dry weight basis) about 70% protein with some starch and small amounts of lipid. The proteins correspond essentially to the prolamins (which account for about 60% of the total grain N) with only traces of non-prolamin components.

Isolated wheat gluten exhibits biomechanical properties which are unique among plant proteins and are not shared, to any extent, by prolamins from other species. These are a combination of elasticity and extensibility (also called viscous flow). In dough the gluten proteins appear to form a dispersed but continuous network, and this confers the cohesive and viscoelastic properties, which allow dough to be processed to form bread, pasta and noodles and a range of other baked goods (cakes, biscuits, pastries, etc.). In addition, gluten is used as a binder in breakfast cereals and pet foods, and can be textured or modified for use in the food industry (e.g. as a meat analogue or foaming agent). It can also be used to produce biodegradable films, e.g. for packaging food.

However, the major interest in wheat gluten relates to its role in breadmaking. In particular, there is considerable interest in understanding and manipulating the level of gluten elasticity, as low elasticity may limit the breadmaking performance. For example, high yielding "general purpose" wheats grown in the UK have lower gluten elasticity than high-quality breadmaking wheats, which have lower yields, while gluten elasticity also varies from year-to-year due to variation in the climatic conditions.

Wheat gluten is a complex mixture of over 50 individual proteins, which can be classified into a number of groups and subgroups. Thus, the monomeric gliadins are classified into α-, β-, γ- and ω-gliadins on the basis of their electrophoretic mobilities at low pH and their N-terminal amino-acid sequences. Similarly, the reduced subunits of glutenin can be classified on the basis of their mobilities on SDS–PAGE into high- (HMW) and low-molecular-weight (LMW) groups, with the latter being subdivided into B, C and D subunits (Fig. 5). However, comparison of amino-acid sequences shows that three broad groups of related proteins are present: the S-rich prolamins comprising the α-, β-, γ-gliadins and the B and C groups of LMW subunits, the S-poor prolamins comprising the D groups of LMW subunits and the ω-gliadins and the HMW prolamins comprising the HMW subunits of glutenin. These groups are shown in Fig. 5 and discussed in detail by Shewry *et al.* (1999).

2.   HMW Subunits and Breadmaking Quality

It has been accepted for many years that the glutenins are responsible for gluten elasticity, but a major breakthrough came about 20 years ago when Payne and co-workers demonstrated that variation in the breadmaking quality of European wheats was associated with allelic variation in the HMW subunits of glutenin (see Payne, 1987). Bread wheats all have six HMW subunit genes, two each (encoding one x-type and one y-type subunit) on the long arms of chromosomes 1A, 1B and 1D. However, gene silencing results in the presence of only 3, 4 or 5 subunits. The work of Payne, and more recent studies in other laboratories (see Shewry *et al.*, 1995), indicates that differences in subunit amount (each subunit accounting for about 2–2.5% of the total grain proteins) and subunit properties (notably between allelic pairs of 1Dx + 1Dy subunits) may both contribute to variation in quality.

Detailed studies have revealed that the individual HMW subunits have an unusual structure, with an extensive central domain comprising repetitive sequences that form a loose spiral structure. The N- and C-terminal domains are non-repetitive and contain cysteine residues, some of which appear to form interchain disulphide bonds. It is, therefore, probable that the HMW subunits form an extensive disulphide-bonded network, providing an "elastic backbone", which interacts with other gluten proteins by disulphide bonds and non-covalent forces (principally hydrogen bonds) as shown in Fig. 6. Although the precise mechanism of gluten elasticity is still not completely understood, recent work suggests that it derives from stretching of chemical bonds rather than being entropic in origin as in protein elastomers from

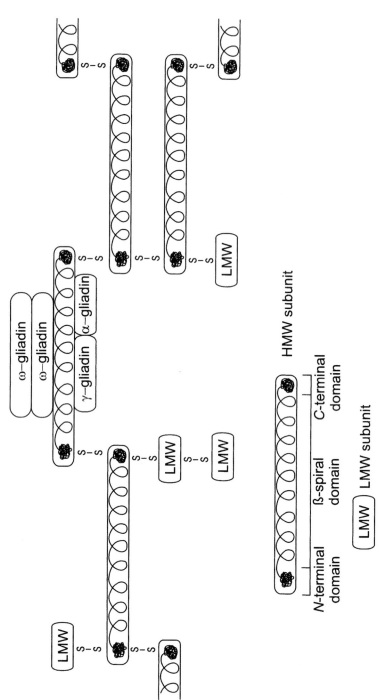

Fig. 6.   A structural model for wheat gluten. S–S, interchain disulphide bonds; LMW, low-molecular-weight subunits.

animals (Tatham and Shewry, 2000), with contributions from the β-spiral structure, interchain disulphide cross-links and interchain hydrogen bonds (Belton, 1999).

3. Manipulation of HMW Subunits

The demonstration of a clear relationship between the number of expressed HMW subunit genes, the total amount of HMW subunit protein and breadmaking quality (Halford *et al.*, 1992) has identified the HMW subunits as a target for attempts to improve wheat quality by genetic engineering. Four laboratories have reported the transformation of wheat with HMW subunit genes, encoding a subunit 1Dy10/1Dx5 hybrid protein (Blechl and Anderson, 1996), subunit 1Ax1 (Altpeter *et al.*, 1996; Alvarez *et al.*, 2000) and subunits 1Ax1 and 1Dx5 (Barro *et al.*, 1997; He *et al.*, 1999; Rooke *et al.*, 1999). In all cases expression levels up to or exceeding those of the endogenous genes were observed.

Increased quality, as measured using a small-scale Mixograph, was also observed when additional subunits were expressed in a bread wheat background expressing only two endogenous subunits (Barro *et al.*, 1997). Similarly, expression of subunits 1Ax1 or 1Dx5 in durum wheat at levels corresponding to those of the endogenous genes gave increased dough strength as measured using the Mixograph (He *et al.*, 1999). However, expression of very high levels of subunit 1Dx5 in either bread wheat expressing five endogenous subunits (Rooke *et al.*, 1999) or durum wheat (He *et al.*, 1999) resulted in dough with unusual mixing properties. However, blending of flour from the transgenic bread wheat line with flour from a normal cultivar gave increased dough strength up to a level of about 50% (Fig. 7).

These results demonstrate that it is possible to modify the properties of wheat gluten by transformation with genes for HMW subunits, either to increase the dough strength or to give new properties. Further fine-tuning may be achieved by constructing and expressing specific HMW subunit mutants or by transformation with genes for other gluten proteins.

D. HIGH-LYSINE PROTEINS AND NUTRITIONAL QUALITY

Animals are only able to synthesize about half of the 20 amino acids commonly found in proteins; the remaining ones are required in the diet and are therefore called "essential". If only one of these "essential" amino acids is limiting in the diet, the remaining amino acids cannot be utilized, and are broken down leading to poor growth and nitrogen loss. This does not pose a problem when humans or monogastic livestock (e.g. pigs, poultry) are fed on a diet that contains various protein sources with different amino acid compositions. Similarly, in ruminants, all 20 amino acids can be synthesized by the microflora in the rumen, which is sufficient to provide for all the

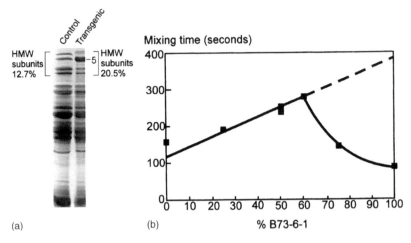

Fig. 7. The manipulation of wheat functionality by transformation. (a) SDS–PAGE of total grain proteins from the control line L88–6 and the transgenic line B73-6-1. Note the increase in the proportion of subunit 1Dx5 in B73-6-1, resulting in an increase in the proportion of total HMW subunits from 12.7% to 20.5% of the total protein. (b) The effect of blending flour of B73-6-1 and the normal cultivar Banks on the mixing time determined with the 2g Mixograph. Increasing the proportion of B73-6-1 up to 60% results in increased mixing time, above which the mixing time falls. (Based on Rooke *et al.*, 1999.)

dietary requirements of the animal. However, the amino-acid composition of cereals can limit the growth of livestock, and perhaps also humans, when fed on a pure cereal diet. Feeding studies show that lysine is the first limiting amino acid in cereals, followed by threonine and, in maize, tryptophan. In all cereals these deficiencies result from low levels of the essential amino acids in the prolamin fractions.

Early attempts to improve the nutritional quality of cereal grain resulted from the discovery in the 1960s and 1970s of naturally occurring high lysine mutants of maize, sorghum and barley, followed by the induction of further mutations in all three species using chemical or physical mutagens; see Shewry *et al.* (1987) and Coleman and Larkins (1999) for reviews. The best known and most widely studied of these mutant high lysine genes are the *opaque-2* gene of maize (Mertz *et al.*, 1964) and the *lys* gene of Hiproly barley (Munck *et al.*, 1970). With the exception of Hiproly, all high lysine mutants so far characterized have decreased proportions of lysine-poor prolamins and compensatory increases in other, more lysine-rich, proteins. In Hiproly there is little or no effect on total prolamins but specific increases in several lysine-rich proteins: β-amylase (5 g% lys), protein Z (a serpin proteinase inhibitor) (7.1 g% lys) and chymotrypsin inhibitors CI-1 (9.5 g% lys) and CI-2 (11.5 g% lys) (Hejgaard and Boisen, 1980). Unfortunately, all of these mutants also have reduced yield, due to pleiotropic effects on starch synthesis. This has limited attempts to develop high-lysine cultivars for commercial production, which has only so far been achieved with *opaque-2* maize. However, this mutation also gives rise to soft endosperm texture, which can make the grain susceptible to damage and pathogen attack. Hard *opaque-2* maize lines have, therefore, been developed using genetic modifiers (Coleman and Larkins, 1999). These lines, called quality protein maize (QPM), contain higher levels of lysine and tryptophan and have gained some level of acceptance in developing countries. Despite considerable effort, it has so far proved impossible to incorporate high-lysine genes into varieties suitable in terms of maximum yield and yield stability for commercial production in developed countries and different strategies need to be adopted. Two such strategies have been proposed to increase the levels of lysine in cereals.

The first is to increase the amounts of free amino acids, by manipulating their biosynthetic pathways to circumvent the feedback regulation that normally occurs. This can be achieved by transforming with feedback-insensitive forms of two key regulatory enzymes, aspartate kinase (AK) and dihydrodipicolinate synthase (DHDPS). Falco *et al.* (1995) showed that a two-fold increase in total seed lysine occurred in canola transformed with a DHDPS gene from *Corynebacterium*, while transformation of soybean with the same gene and an AK gene from *E. coli* resulted in a five-fold increase.

More extensive studies of free lysine synthesis in seeds of transgenic tobacco, soybean, canola (oilseed rape) and maize have been reported in a recent patent (Falco, 1998). Expression of the *Corynebacterium* DHDPS in maize under control of the maize globulin 1 gene promoter gave increases in free lysine from 1.4% of total free amino acids to 15–27% in three independent lines. This represented increases in total grain lysine from 2.3% in control seeds to 3.6%, 5.1% and 5.3% in the three transgenic lines. However, no increases in lysine were observed when the same gene was expressed under control of the endosperm-specific glutelin 2 gene promoter. This was probably due to increased lysine catabolism in the grain, and down-regulation of lysine catabolism would therefore be required to achieve overaccumulation of lysine in the corn endosperm. In contrast, Brinch-Pedersen *et al.* (1996) failed to detect any increase in total grain lysine when AK or DHDPS genes from *E. coli* were expressed in transgenic barley under control of the "constitutive" CaMV 35S promoter.

The second approach is to transform with genes for specific lysine-rich proteins. An obvious candidate protein for this is the barley chymotrypsin inhibitor CI-2. The major form of CI-2 comprises only 84 residues including seven lysines (i.e. 8.3 mol% lys) (Williamson *et al.*, 1987) and the combined activities of CI-1 and CI-2 are increased by up to almost 10-fold in Hiproly and derived lines with no apparent adverse effects on grain development or digestibility in animal diets (Hejgaard and Boisen, 1980). Although it would be possible to express the wild type CI-2 protein in transgenic plants, our detailed understanding of the structure and folding pathway of CI-2 (see Campbell, 1992) has facilitated the use of protein engineering to increase the lysine content even further. Recently, Roesler and Rao (1999) have reported the expression in *E. coli* of five mutant forms of CI-2 containing 20–25 mol% lysine. All had similar conformations (as determined by circular dichroism spectroscopy) as the wild-type protein but lower thermodynamic stabilities. However, at least one mutant was considered to be potentially suitable for expression in transgenic plants.

An alternative approach to transforming with genes for wild-type or mutant high-lysine proteins is the *ab initio* design of proteins rich in essential amino acids. Keeler *et al.* (1997) designed proteins with an α-helical coiled coil structure containing up to 43 mol% lysine and expressed them in *E. coli* using synthetic oligonucleotides. One protein containing 31 mol% lysine and 20 mol% methionine was expressed in seeds of tobacco under control of the bean phaseolin and soybean β-conglycinin gene promoters. The primary transformants contained increases in lysine varying from zero up to 0.8 mol% above a level of 2.56 mol% in the wild-type seeds. These increases were also inherited through three generations. In contrast, most of the lines showed little or no increase in methionine content.

E.   HORDEIN, β-AMYLASE AND BARLEY MALTING QUALITY

Hordeins, the prolamins of barley, can be classified into four groups, which are related to the gluten proteins of wheat. The two major groups are the B hordeins (polymeric S-rich prolamins) and C hordeins (monomeric S-poor prolamins), which account for about 70–80% and 10–20% of the total fraction, respectively. In addition, the γ-hordeins (S-rich prolamins, which occur as monomers and polymers) and D hordeins (polymeric HMW prolamins) each account for less than 5% of the total fraction.

Work carried out as long ago as the 1930s demonstrated that high levels of total grain protein were disadvantageous for malting quality, owing to a negative correlation with hot water extract (Bishop, 1930a). Furthermore, this effect varied with cultivar and increases in respiration and rootlet growth were shown to contribute to malting losses. The major effect of high grain nitrogen is an increase in the proportion of hordeins (Bishop, 1930b; Kirkman *et al.*, 1982) and this is now generally considered to account for the inverse correlation between grain nitrogen and malting performance. However, more recent work indicates that differences in the amounts and properties of the individual hordein groups and proteins may also be important.

Initial attempts to relate malting performance to differences in the hordein composition revealed by SDS–PAGE or gel electrophoresis at low pH proved inconclusive (Baxter and Wainwright, 1979a: Shewry *et al.*, 1980; Riggs *et al.*, 1983). However, other studies indicated that the ability of the hordeins to form high-M_r polymers was important and it was proposed that such polymers could entrap starch granules and hinder digestion during malting and brewing (Baxter and Wainwright, 1979b; Shewry *et al.*, 1981).

Further support for the importance of polymeric hordeins came from the analysis of the gel protein fraction. This fraction can be prepared by stirring barley meal with 1.5% SDS followed by centrifugation. It consists of polymeric B and D hordeins, being particularly rich in the latter, and the amount (measured by weight or as N) is inversely correlated with hot water extract (Smith and Lister, 1983; Skerritt and Janes, 1992). Furthermore, Howard *et al.* (1996) reported that the amount of D hordein (expressed as mg/g flour) present in three varieties of barley grown under five nutrient regimes was negatively correlated with malt extract, the effect being independent of cultivar. Based on these studies it can be suggested that the gel protein represents a disperse polymeric protein network, analogous to gluten in wheat, which physically limits modification during malting. However, a recently reported study casts doubt on this hypothesis.

This study compared the amount of gel protein in a series of near-isogenic pairs of lines that differed in the absence or presence of D hordein. Statistical analysis showed a strong correlation between the absence or presence of D hordein and the amount of gel protein, but no relationship was found between these parameters and hot water extract or any other malting parameters (Brennan *et al.*, 1998). The relationship between D hordein, gel protein, hordein polymers and malting performance therefore remains unresolved.

Hordein is only one, and by no means the most important, of a number of factors that influence the malting performance of barley. These can be broadly divided into two groups: those that operate during grain development and influence the structure and composition of the mature grain, and those that operate during germination. Several are discussed in detail in other parts of this chapter: endosperm texture, cell wall β-glucans and surface active proteins. However, among the most important determinants of malting quality is the amount and distribution of degradative enzymes, which may be stored in the mature grain or synthesised *de novo* during germination.

Endosperm breakdown during malting and brewing depends on the combined activities of three main classes of enzyme: proteinases, cell wall degrading enzymes (notably β-glucanases) and starch degrading enzymes. The vast majority of these enzymes are synthesized in the aleurone cells and the scutellar epithelium during germination (see Briggs, 1992 for a detailed discussion), the major exception being β-amylase.

β-Amylase ((1→4)-α-D-glucan maltohydrolase) is the second most important of four activities that contribute to diastatic power, with α-amylase being the major activity and α-glucosidase and limit dextrinase being less active. Diastatic power is particularly important when brewing with high levels of starch adjunct and may be limiting under such conditions (Evans *et al.*, 1997).

β-Amylase accumulates in the developing barley grain, being particularly concentrated in the aleurone layer and starchy endosperm. The amount also appears to be regulated by nitrogen availability, in the same way as hordein storage proteins (Giese and Hejgaard, 1984). β-Amylase is synthesized without a signal peptide, presumably in the cytosol, but becomes partially associated with the periphery of the starch granules during grain desiccation (Hara-Nishimura *et al.*, 1986). The mature grain contains free and bound forms of the protein, the latter being only extracted in the presence of a reducing agent or after partial proteolytic digestion with papain. The mature protein consists of 535 amino acids with an M_r of about 60 000 (Kreis *et al.*, 1987) but release of the bound form during seed germination results from partial proteolysis at the C-terminus, resulting in additional enzymatically active forms with M_r of about 58 000, 56 000 and 54 000 (Lundgaard and Svensson, 1987).

Although it is possible to select for lines with high levels of β-amylase, a more important target is to increase the stability of the enzymes to heating. This is because the activity of β-amylase is reduced by kilning with most of the remaining activity being lost during a typical mash at 65°C (Palmer, 1989; Bamforth and Quain, 1989). This contrasts with α-amylases, which remain active under these conditions. Kihara *et al.* (1998) demonstrated that modern varieties of barley do differ in the extent to which their β-amylase enzymes are stable to heating. They divided modern varieties into three groups. The group A varieties, all of which were Japanese, had 60–63% of β-amylase activity remaining after heating at 57.5°C for 30 minutes. In contrast, only 26–39% activity remained in group B varieties and 2–5% in group C, both groups containing varieties from Europe, North America and Australia. However, the enzymes from typical varieties of all three groups were totally inactivated when heated at 62.5°C or 65°C for 30 minutes.

Higher levels of thermostability can be achieved by protein engineering, with mutagenesis of seven single residues in barley β-amylase giving increases in T_{50} (the temperature at which 50% of the initial activity is lost over 30 minutes) ranging from 0.8°C to 3.2°C. Combining all seven mutations in one protein resulted in an increase in T_{50} of 11.6°C, from 57.4°C to 69°C (Yoshigi *et al.*, 1995). This increase would certainly be expected to result in greater stability of the enzyme during malting and brewing. Alternatively, heat-stable forms of β-amylase from thermophilic organisms such as *Clostridium thermosulphurogenes* (Kitamoto *et al.*, 1988) could also be expressed in transgenic barley. Kihara *et al.* (1997) have reported the expression of thermostable β-amylase in seeds of transgenic barley, but the origin of the enzyme and the impact on malting quality were not reported.

F. THE ROLE OF PROTEINS IN DETERMINING GRAIN HARDNESS AND VITREOUSNESS

1. Grain Hardness in Wheat

Hardness, or kernel texture, has been described as the "most important single characteristic that affects the functionality of a common wheat" (Pomeranz and Williams, 1990). It affects a range of characters including the milling (tempering, yield, size, shape and density of flour particles) and end-use properties (for breadmaking, noodle making, cakes and biscuits) and is the primary character used to separate wheat into groups.

An important functional difference between hard and soft wheats is their water absorption. Flours milled from hard wheats have a higher baking absorption, giving higher quality and increased profit. This difference is assumed to result from more starch damage during milling of hard wheats, implying that the starch granules are more tightly bound into the protein matrix.

Hardness reflects the ease with which the endosperm can be "separated into fragments" ("Oxford English Dictionary", quoted by Pomeranz and Williams, 1990) and can be readily measured as the amount of energy required for milling. It is genetically determined, and the major determinant has been mapped to a single genetic locus (designated *Ha*) on the short arm of chromosome 5D (Law *et al.*, 1978). This locus is absent from tetraploid durum wheats with the result that the grain is ultrahard.

It is important to distinguish hardness from vitreousness, in which the endosperm appears glassy. Vitreousness is not under strong genetic control but can occur in all varieties, particularly when grown with high nitrogen fertilization and at high temperature (Pomeranz and Williams, 1990).

Similarly, hardness is not linked genetically to the major determinants of breadmaking quality, which are the gluten proteins. Although most breadmaking wheats are hard, this has almost certainly resulted from concurrent selection by plant breeders for hardness and strong gluten.

The most widely accepted theory for the origin of hardness is that it is determined by the strength of bonding between the cell contents, principally between the starch granules and the matrix proteins but possibly also between the matrix proteins and the cell walls. As a result, the flour particles milled from hard wheats consist mainly of whole cells or groups of cells, while in soft wheats the cells are ruptured to release the contents including starch granules (Simmonds, 1974). Differences in adhesion between the starch granule and matrix proteins in hard and soft wheats are supported by microscopical and biochemical studies. In particular, starch granules prepared from hard wheats under non-aqueous conditions show a much higher level of adherent protein than do granules from soft wheats (Barlow *et al.*, 1973).

The major breakthrough in understanding the biochemical basis for hardness came with the demonstration that the surface of water-washed starch granules from soft wheat contained an M_r 15 000 protein that was not present on water-washed starches from hard bread wheat or durum wheat (Greenwell and Schofield, 1986). This protein, which was subsequently called "friabilin", was proposed to act as a "non-stick" agent preventing adhesion between the starch and protein.

Detailed analyses have shown that friabilin is a mixture of protein, that include two major basic components that are identical to puroindolines a and b, and previously characterized inhibitors of α-amylase (Gautier *et al.*, 1994; Morris *et al.*, 1994; Oda and Schofield, 1997). Rahman *et al.* (1994) isolated cDNA clones for a group of minor components related to puroindolines and subsequently showed that they were encoded by structural genes at a locus (*GSP-1*), which is tightly linked to the *Ha* locus (Jolly *et al.*, 1996). However, most attention has focused on the puroindolines.

The puroindolines (pins) were initially isolated from wheat flour using a Triton X114 phase partitioning procedure designed to extract proteins that are bound to membrane lipids (Blochet *et al.*, 1993). Two major isoforms are present (pin a and pin b), which are 55% similar in amino acid sequence (Gautier *et al.*, 1994) and contain an unusual tryptophan-rich motif, which includes five tryptophans in pin a (WRWWKWWK) but only three in pin b (WPTKWWK). Detailed sequence comparisons show that the pins share a conserved skeleton of cysteine residues with the α-amylase inhibitors, lipid binding proteins (see below) and 2S albumins (Gautier *et al.*, 1994), demonstrating that they belong to the "prolamin superfamily" of seed proteins.

Pins are located in the aleurone and starchy endosperm cells, being present in the protein matrix and at the interface between the matrix and the starch granule surface in the latter (Dubreil *et al.*, 1998a), with pin a being the major isoform (Blochet *et al.*, 1993; Dubreil *et al.*, 1998a).

Giroux and Morris (1997, 1998) have compared the amounts and amino-acid sequences of pin a and pin b in hard and soft wheats using Northern blotting and polymerase chain reaction (PCR) amplification. Their results suggested that two mutations might independently result in grain hardness. The first mutation leads to the absence of pin a from hard varieties while the second leads to a glycine to serine substitution at position 46 in pin b. This position is adjacent to the tryptophan-rich motif [39]W–[45]K, which is the putative starch-binding site. Thus, mutation could affect the binding of the protein to the starch granule surface. This provides an attractive hypothesis but has not so far been confirmed by studies in other laboratories (Turnball *et al.*, 2000). Several workers have also reported that there is little difference between the total amounts of "friabilin" components in flours of hard and soft wheats (Rahman *et al.*, 1994; Greenblatt *et al.*, 1995), implying that binding of the "friabilin" to the starch granules is a more important determinant of hardness than the total amount.

It has long been known that soft wheats contain higher levels of free polar lipids than hard wheats and Morrison *et al.* (1989) showed that this character was controlled by a locus (*Fpl-1*), which is tightly linked (or identical) to *Ha* on chromosome 5D. Furthermore, Greenblatt *et al.* (1995) demonstrated that the presence of "friabilin" on starch granules is associated with higher levels of bound phospholipids and glycolipids. This property appears to result from a specific interaction of pins with polar lipids (Dubreil *et al.*, 1997) but it is not known whether the bound lipid contributes to the strength of binding to starch.

2. Grain Hardness in Other Cereals

Starch granule-associated proteins with similar M_r to "friabilins" are also present in rye, where they are encoded by a locus on chromosome 5R (Greenwell, 1992), oats, barley (Jagtap *et al.*, 1993; Darlington *et al.*, 2000) and wild species of *Triticum* and *Aegilops*. The latter include species related to the progenitors of the A and B genomes of durum and bread wheats. In barley "friabilin" appears to show differences in binding patterns in hard and soft types, which may influence malting performance, with hard-textured varieties generally malting less well than soft-textured types (Brennan *et al.*, 1996; Darlington *et al.*, 2000). Protein analyses and PCR-based cloning demonstrated that barley "friabilins" include pin homologues (called hordoindoline) but no consistent differences are observed between the sequences of these in soft varieties with good malting performance and hard varieties with poor malting performance (H. Darlington and P. R. Shewry, unpublished results).

3. Manipulation of Hardness in Wheat and Related Species

Manipulating grain texture to produce a wider range of properties could improve the end-use quality of wheat and barley. Increasing the hardness of both species could improve the feed quality for both ruminant and monogastric livestock. Whole grains are usually used for feeding ruminants and, in this case, stronger binding of the matrix proteins could limit the loss of starch by digestion in the rumen. In contrast, milled grain is usually used to feed monogastrics (pigs and poultry) and harder grains would have a higher level of starch damage during milling, resulting in greater digestibility and more efficient utilization by the animal.

In the case of barley, soft-textured varieties may be more rapidly and evenly modified during malting (Brennan *et al.*, 1996, 1997). A wider range of textural characteristics would also facilitate the wider use of bread and pasta wheats in the food industry, particularly pasta wheats, which are uniformly "ultra hard".

Genetic engineering could be used in two ways: to explore the molecular basis for grain texture, and to produce modified lines as raw material for plant breeders. Pasta wheat provides an ideal model system for the former, owing to the absence of pins and other proteins associated with the *Ha* locus on chromosome 5D. Transformation of pasta wheat with genes for pin a, pin b and other starch granule-associated proteins would provide information on their relative importance, while structure:function relationships could be explored using mutant genes. The latter would be facilitated by detailed structural studies, such as those in progress in the laboratory of Didier Marion at INRA-Nantes (see Shewry *et al.*, 2000). In particular, it would be important to determine how pins interact with starch at the molecular level and how the serine → glycine mutation in pin b affects this. Similarly, the role of polar lipids needs to be elucidated.

Mutant pin genes could in the future be used to develop new breeding material, together with up- or down-regulation of endogenous genes and transformation with genes from other species. For example, genes from bread wheat could be used to develop soft-textured types of durum wheat, while genes from wild species could be used to extend the range of properties of bread wheat.

4.   Vitreousness of Wheat, Barley and Other Cereals

Whereas grain hardness can only be assessed by mechanical testing, clear visual differences can be observed between vitreous (steely) and floury (mealy) grains of wheat and barley. The molecular basis for vitreousness is not known, but vitreous grain appear to contain higher levels of protein, which is also more tightly packed, the absence of air spaces being proposed to account for the different appearance (Hoseney, 1992). Furthermore, single grains may contain both vitreous and floury areas. In barley, steely endosperms are less rapidly and uniformly modified during malting and floury grains are preferred. However, the absence of any clear genetic control means this character would be most readily manipulated by control of environmental and nutritional factors.

Vitreous and floury grains and areas within individual grains also occur in maize and sorghum, where they affect cooking quality and resistance to pathogenic fungi (Chandrashekar and Mazhar, 1999). Vitreous endosperms also appear to have higher protein contents, and are enriched in one specific group of prolamins, the γ-zeins and γ-kafirins in maize and sorghum, respectively (Chandrashekar and Mazhar, 1999). Convincing evidence for a role of the γ-zeins in determining vitreousness comes from analysis of *opaque-2* lines of maize. Normal *opaque-2* lines of maize are generally soft, but the introduction of genetic modifiers to give QPM results in vitreous grains with a 2–3 increase in the content of γ-zein (Wallace *et al.*, 1990). Pratt *et al.* (1995) also demonstrated a relationship between the amount of γ-zeins and grain hardness. Thus, both environmental and genetic factors affect vitreousness in maize and sorghum, with the latter being amenable to manipulation in plant breeding. Transformation could possibly also be used to manipulate the expression of γ-prolamins in these species, but this approach has not so far been used.

G. LIPID-BINDING PROTEINS

The puroindolines (pins) discussed above are one of two major groups of lipid-binding proteins present in cereal starchy endosperms. The second is the non-specific lipid transfer proteins (nsLTPs). The first member of this group of proteins to be described was from grain of Indian finger millet (ragi) (Campos and Richardson, 1984). This was shown to inhibit α-amylase and a related protein encoded by a cDNA from barley aleurone tissue was therefore called "probable amylase/proteinase inhibitor" (PAPI) (Svensson *et al.*, 1986; Mundy and Rogers, 1986). It was not until 1989 that Bernhardt and Somerville (1989) noted that both the ragi inhibitor and PAPI belonged to a previously characterized family of non-specific lipid transfer proteins, with homologues characterized from spinach, maize and castor bean. Although PAPI was subsequently shown to transfer phosphatidyl choline from liposomes to potato mitochondria *in vitro* (Breu *et al.*, 1989), its *in vivo* function is still a matter of debate.

More recently, Marion and colleagues have shown that nsLTPs can be extracted from wheat grain using the Triton X114 phase partition method discussed above (Désmoreaux *et al.*, 1992). nsLTPs have now been isolated from a number of species and several three-dimensional structures have been determined (see Kader, 1996; Lerche and Poulsen, 1998; Charvolin *et al.*, 1999; Douliez *et al.*, 2000). The nsLTPs can be divided into two classes of M_r about 9000 (nsLTP1) and 7000 (nsLTP2), with those in most cereal grains falling into the former class (see Douliez *et al.*, 2000). Their three-dimensional structures show the presence of a large hydrophobic cavity or tunnel, which is assumed to play a role in lipid binding.

The nsLTPs of barley and wheat are confined to the aleurone (Mundy and Rogers, 1986; Dubreil *et al.*, 1998a), casting doubt on their role in the intracellular transport of lipids. Similarly, nsLTP are also concentrated in the epidermal layers of other tissues, and may be secreted into the cell walls or present in surface waxes. It has, therefore, been proposed that they play a role in the transport of hydrophobic monomers, which form protective layers of wax, cutin and suberin (as discussed by Douliez *et al.*, 2000). Furthermore, nsLTPs (including those from seeds of wheat and maize) inhibit the growth of fungal or bacterial pathogens, either alone or additively or synergistically with thionins (see Shewry and Lucas, 1997).

The ability of nsLTPs and pins to bind lipids means that they have current and potential applications in food systems. Foam stability is an important component of beer quality and is adversely affected by lipids. Sørensen *et al.* (1993) showed that barley nsLTP (i.e. PAPI) was a major component present in the protein fraction recovered from beer foam. Furthermore, it appears to be modified during malting and brewing, resulting in changes in pI, M_r and immunoreactivity and in an increased efficiency in stabilizing foam (Sørensen *et al.*, 1993; Bech *et al.*, 1995). This modification presumably includes denaturation during wort boiling and the denatured protein is also a competitive inhibitor of the major malt cysteine endoproteinase, EP-B (Davy *et al.*, 1999).

Whereas nsLTPs only form stable foams after denaturation, the pins are highly active in their native form (Dubreil *et al.*, 1997). The foams formed by pins are highly stable in the presence of neutral and polar lipids and the latter may even enhance foam stability (Wilde *et al.*, 1993; Dubreil *et al.*, 1997). These properties can be exploited to stabilize beer foams in the presence of lipids (Clark *et al.*, 1994) and to improve the crumb structure of bread (Dubreil *et al.*, 1998b).

In addition to these traditional applications of barley and wheat, surface-active proteins with the ability to stabilize foams and emulsions have a wide range of other uses in the food and personal care (cosmetics, lotions, etc.) industries. Cereal grains could, therefore, prove to be valuable sources of such proteins in the future, replacing animal-derived products.

VI.   OTHER GRAIN COMPONENTS

Cereal grains contain many other components that contribute to their nutritional and functional properties, including lipids, minerals and pigments.

A.   CEREAL LIPIDS

Lipids only account for about 2–3% of the total grain weight in barley, wheat, rye and oats and are concentrated in the embryo. Consequently, white wheat flour contains only about 1.8% lipid (Fujino *et al.*, 1996). In maize the oil content is higher, an average of about 4.4% (Weber, 1987), and is again concentrated in the embryo, chiefly in the scutellum (Weber, 1987). Because of this a higher oil content may be achieved in maize by increasing the embryo size, decreasing the endosperm size, or increasing the oil content of the embryo. Thus the higher oil contents of several high lysine mutants of maize (*brittle-2, floury-2, sugary-2, opaque-2*) all appear to result from changes in the ratio of the embryo:endosperm (see Weber, 1987). In contrast, endosperms of the high lysine barley mutant Risø1508 contain about 3.6% lipid compared with about 1.8% in the parental variety Bomi

(Shewry *et al.*, 1979). However, this increase is associated with reduced accumulation of starch resulting in lower grain yield (see above).

Oats has the highest oil content of the cultivated cereals, with means reported as 5.2% dry weight for samples grown in Germany, UK and USA, 5.2% for samples from Wales, and 5.7% for samples from Sweden (Welch, 1995). The oil levels are higher in groats (i.e. grain processed to remove the hull), ranging from about 5–9%. Analysis of hand dissected groats shows that all parts contain high amounts of oil, with 7, 9.2, 15.3 and 24.0% in the starchy endosperm, bran, embryonic axis and scutellum, respectively (Youngs, 1972, 1986).

High oil corn gives higher body weight gains in pigs and poultry and a better egg-to-feed ratio in chickens (see Weber, 1987). Selecting or transforming corn for increased oil content is therefore a commercially viable target. However, damage of high oil corn could result in rancidity, which occurs in damaged oats owing to the action of lipase followed by lipoxygenase and lipoperoxidase. Down-regulation of lipoxygenase activity might therefore be valuable in oats or in other cereals engineered to have high oil contents, as discussed below for durum wheat. Finally, the wide consumption of wheat in bread and other cereals in breakfast goods means that cereals could be a good mechanism to deliver oils with enhanced nutritional properties, such as those containing long-chain polyunsaturated fatty acids (Newton and Snyder, 1997).

B.   PHYTIC ACID AND MINERAL AVAILABILITY

About 60–80% of the total phosphate in cereal grains is present as phytin, *myo*-inositol 1,2,3,4,5,6-hexabisphosphate (IP_6) (Raboy and Gerbasi, 1996). In wheat and rice (and probably also in barley) about 90% of the phytin is located in the aleurone layer with the remaining 10% being in the embryo and scutellum. In contrast, about 90% of the phytin of maize is in the embryo but only 10% in the aleurone (O'Dell *et al.*, 1972). In the barley aleurone phytin is clearly deposited as inclusions in aleurone grains.

Phytin is anti-nutritional for humans and monogastric animals as the phosphate is poorly utilized and it also chelates other essential minerals and forms complexes with proteins. It can, therefore, result in anaemia in women, particularly those living in developing countries. The disposal of phosphate-rich slurry can also result in environmental pollution in countries with intensive livestock production.

Two approaches have been adopted to produce low phytase barley and wheat. Rasmussen and Hatzack (1998) mutagenized barley with sodium azide and identified two mutant phenotypes. Type A plants contained very low phytin with high free phosphate and traces of additional phosphate-containing compounds. In contrast, seeds of type B plants contained an average of about 50% of the phytin present in seeds of wild-type plants, with lower levels of free phosphate and no additional phosphate-containing compounds. Genetic analyses showed that at least three recessive alleles caused the type A phenotype with a separate recessive gene for the B type phenotype.

Brinch-Pedersen et al. (2000) have attempted to improve the nutritional value of wheat by expressing an Aspergillus niger phytase gene using the "constitutive" maize ubiquitin-1 promoter, following success reported for expression of heterologous phytase in seeds of dicotyledonous plants. Analyses of the mature grains showed increases of up to four-fold in the total phytase activity, but no differences in inositol composition were observed. Nevertheless, this work demonstrates the feasibility of using genetic engineering to give significant increases in grain phytase.

C.   CAROTENOIDS AND LIPOXYGENASE

A bright yellow colour is a quality attribute in durum wheat and results from the presence of lutein, a carotenoid pigment. Although the colour of durum wheat semolina is affected by genetic and environmental factors (Irvine and Winkler, 1950; Irving and Anderson, 1953), bleaching also occurs during processing, catalysed by enzymes, notably lipoxygenase (LOX). Three LOX isoenzymes have been characterized in wheat (Hsieh and McDonald, 1984), two of which (LOX2 and LOX3) were shown to be active at the pH present in dough and were therefore considered to be likely to play a role in the bleaching of lutein during processing (Hsieh and McDonald, 1984; McDonald, 1979). However, Manna et al. (1998) showed a negative correlations between the level of LOX-1 mRNA and both the semolina β-carotene content and the semolina yellow index in a range of durum wheat varieties. Reduction in bleaching may therefore be achieved by selection for low LOX activity among cultivars or by down-regulation of the expression of one or more LOX isoenzymes in transgenic plants.

VII.   CONCLUSIONS

The application of sophisticated analytical, molecular and biophysical techniques over the past two decades has provided us with a vastly improved understanding of the structures of the major components of the mature cereal grain, their biosynthetic pathways and their role in determining end-use properties. Although our knowledge is still far from complete, for example, we know little or nothing of the pathway of β-glucan synthesis, the development of routine transformation and gene tagging systems will provide new opportunity to identify key genes and explore their functions.

The present chapter has focused on understanding and improving grain quality for traditional end uses, as food and feed. There is no doubt that these will remain the major end uses of cereals, with a greater emphasis on enhancing the properties for human nutrition (i.e. vitamin-enriched, high fibre, increased antioxidants, long-chain polyunsaturated fatty acids) and developing high-value ingredients with novel functional properties for food processing. However, it is probable that novel types of cereals will also be developed for non-food uses, including high-value/low-volume products, such as pharmaceuticals and cosmetics, and low-value/high-volume products, such as structural and packaging materials (plastics, films, etc.). This should result in dramatic changes in the patterns of cereal production and utilization over the next few decades.

ACKNOWLEDGEMENT

IACR receives grant-aided support from the Biotechnology and Biological Sciences Research Council of the United Kingdom.

REFERENCES

Ainsworth, C., Clark, J. and Balsdon, J. (1993a). Expression, organisation and structure of the genes encoding the *waxy* protein (granule-bound starch synthase) in wheat. *Plant Molecular Biology* **22**, 67–82.

Ainsworth, C., Tarvis, M. and Clark, J. (1993b). Isolation and analysis of a cDNA clone encoding the small subunit of ADP-glucose pyrophosphorylase from wheat. *Plant Molecular Biology* **23**, 23–33.

Ainsworth, C., Hosein, F., Tarvis, M., Weir, F., Burrell, M., Devos, K. M. and Gale, M. D. (1995). Adenosine diphosphate glucose pyrophosphorylase genes in wheat: differential expression and gene mapping. *Planta* **197**, 1–10.

Altpeter, E., Vasil, V., Srivastava, V. and Vasil, I. K. (1996). Integration and expression of the high-molecular-weight glutenin subunit 1Ax1 gene into wheat. *Nature Biotechnology* **14**, 1155–1159.

Alvarez, M. L., Guelman, S., Halford, N. G., Lustig, S., Reggiardo, M. I., Ryabushkina, N., Shewry, P. R., Stein, J. and Vallejos, R. H. (2000). Silencing of HMW glutenins in transgenic wheat expressing extra HMW subunits. *Theoretical and Applied Genetics* **100**, 319–327.

Andersson, L., Fredriksson, H., Bergh, M. O., Andersson, R. and Aman, P. (1999). Characterisation of starch from inner and peripheral parts of normal and waxy barley kernels. *Journal of Cereal Science* **30**, 165–171.

Antoniou, T., Marquardt, R. R. and Cansfield, P. E. (1981). Isolation, partial characterization and antinutritional activity of a factor (pentosans) in rye grain. *Journal of Agricultural and Food Chemistry* **29**, 1240.

Araki, E., Miura, H. and Sawada, S. (1999). Identification of genetic loci affecting amylose content and agronomic traits on chromosome 4A of wheat. *Theoretical and Applied Genetics* **98**, 977–984.

Asp, N.-G., Björk, I. and Nyman, M. (1993). Physiological effects of cereal dietary fibre. *Carbohydrate Polymers* **21**, 183–187.

Bacic, A. and Stone, B. A. (1980). A $(1{\rightarrow}3)$- and $(1{\rightarrow}4)$-linked β-D-glucan in the endosperm cell walls of wheat. *Carbohydrate Research* **82**, 372–377.

Baga, M., Repellin, A., Demeke, T., Caswell, K., Leung, N., Abdel-aal, E. S.M., Hucl, P. and Chibbar, R. N. (1999). Wheat starch modification through biotechnology. *Starch* **51**, 111–116.

Ball, S., Guan, H. P., James, M., Myers, A., Keeling, P., Mouille, G., Buléon, A., Colonna, P. and Preiss, J. (1996). From glycogen to amylopectin: a model for the biogenesis of the plant starch granule. *Cell* **86**, 349–352.

Bamforth, C. W. and Quain, D. E. (1989). Enzymes in brewing and distilling. *In* "Cereal Science and Technology" (G. H. Palmer, ed.), pp. 326–366. Aberdeen University Press, Aberdeen.

Barlow, K. K., Buttrose, M. S., Simmonds, D. H. and Vesk, M. (1973). The nature of starch–protein interface in wheat endosperm. *Cereal Chemistry* **50**, 443–454.

Barro, F., Rooke, L., Békés, F., Gras, P., Tatham, A. S., Fido, R., Lazzeri, P. A., Shewry, P. R. and Barceó, P. (1997). Transformation of wheat with high molecular weight subunit genes results in improved functional properties. *Nature Biotechnology* **15**,1295–1299.

Baruch, D. W., Meridith, P., Jenkins, L. D. and Simmons, L. D. (1979). Starch granules of developing wheat kernels. *Cereal Chemistry* **56**, 554–558.

Baruch, D. W., Jenkins, L. D., Dengate, H. N. and Meridith, P. (1982). Nonlinear model of wheat starch granule distribution at several stages of development. *Cereal Chemistry* **60**, 32–35.

Batey, I. L., Wrigley, C. W. and Gras, P. (1989). Novel approaches to grain quality and process control. *In* "Wheat is Unique" (Y. Pomeranz, ed.), pp. 161–176. American Association of Cereal Chemists, St Paul, MN.

Baxter, E. D. and T. Wainwright. (1979a). Hordein and malting quality. *Journal of American Society of Brewing Chemists* **37**, 8–12.

Baxter, E. D. and T. Wainwright. (1979b). The importance in malting and mashing of hordein proteins with a relatively high sulphur content. *In* "Proceedings of European Brewing Convention Congress", Berlin, pp. 131–143.

Bech, L. M., Vaag, P., Heinemann, B. and Breddam, K. (1995). Throughout the brewing process barley lipid transfer protein 1 (LTP1) is transformed into a more foam-promoting form. *In* "Proceedings of the European Brewing Convention", Brussels, pp. 561–568.

Bechtel, D. B., Zayas, I., Kaleikau, L. and Pomeranz, Y. (1990). Size-distribution of wheat starch granules during endopserm development. *Cereal Chemistry* **67**, 59–63.

Belton, P. S. (1999). On the elasticity of wheat gluten. *Journal of Cereal Science* **29**, 103–107.

Bennett, M. D., Smith, J. B. and Barclay, I. (1975). Early seed development in the Triticeae. *Philosophical Transactions of the Royal Society of London, Series B* **272**, 199–227.

Bernhardt, W. R. and C.R. Somerville. (1989). Coidentity of putative amylase inhibitors from barley and finger millet with phospholipid transfer proteins inferred from amino acid sequence homology. *Archives of Biochemistry and Biophysics* **269**, 695–697.

Betschart, A. A. (1988). Nutritional quality of wheat and wheat foods. *In* "Wheat: Chemistry and Technology" (Y. Pomeranz, ed.), pp. 91–130. American Association of Cereal Chemists, St Paul, MN.

Bhattacharyya, M., Smith, A. M., Ellis, T. H. N., Hedley, C. and Martin, C. (1990). The wrinkled-seed character of pea described by Mendel is caused by a transposon-like insertion in a gene encoding starch-branching enzyme. *Cell* **60**, 115–122.

Bhattacharya, M., Jafari-Shabestari, J., Qualset, C. O. and Corke, H. (1997). Diversity of starch pasting properties in Iranian hexaploid wheat landraces. *Cereal Chemistry* **74**, 417–423.

Biliaderis, C. G. (1991). The structure and interactions of starch with food constituents. *Canadian Journal of Physiology and Pharmacology* **69**, 60–78.

Bishop, L. R. (1930a). The Institute of Brewing research scheme. 1. The prediction of extract. *Journal of the Institute of Brewing* **36**, 421–434.

Bishop, L. R. (1930b). The nitrogen content and quality of barley. *Journal of the Institute of Brewing* **36**, 352–369.

Blechl, A. E. and Anderson, O. D. (1996). Expression of a novel high-molecular-weight glutenin subunit gene in transgenic wheat. *Nature Biotechnology* **14**, 875–879.

Blochet, J.-E., Chevalier, C., Forest, E., Pebay-Peyroula, E., Gautier, M.-F., Joudrier, P., Pézolet, M. and Marion, D. (1993). Complete amino acid sequence of puroindoline, a new basic and cystine-rich protein with a unique tryptophan-rich domain, isolated from wheat endosperm by Triton X-114 phase partitioning. *FEBS Letters* **329**, 336–340.

Blumenthal, C., Rawson, H. M., and Wrigley, C. W. (1996). Changes in wheat grain quality due to doubling the level of atmospheric CO_2. *Cereal Chemistry* **73**, 762.

Blumenthal, C., Wrigley, C. W., Batey, I. L. and Barlow, E. W. R. (1994). The heat-shock response relevant to molecular and structural changes in wheat yield and quality. *Australian Journal of Plant Physiology* **21**, 901–909.

Bowler, P., Williams M. R. and Angold, R. E. (1980). A hypothesis for the morphological changes which occur on heating lenticular starch in water. *Starch* **32**, 186.

Boyer, C. D. and Preiss, J. (1981). Evidence for independent genetic control of the multiple forms of maize endosperm branching enzymes and starch synthases. *Plant Physiology* **67**, 1141–1145.

Boyer, C. D., Damewood, P. A. and Simpson, E. K.G. (1980). Effect of gene dosage at high amylose loci on the properties of the amylopectin fractions of the starches. *Starch* **32**, 217–222.

Brennan, C. S., Harris, N., Smith, D. and Shewry, P. R. (1996). Structural differences in mature endosperms of good and poor malting barley cultivars. *Journal of Cereal Science* **24**, 171–177.

Brennan, C. S., Amor, M. A., Harris, N., Smith, D., Cantrell, I. and Shewry, P. R. (1997). Cultivar differences in modification patterns of protein and carbohydrate reserves during malting of barley. *Journal of Cereal Science* **26**, 83–93.

Brennan, C. S., Smith, D. B., Harris, N. and Shewry, P. R. (1998). The production and characterisation of *Hor* 3 null lines of barley provides new information on the relationship of D hordein to malting performance. *Journal of Cereal Science* **28**, 291–299.

Breu, V., Guerbette, F., Kader, J. C., Kannangara, C. G., Svensson, B. and Von Wettstein-Knowles, P. (1989). A 10 kD barley basic protein transfers phosphatidylcholine from liposomes to mitochondria. *Carlsberg Research Communications* **54**, 81–84.

Briarty, L. G., Hughes, C. E. and Evers, A. D. (1979). The developing endosperm of wheat – a stereological analysis. *Annals of Botany* **44**, 641–658.

Briggs, D. E. (1987). Endosperm breakdown and its regulation in germinating barley. *In* "Brewing Science", Vol. 3 (J. R. A. Pollock, eds), pp. 491–532. Academic Press, London.

Briggs, D. E. (1992). Barley germination: biochemical changes and hormonal control. *In* "Barley: Genetics, Biochemistry, Molecular Biology and Biotechnology" (P. R. Shewry, ed.), pp. 369–401. CAB International, Wallingford.

Brinch-Pedersen, H., Galili, G., Knudsen, S. and Holm, P. B. (1996). Engineering of the aspartate family biosynthetic pathway in barley (*Hordeum vulgare* L.) by transformation with heterologous genes encoding feed-back-insensitive aspartate kinase and dihydrodipicolinate synthase. *Plant Molecular Biology* **32**, 611–620.

Brinch-Pedersen, H., Olesen, A., Rasmussen, S. K. and Holm, P. B. (2000). Generation of transgenic wheat (*Triticum aestivum* L.) for constitutive accumulation of an *Aspergillus* phytase. *Molecular Breeding* **6**, 195–206.

Brocklehurst, P. A. (1977). Factors controlling grain weight in wheat. *Nature* **266**, 348–349.

Brocklehurst, P. A. and Evers, A. D. (1977). The size distribution of starch granules in endosperm of different sized kernels of the wheat cultivar Maris Huntsman. *Journal of the Science of Food and Agriculture* **28**, 1084–1089.

Brooks, A., Jenner, C. F., and Aspinall, D. (1982). Effects of water deficit on endosperm starch granules and on grain physiology of wheat and barley. *Australian Journal of Plant Physiology* **9**, 423–436.

Brown, I. (1996). Complex carbohydrates and resistant starch. *Nutrition Reviews* **54**, 115.

Buleon, A., Colonna, P., Planchot, V. and Ball, S. (1998). Starch granules – structure and biosynthesis. *International Journal of Biological Macromolecules* **23**, 85–112.

Buliga, G. S., Brant, D. A. and Fincher, G. B. (1986). The sequence statistics and solution configuration of barley (1→3, 1→4)-β-D-glucan. *Carbohydrate Research* **157**, 139–156.

Burton, R. A., Zhang, X. Q., Hrmova, M. and Fincher, G. B. (1999). A single limit dextrinase gene is expressed both in the developing endosperm and in germinated grains of barley. *Plant Physiology* **119**, 859–871.

Buttrose, M. S. (1960). Submicroscopic development and structure of starch granules in cereal endosperms. *Journal of Ultrastructure Research* **4**, 231–257.

Buttrose, M. S. (1963). Ultrastructure of the developing wheat endosperm. *Australian Journal of Biological Science* **16**, 305–317.

Campbell, A. F. (1992). Protein engineering of the barley chymotrypsin inhibitor 2. *In* "Plant Protein Engineering" (P. R. Shewry and S. Gutteridge, eds.), pp. 257–268. Edward Arnold, Sevenoaks.

Campos, F. A.P. and Richardson, M. (1984). The complete amino acid sequence of the α-amylase inhibitor 1–2 from seeds or ragi (Indian finger millet, *Eleusine coracana* Gaern.). *FEBS Letters* **167**, 212–225.

Cao, H. P., Imparl-Radosevich, J., Guan, H. P., Keeling, P. L., James, M. G. and Myers, A. M. (1999). Identification of the soluble starch synthase activities of maize endosperm. *Plant Physiology* **120**, 205–215.

Carbonero, P. and Garcia-Olmedo, F. (1999). A multigene family of trypsin/α-amylase inhibitors from cereals. *In* "Seed Proteins" (P. R. Shewry and R. Casey, eds), pp. 617–633. Kluwer Academic Publishers, Dordrecht.

Casey, R. (1999). Distribution and some properties of seed globulins. *In* "Seed Proteins" (P. R. Shewry and R. Casey, eds), pp 159–169. Kluwer Academic Publishers, Dordrecht.

Chandrashekar, A. and Mazhar, H. (1999). The biochemical basis and implacations of grain strength in sorghum and maize. *Journal of Cereal Science* **30**, 193–207.

Charvolin, D., Douliez, J. P., Marion, D., Cohen-Addad, C. and Pebay-Peyroula, E. (1999). The crystal structure of a wheat non-specific lipid transfer protein (nsLTP1) complexed with two phospholipid molecules at 2.1 Å resolution. *European Journal of Biochemistry* **264**, 562–568.

Choct, M. and Annison, G. (1992). The inhibition of nutrient digestion by wheat pentosans. *British Journal of Nutrition* **67**, 123–132.

Chojecki, A. J. S., Baylis, M. W. and Gale, M. D. (1986a). Cell production and DNA accumulation in the wheat endosperm, and their association with grain weight. *Annals of Botany* **58**, 809–817.

Chojecki, A. J. S., Gale, M. D. and Baylis, M. W. (1986b). The number and sizes of starch granules in the wheat endosperm, and their association with grain weight. *Annals of Botany* **58**, 819–831.

Ciacco, C. F. and D'Appolonia, B. L. (1977). Characterisation of starches from various tubers and their use in bread. *Cereal Chemistry* **54**, 1096.

Clark, D. C., Wilde, P. J. and Marion, D. (1994). The effect of lipid binding protein on the foaming properties of bear containing lipid. *Journal of the Institute of Brewing* **100**, 23–25.

Cochrane, M. P. and Duffus, C. M. (1981). Endosperm cell number in barley. *Nature* **289**, 399–401.

Cochrane, M. P. and Duffus, C. M. (1983). Endosperm cell number in cultivars of barley differing in grain weight. *Annals of Applied Biology* **102**, 177–181.

Coleman, C. E. and Larkins, B. A. (1999). Prolamins of maize. *In* "Seed Proteins" (P. R. Shewry and R. Casey, eds), pp. 109–139. Kluwer Academic Publishers, Dordrecht.

Colleoni, C., Dauvillee, D., Mouille, G., Buleon, A., Gallant, D., Bouchet, B., Morell, M., Samuel, M., Delrue, B., Dhulst, C., Bliard, C., Nuzillard, J. M. and Ball, S. (1999a). Genetic and biochemical evidence for the involvement of α-1,4 glucanotransferases in amylopectin synthesis. *Plant Physiology* **120**, 993–1003.

Colleoni, C., Dauvillee, D., Mouille, G., Morell, M., Samuel, M., Slomiany, M. C., Lienard, L., Wattebled, F., Dhulst, C. and Ball, S. (1999b). Biochemical characterization of the *Chlamydomonas reinhardtii* α-1,4 glucanotransferase supports a direct function in amylopectin biosynthesis. *Plant Physiology* **120**, 1005–1013.

Courtin, C. M. and Delcour, J. A. (1998). Physicochemical and bread-making properties of low molecular weight wheat-derived arabinoxylans. *Journal of Agricultural and Food Chemistry* **46**, 4066–4073.

Craig, S. A. S. (1989). Modification of wheat starch. *In* "Wheat is Unique" (Y. Pomeranz, ed.), pp. 235–250. American Association of Cereal Chemists, St Paul, MN.

Crosbie, G. B. (1991). The relationship between starch swelling properties, paste viscosity and boiled noodle quality in wheat flours. *Journal of Cereal Science* **13**, 145–150.

Crosbie, G. B. and Lambe, W. J. (1993). The application of the flour swelling volume test for potential noodle quality to wheat breeding lines affected by sprouting. *Journal of Cereal Science* **18**, 267–276.

Crosbie, G. B., Lambe, W. J., Tsutsui, H. and Gilmour, R. F. (1992). Further evaluation of the flour swelling volume test for identifying wheats potentially suitable for Japanese noodles. *Journal of Cereal Science* **15**, 271–280.

D'Appolonia, B. L. and Gilles, K. A. (1971). Effect of various starches in baking. *Cereal Chemistry* **48**, 625–636.

Darlington, H. F., Tesci, L., Harris, N., Griggs, D., Cantrell, I. and Shewry, P. R. (2000). Starch granule associated proteins in barley and wheat. *Journal of Cereal Science* **32**, 21–29.

Davy, A., Svendsen, Ib., Bech, L., Simpson, D. and Cameron-Mills, V. (1999). LTP does not represent a cysteine endoprotease inhibitor in barley grains. *Journal of Cereal Science* **30**, 237–244.

De Boer, E. D. (1991). Chemically modified derivatives of starch from a new genetic variety of corn. *Cereal Foods World* **36**, 631.

Delcour, J. A., Vanhamel, S. and De Geest, C. (1989). Physico-chemical and functional properties of rye nonstarch polysaccharides. I. Colorimetric analysis of pentosans and their relative monosaccharide compositions in fractionated (milled) rye products. *Cereal Chemistry* **66**, 107–111.

DeMason, D. A. (1997). Endosperm structure and development. *In* "Cellular and Molecular Biology of Plant Seed Development" (B. A. Larkins and I. K. Vasil, eds), pp. 73–115. Kluwer Academic Publishers, Dordrecht.

Demeke, T., Hucl, P., Nair, R. B., Nakamura, T. and Chibbar, R. N. (1997). Evaluation of Canadian and other wheats for waxy proteins. *Cereal Chemistry* **74**, 442–444.

Demeke, T., Huel, P., Abdel-aal, E. S. M., Baga, M. and Chibbar, R. N. (1999). Biochemical characterization of the wheat waxy A protein and its effect on starch properties. *Cereal Chemistry* **76**, 694–698.

Denyer, K., Hylton, C. M., Jenner, C. F. and Smith, A. M. (1995). Identification of multiple isoforms of soluble and granule-bound starch synthase in developing wheat endosperm. *Planta* **196**, 256–265.

Denyer, K., Dunlap, F., Thorbjornsen, T., Keeling, P. and Smith, A. M. (1996). The major form of ADP-glucose pyrophosphorylase in maize endosperm is extra-plastidial. *Plant Physiology* **112**, 779–785.

Désormeaux, A., Blochet, J.-E., Pézolet, M. and Marion, D. (1992). Amino acid sequence of a non-specific wheat phospholipid binding protein and its conformation revealed by infrared and Raman spectroscopy, role of disulphide bridges and phospholipids in stabilizing the α-helix. *Biochimica et Biophysica Acta* **1121**, 137–152.

Doan, D. N.P., Rudi, H. and Olsen, O. A. (1999). The allosterically unregulated isoform of ADP-glucose pyrophosphorylase from barley endosperm is the most likely source of ADP-glucose incorporated into endosperm starch. *Plant Physiology* **121**, 965–975.

Douliez, J.-P., Michon, T., Elmorjani, K. and Marion, D. (2000). Structure, biological and technological functions of lipid transfer proteins and indolines, the major lipid binding proteins from cereal kernels. *Journal of Cereal Science* **32**, 1–20.

Dubreil, L., Compoint, J.-P. and Marion, D. (1997). The interaction of puroindolines with wheat polar lipids determines their foaming properties. *Journal of Agrictural and Food Chemistry* **45**, 108–116.

Dubreil, L., Gaborit, T., Bouchet, B., Gallant, D. J., Broekaert, W. F., Quillien, L. and Marion, D. (1998a). Spatial and temporal distribution of the major isoforms of puroindolines (puroindoline-a and puroindoline-b) and non specific lipid transfer protein (ns-LTP1e$_1$) of *Triticum aestivum* seeds. Relationships with their *in vitro* antifungal properties. *Plant Science* **138**, 121–135.

Dubreil, L., Méliande, S., Chiron, H., Compoint, J. P., Quillien, L., Branlard, G. *et al.* (1998b). The effect of puroindolines on the breadmaking properties of wheat flour. *Cereal Chemistry* **75**, 222–229.

Duffus, C. M. and Binnie, J. (1990). Sucrose relationships during endosperm and embryo development in wheat. *Plant Physiology and Biochemistry* **28**, 161–165.

Egli, D. B. (1998). "Seed Biology and the Yield of Grain Crops". CAB International, Wallingford.

Entwhistle, G., Tyson, R. H. and Ap Rees, T. (1988). Isolation of amyloplasts from wheat endosperm. *Phytochemistry* **27**, 993–996.

Esposito, S., Bowsher, C. G., Emes, M. J. and Tetlow, I. J. (1999). Phosphoglucomutase activity during development of wheat grains. *Journal of Plant Physiology* **154**, 24–29.

Evans, D. E., MacLeod, L. C., Eglinton, J. K., Gibson, C. E., Zhang, X., Wallace, W., Skerritt, J. H. and Lance, R. C. M. (1997). Measurement of *beta*-amylase in malting barley (*Hordeum vulgare* L.) I. Development of a quantitative ELISA for *beta*-amylase. *Journal of Cereal Science* **26**, 229–239.

Evers, A. D. (1971). Scanning electron microscopy of wheat starch. III. Granule development in endosperm. *Starch* **23**, 157–162.

Evers, A. D. (1973). The size distribution among starch granules in wheat endosperm. *Starch* **9**, 303–304.

Evers, A. D. and Bechtel, D. B. (1988). Microscopic structure of the wheat grain. *In* "Wheat: Chemistry and Technology", Vol. 1 (Y. Pomeranz, ed.), pp. 47–95. American Association of Cereal Chemists Inc., St Paul, MN.

Evers, A. D. and Lindley, J. (1977). The particle-size distribution in wheat endosperm starch. *Journal of the Science of Food and Agriculture* **28**, 98–102.

Evers, A. D., Blakeney, A. B. and O'Brien, L. (1999). Cereal structure and composition. *Australian Journal of Agricultural Research* **50**, 629–650.

Every, D., Gerrard, J. A., Gilpin, M. J., Ross, M. and Newberry, M. P. (1998). Staling in starch bread: the effect of gluten additions on specific loaf volume and firming rate. *Starch* **50**, 443–446.

Falco, S. C. (1998). Chimeric genes and methods for increasing the lysine and threonine content of the seeds of plants. U.S. Patent 5,773,691.

Falco, S. C., Guida, T., Locke, M., Mauvais, J., Sanders, C., Ward, R. T. and Webber, P. (1995). Transgenic canola and soybean seeds with increased lysine. *Bio/Technology* **13**, 577–582.

Fennema, O. R. ed. (1985). "Food Chemistry", 2nd edn, pp. 875. Marcel Dekker, New York.

Filpse, E., Suurs, L. and Visser, R. G. F. (1996). Introduction of sense and antisense cDNA for branching enzyme in the amylose-free potato mutant leads to physico-chemical changes in the starch. *Planta* **198**, 340.

Fincher, G. B. (1975). Morphology and chemical composition of barley endosperm cell walls. *Journal of the Institute of Brewing* **81**, 116–122.

Fincher, G. B. (1992). Cell wall metabolism in barley. *In* "Barley: Genetics, Biochemistry, Molecular Biology and Biotechnology" (P. R. Shewry, ed.), pp. 413–437. CAB International, Wallingford.

Fincher, G. B. and Stone, B. A. (1986). Cell walls and their components in cereal technology. *Advances in Cereal Science and Technology* **8**, 207–295.

Fisher, D. K., Boyer, C. D. and Hannah, L. C. (1993). Starch branching enzyme II from maize endosperm. *Plant Physiology* **102**, 1045–1046.

Fisher, D. K., Kim, K.-N., Gao, M., Boyer, C. D. and Guiltinan, M. J. (1995). A cDNA encoding starch branching enzyme I from maize endosperm. *Plant Physiology* **108**, 1313–1314.

Fisher, D. K., Gao, M., Kim, K. N., Boyer, C. D. and Guiltinan, M. J. (1996a). Two closely related cDNAs encoding starch branching enzyme from *Arabidopsis thaliana*. *Plant Molecular Biology* **30**, 97–108.

Fisher, D. K., Gao, M., Kim, K.-N., Boyer, C. D. and Guiltinan, M. J. (1996b). Allelic analysis of the maize *amylose-extender* locus suggests that independent genes encode starch-branching enzymes IIa and IIb. *Plant Physiology* **110**, 611–619.

Fredriksson, H., Silverio, J., Andersson, R., Eliasson, A. C. and Aman, P. (1998). The influence of amylose and amylopectin characteristics on gelatinization and retrogradation properties of different starches. *Carbohydrate Polymers* **35**, 119–134.

Fujino, Y., Kuwata, J., Mano, Y. and Ohnishi, M. (1996). Other grain components. In "Cereal Grain Quality" (R. J. Henry and P. S. Kettlewell, eds), pp. 289–317. Chapman and Hall, London.

Fujita, N. and Taira, T. (1998). A 56-kda protein is a novel granule-bound starch synthase existing in the pericarps, aleurone layers, and embryos of immature seed in diploid wheat (*Triticum monococcum L.*). *Planta* **207**, 125–132.

Fujita, N., Wadano, A. and Taira, T. (1996). Comparison of the primary structure of waxy proteins (granule-bound starch synthase) between polyploid wheats and related diploid species. *Biochemical Genetics* **34**, 403–403.

Fujita, N., Kubo, A., Francisco, P. B., Nakakita, M., Harada, K., Minaka, N. and Nakamura, Y. (1999). Purification, characterization, and cDNA structure of isoamylase from developing endosperm of rice. *Planta* **208**, 283–293.

Gaines, R. L., Bechtel, D. B. and Pomeranz, Y. (1985). Endosperm structural and biochemical differences between a high-protein amphiploid wheat and its progenitors. *Cereal Chemistry* **62**, 25–31.

Gallant, D. J., Bouchet, B. and Baldwin, P. M. (1997). Microscopy of starch: evidence of a new level of granule organisation. *Carbohydrate Polymers* **32**, 177–191.

Gao, M., Fisher, D. K., Kim, K. N., Shannon, J. C. and Guiltinan, M. J. (1996). Evolutionary conservation and expression patterns of maize starch branching enzyme I and IIb genes suggests isoform specialization. *Plant Molecular Biology* **30**, 1223–1232.

Gao, M., Fisher, D. K., Kim, K. N., Shannon, J. C. and Guiltinan, M. J. (1997). Independent genetic control of maize starch-branching enzymes IIa and IIb – isolation and characterization of a *Sbe2a* cDNA. *Plant Physiology* **114**, 69–78.

Gao, M., Wanat, J., Stinard, P. S., James, M. G. and Myers, A. M. (1998). Characterization of *dull1*, a maize gene coding for a novel starch synthase. *Plant Cell* **10**, 399–412.

Gao, X., Francis, D., Ormrod, J. C. and Bennett, M. E. (1992). Changes in cell number and cell division activity during endosperm development in allohexaploid wheat, *Triticum aestivum L. Journal of Experimental Botany* **43**, 1603–1609.

García-Olmedo, F. (1999). Thionins. In "Seed Proteins" (P.R. Shewry and R. Casey, eds), pp. 709–726. Kluwer Academic Publishers, Dordrecht.

Gautier, M.-F., Aleman, M.-E., Guirao, A., Marion, D. and Joudrier, P. (1994). *Triticum aestivum* puroindolines, two basic cystine-rich seed proteins: cDNA sequence analysis and developmental gene expression. *Plant Molecular Biology* **25**, 43–57.

Gidley, M. J. and Bociek, S. M. (1988). ^{13}C CP/MAS NMR studies of amylose inclusion complex. *Journal of the American Chemical Society* **110**, 3820.

Giese, H. and Hejgaard, J. (1984). Synthesis of salt-soluble proteins in barley. Pulse-labelling study of grain filling in liquid-cultured detached spikes. *Planta* **161**, 172–177.

Girhammar, U. and Nair, B. M. (1992). Isolation, separation and characterization of water soluble non-starch polysacharides from wheat and rye. *Food Hydrocolloids* **6**, 285–299.

Giroux, M. J. and Morris, C. F. (1997). A glycine to serine change in puroindoline b is associated with wheat grain hardness and low levels of starch-surface friabilin. *Theoretical and Applied Genetics* **95**, 857–864.

Giroux, M. J. and Morris, C. F. (1998). Wheat grain hardness results from highly conserved mutations in the friabilin components puroindoline a and b. *Proceedings of the National Academy of Sciences (USA)* **95**, 6262–6266.

Glitsø, L. V. and Bech Knudsen, K. E. (1999). Milling of whole grain rye to obtain fractions with different dietary fibre characteristics. *Journal of Cereal Science* **29**, 89–97.

Greenblatt, G. A., Bettge, A. D. and Morris, C. F. (1995). Relationship between endosperm texture and the occurrence of "friabilin" and bound polar lipids in wheat starch. *Cereal Chemistry* **72**, 172–176.

Greene, T. W. and Hannah, L. C. (1998). Enhanced stability of maize endosperm ADP-glucose pyrophosphorylase is gained through mutants that alter subunit interactions. *Proceedings of the National Academy of Sciences (USA)* **95**, 13342–13347.

Greenwell, P. (1992). Genes, molecules and milling quality. *Chorleywood Digest*, December 1992, issue 122, pp. 133–136.

Greenwell, P. and Schofield, J. D. (1986). A starch granule protein associated with endosperm softness in wheat. *Cereal Chemistry* **63**, 378–380.

Gregory, A. C. E., O'Connell, A. P. and Bolwell, G. P. (1998). Xylans. *Biotechnology and Genetic Engineering Reviews* **15**, 439–455.

Halford, N. G., Field, J. M., Blair, H., Urwin, P., Moore, K., Robert, L., Thompson, R., Flavell, R. B., Tatham, A. S. and Shewry, P. R. (1992). Analysis of HMW glutenin subunits encoded by chromosome 1A of bread wheat (*Triticum aestivum* L.) indicates quantitative effects on grain quality. *Theoretical and Applied Genetics* **83**, 373–378.

Hara-Nishimura, I., Nishimura, M. and Daussant, J. (1986). Conversion of free β-amylase to bound β-amylase on starch granules in the barley endosperm during desiccation phase of seed development. *Protoplasma* **134**, 149–153.

Harn, C., Knight, M., Ramakrishnan, A., Guan, H. P., Keeling, P. L. and Wasserman, B. P. (1998). Isolation and characterization of the ZSSIIa and ZSSIIb starch synthase cDNA clones from maize endosperm. *Plant Molecular Biology* **37**, 639–649.

Harris, R. H. and Sibbitt, L. D. (1941). The comparative baking qualities of starches prepared from different wheat varieties. *Cereal Chemistry* **18**, 585–604.

Hashimoto, S., Shogren, M. D. and Pomeranz, Y. (1987). Cereal pentosans: their estimation and significance. I. Pentosans in wheat and milled wheat products. *Cereal Chemistry* **64**, 30–34.

He, G., Rooke, L., Steele, S., Békés, F., Gras, P., Tatham, A. S., Fido, R., Barcelo, P., Shewry, P. R. and Lazzeri, P. (1999). Transformation of pasta wheat (*Triticum durum* L.) with HMW glutenin subunit genes and modification of dough functionality. *Molecular Breeding*, **5**, 377–386.

Hejgaard, J. and Boisen. S. (1980). High lysine proteins in Hiproly barley breeding: Identification, nutritional significance and new screening methods. *Hereditas* **93**, 311–320.

Henry, R. J. (1987). Pentosan and $(1\rightarrow3),(1\rightarrow4)$-β-glucan concentrations in endosperm and whole grain of wheat, barley, oats and rye. *Journal of Cereal Science* **6**, 253–258.

Hermansson, A.-M. and Svegmark, K. (1996). Developments in the understanding of starch functionality. *Trends in Food Science and Technology* **7**, 345–353.

Hizukuri, S. (1988). Recent advances in molecular structures of starch. *Journal of the Japanese Society of Starch Science* **35**, 185.

Hizukuri, S. and Maehara, Y. (1990). Fine structure of wheat amylopectin: the mode of A to B chain binding. *Carbohydrate Research* **206**, 145–159.

Hong, B. H., Rubenthaler, G. L. and Allan, R. E. (1989). Wheat pentosans. I. Cultivar variation and relationship to kernel hardness. *Cereal Chemistry* **66**, 369–373.

Hoseney, R. C. (1992). "Principles of Cereal Science and Technology". American Association of Cereal Chemists, St Paul, MN, pp. 327.

Howard, K. A., Gayler, K. R., Eagles, H. and Halloran, G. M. (1996). The relationship between D hordein and malting quality in barley. *Journal of Cereal Science* **24**, 47–56.

Hsieh, C. C. and McDonald, C. E. (1984). Isolation of lipoxygenase isoenzymes from flour of durum wheat endosperm. *Cereal Chemistry* **61**, 392–398.

Hucl, P. and Chibbar, R. N. (1996). Variation for starch concentration in spring wheat and its repeatability relative to protein concentration. *Cereal Chemistry* **73**, 756–756.

Irvine, G. N. and Anderson, J. A. (1953). Variation in principal quality factors of durum wheat with a quality prediction test for wheat or semolina. *Cereal Chemistry* **30**, 334–342.

Irvine, G. N. and Winkler, C. A. (1950). Factors affecting the color of macaroni. II: Kinetic studies of pigment destruction during making. *Cereal Chemistry* **27**, 205–218.

Jagtap, S. S., Beardsley, A., Forrest, J. M. S. and Ellis, R. P. (1993). Protein composition and grain quality in barley. *Aspects of Applied Biology* **36**, 51–59.

James, M. G., Robertson, D. S. and Myers, A. M. (1995). Characterization of the maize gene *sugary1*, a determinant of starch composition in kernels. *Plant Cell* **7**, 417–429.

Jane, J., Chen, Y. Y., Lee, L. F., McPherson, A. E., Wong, K. S., Radosavljevic, M. and Kasemsuwan, T. (1999). Effects of amylopectin branch chain length and amylose content on the gelatinization and pasting properties of starch. *Cereal Chemistry* **76**, 629–637.

Jenkins, D. J. A., Vuksan, V., Kendall, C. W. C., Wursch, P., Jeffcoat, R., Waring, S., Mehling, C. C., Vidgen, E., Augustin, L. S. A. and Wong, E. (1998). Physiological effects of resistant starches on fecal bulk, short chain fatty acids, blood lipids and glycemic index. *Journal of the American College of Nutrition* **17**, 609–616.

Jenner, C. F. (1994). Starch synthesis in the kernel of wheat under high temperature conditions. *Australian Journal of Plant Physiology* **21**, 791–806.

Jenner, C. F. and Hawker, J. S. (1993). High temperature affects the activity of enzymes in the committed pathway of starch synthesis in developing wheat endosperm. *Australian Journal of Plant Physiology* **20**, 197–209.

Jenner, C. F., Ugalde, T. D. and Aspinall, D. (1991). The physiology of starch and protein deposition in the endosperm of wheat. *Australian Journal of Plant Physiology* **18**, 211–226.

Jenner, C. F., Siwek, K. and Hawker, J. S. (1993). The synthesis of [^{14}C]starch from [^{14}C]sucrose in isolated wheat grains is dependent upon the activity of soluble starch synthase. *Australian Journal of Plant Physiology* **20**, 329–335.

Jenner, C. F., Denyer, K. and Guerin, J. (1995). Thermal characteristics of soluble starch synthase from wheat endosperm. *Australian Journal of Plant Physiology* **22**, 703–709.

Jobling, S. A., Schwall, G. P., Westcott, R. J., Sidebottom, C. M., Debet, M., Gidley, M. J., Jeffcoat, R. and Safford, R. (1999). A minor form of starch branching enzyme in potato (*Solanum tuberosum* L.) tubers has a major effect on starch structure: cloning and characterisation of multiple forms of SBE A. *Plant Journal* **18**, 163–171.

Jolly, C. J., Glen, G. M. and Rahman, S. (1996). *GSP-1* genes are linked to the grain hardness locus (*Ha*) on wheat chromosome 5D. *Proceedings of the National Academy of Sciences (USA)* **93**, 2408–2413.

Jones, R. J., Roessler, J. and Ouattar, S. (1985). Thermal environment during endosperm cell division in maize: effects on number of endosperm cells and starch granules. *Crop Science* **25**, 830–834.

Kader, J. C. (1996). Lipid transfer proteins in plants. *Annual Review of Plant Physiology and Plant Molecular Biology* **47**, 627–654.

Kahlon, T. S. and Chow, F. I. (1997). Hypocholesterolemic effects of oat, rice and barley dietary fibers and fractions. *Cereal Foods World* **42**, 86–92.

Karlsson, R., Olered, R. and Eliasson, A.-C. (1983). Changes in starch granule size distribution and starch gelatinisation properties during development and maturation of wheat, barley and rye. *Starch* **35**, 335–340.

Keeler, S. J., Maloney, C. L., Webber, P. Y., Patterson, C., Hirata, L. T., Falco, S. C. and Rice, J. A. (1997). Expression of *de novo* high-lysine α-helical coiled-coil proteins may significantly increase the accumulated levels of lysine in mature seeds of transgenic tobacco plants. *Plant Molecular Biology* **34**, 15–29.

Keeling, P. L., Wood, J. R., Tyson, R. H. and Bridges, I. G. (1988). Starch biosynthesis in developing wheat grain. Evidence against the direct involvement of triose phosphates in the metabolic pathway. *Plant Physiology* **87**, 311–319.

Keeling, P. L., Bacon, P. J. and Holt, D. C. (1993). Elevated temperature reduces starch deposition in wheat endosperm by reducing the activity of soluble starch synthase. *Planta* **191**, 342–348.

Keeling, P. L., Banisadr, R., Barone, L., Wasserman, B. P. and Singletary, G. W. (1994). Effect of temperature on enzymes in the pathway of starch biosynthesis in developing wheat and maize grain. *Australian Journal of Plant Physiology* **21**, 807–827.

Kihara, M., Okada, Y., Kuroda, H., Saeki, K. and Ito, K. (1997). Generation of fertile transgenic barley synthesizing thermostable β-amylase. *Journal of the Institute of Brewing* **103**, 153.

Kihara, M., Kaneko, T. and Ito, K. (1998). Genetic variation of β-amylase thermostability among varieties of barley, *Hordeum vulgare* L. and relation to malting quality. *Plant Breeding* **117**, 425–428.

Kim, K. N., Fisher, D. K., Gao, M. and Guiltinan, M. J. (1998). Molecular cloning and characterization of the amylose-extender gene encoding starch branching enzyme IIB in maize. *Plant Molecular Biology* **38**, 945–956.

Kiribuchi-Otobe, C., Nagamine, T. and Yamaguchi, I. (1997). Production of hexaploid wheats with waxy endosperm character. *Cereal Chemistry* **74**, 72–72.

Kirkman, M. A., Shewry, P. R. and Miflin, B. J. (1982). The effect of nitrogen nutrition on the lysine content and protein composition of barley seeds. *Journal of the Science of Food and Agriculture* **33**, 115–127.

Kitamoto, N., Yamagata, H., Kato, T., Tsukagoshi, N. and Udaka, S. (1988). Cloning and sequencing of the gene encoding thermophilic β-amylase of *Clostridium thermosulfurogenes*. *Journal of Bacteriology* **170**, 5848–5854.

Knight, M. E., Harn, C., Lilley, C. E. R., Guan, H. P., Singletary, G. W., Muforster, C. M., Wasserman, B. P. and Keeling, P. L. (1998). Molecular cloning of starch synthase I from maize (W64) endosperm and expression in *Escherichia coli*. *Plant Journal* **14**, 613–622.

Kreis, M., Forde, B. G., Rahman, S., Miflin, B. J. and Shewry, P. R. (1985a). Molecular evolution of the seed storage proteins of barley, rye and wheat. *Journal of Molecular Biology* **183**, 499–502.

Morrison, W. R. (1988b). Lipids in cereal starches: a review. *Journal of Cereal Science* **8**, 1–15.

Morrison, W. R. (1989). Uniqueness of wheat starch. *In* "Wheat is Unique" (Y. Pomeranz, ed.), pp. 193–214. American Association of Cereal Chemists, St Paul.

Morrison, W. R. and Scott, D. C. (1986). Measurement of the dimensions of wheat starch granule populations using a Coulter Counter with 100-channel analyzer. *Journal of Cereal Science* **4**, 13–21.

Morrison, W. R., Tester, R. F., Snape, C. E., Law, R. and Gidley, M. J. (1993). Swelling and gelatinization of cereal starches. IV. Some effects of lipid-complexed amylose and free amylose in waxy and normal barley starches. *Cereal Chemistry* **70**, 385–391.

Morrison, W. R., Law, C. N., Wylie, L. J., Coventry, A. M. and Seekings, J. (1989). The effect of group 5 chromosomes on the free polar lipids and breadmaking quality of wheat. *Journal of Cereal Science* **9**, 41–51.

Mouille, G., Maddelein, M.-L. and Ball, S. (1996). Preamylopectin processing: a mandatory step for starch biosynthesis in plants. *Plant Cell* **8**, 1353–1366.

Muench, D. G., Ogawa, M. and Okita, T. W. (1999). The prolamins of rice. *In* "Seed Proteins" (P. R. Shewry and R. Casey, eds), pp. 93–108. Kluwer Academic Publishers, Dordrecht.

Munck, L., Karlsson, K. E., Hagberg, A. and Eggum, B. O. (1970). Gene for improved nutritional value in barley seed protein. *Science* **168**, 985–987.

Mundy, J. and Rogers, J. C. (1986). Selective expression of a probable amylase-protease inhibitor in barley aleurone cells: Comparison to the barley amylase-subtilisin inhibitor. *Planta* **169**, 51–63.

Murai, J., Taira, T. and Ohta, D. (1999). Isolation and characterization of the three waxy genes encoding the granule-bound starch synthase in hexaploid wheat. *Gene* **234**, 71–79.

Myers, A. M., Morell, M. K., James, M. G. and Ball, S. G. (2000). Recent progress toward understanding biosynthesis of the amylopectin crystal. *Plant Physiology* **122**, 989–997.

Nagamine, T. and Komae, K. (1996). Improvement of a method for chain-length distribution analysis of wheat amylopectin. *Journal of Chromatography A* **732**, 255–259.

Nair, R. B., Baga, M., Scoles, G. J., Kartha, K. K. and Chibbar, R. N. (1997). Isolation, characterization and expression analysis of a starch branching enzyme II cDNA from wheat. *Plant Science* **122**, 153–163.

Nakamura, T., Yamamori, M., Hirano, H., Hidaka, S. and Nagamine, T. (1995). Production of waxy (amylose-free) wheats. *Molecular and General Genetics* **248**, 253–259.

Nakamura, Y., Umemoto, T. and Satoh, H. (1996). Changes in structure of starch and enzyme activities affected by sugary mutations in developing rice endosperm. Possible role of starch debranching enzyme (R-enzyme) in amylopectin biosynthesis. *Physiologia Plantarum* **97**, 491–498.

Nakamura, Y., Kubo, A., Shimamune, T., Matsuda, T., Harada, K. and Satoh, H. (1997). Correlation between activities of starch debranching enzyme and α-polyglucan structure in endosperms of *sugary-1* mutants of rice. *Plant Journal* **12**, 143–153.

Nelson, O. and Pan, D. (1995). Starch synthesis in maize endosperms. *Annual Review of Plant Physiology and Plant Molecular Biology* **46**, 475–496.

Newman, C. W. and Newman, R. K. (1992). Nutritional aspects of barley seed structure and composition. *In* "Barley: Genetics, Biochemistry, Molecular Biology and

Biotechnology" (P. R. Shewry, ed.), pp. 351–368. CAB International, Wallingford.

Newton, I. and Snyder, D. (1997). Nutritional aspects of long-chain omega-3 fatty acids and their use in bread enrichment. *Cereal Foods World* **42**, 126–131.

Newton, J., Stark, J. R. and Riffkin, H. L. (1994). The use of maize and wheat in the production of alcohol. *In* "Proceedings of the Fourth Conference on Malting, Brewing and Distilling". Institute of Brewing, London.

Niemietz, C. and Jenner, C. F. (1993). Mechanisms of sugar uptake into endosperm and aleurone protoplasts isolated from developing wheat grains. *Australian Journal of Plant Physiology* **20**, 371–378.

Oda, S. and Schofield, J. D. (1997). Characterisation of friabilin polypeptides. *Journal of Cereal Science* **26**, 29–36.

O'Dell, B. L., de Boland, A. R. and Koirtyohann, S. R. (1972). Distribution of phytate and nutritionally important elements among the morphological components of cereal grains. *Journal of Agriculture and Food Chemistry* **20**, 718–721.

Olive, M. R., Ellis, R. J. and Schuch, W. W. (1989). Isolation and nucleotide sequences of cDNA clones encoding ADP-glucose pyrophosphorylase polypeptides from wheat leaf and endosperm. *Plant Molecular Biology* **12**, 525–538.

Oliveira, A. B., Rasmusson, D. C. and Fulcher, R. G. (1994). Genetic aspects of starch granule traits in barley. *Crop Science* **34**, 1176–1180.

Olsen, O.-A. (1998). Endosperm developments. *Plant Cell* **10**, 485–488.

Osborn, R. W. and Broekaert, W. F. (1999). Antifungal proteins. *In* "Seed Proteins" (P. R. Shewry and R. Casey, eds), pp. 727–751. Kluwer Academic Publishers, Dordrecht.

Osborne, T. B. (1924). "The Vegetable Proteins". Longmans, Green & Co., London.

Palmer, G. H. (1989). Cereals in malting and brewing. *In* "Cereal Science and Technology" (G.H. Palmer, ed.), pp. 61–242. Aberdeen University Press, Aberdeen.

Pan, D. and Nelson, O. E. (1984). A debranching enzyme deficiency in endosperms of the *Sugary-1* mutants of maize. *Plant Physiology* **74**, 324–328.

Panozzo, J. F. and Eagles, H. A. (1998). Cultivar and environmental effects on quality characters in wheat starch. *Australian Journal of Agricultural Research* **49**, 757–766.

Parker, M. L. (1985). The relationship between A-type and B-type starch granules in the developing endosperm of wheat. *Journal of Cereal Science* **3**, 271–278.

Payne, P. I. (1987). Genetics of wheat storage proteins and the effect of allelic variation on breadmaking quality. *Annual Review of Plant Physiology* **38**, 141–153.

Peng, M., Gao, M., Abdel-aal, E. S. M., Hucl, P. and Chibbar, R. N. (1999). Separation and characterization of A- and B-type starch granules in wheat endosperm. *Cereal Chemistry* **76**, 375–379.

Peumans, W. J. and Van Damme, E. J. M. (1999). Seed lectins. *In* "Seed Proteins" (P. R. Shewry and R. Casey, eds), pp. 657–683. Kluwer Academic Publishers, Dordrecht.

Pollock, C. and Preiss, J. (1980). The citrate-stimulated starch synthase of starchy maize kernels: purification and properties. *Archives of Biochemistry and Biophysics* **204**, 578–588.

Pomeranz, Y. (1988). Chemical composition of kernel structures. *In* "Wheat: Chemistry and Technology", Vol. 1 (Y. Pomeranz, ed.), pp. 97–158. American Association of Cereal Chemists Inc., St Paul, MN.

Pomeranz, Y. and Williams, P. C. (1990). Wheat hardness: its genetic structural, and biochemical background, measurement, and significance. *In* "Advances in

Cereal Science and Technology", Vol. 10 (Y. Pomeranz, ed.), pp. 471–544. American Association of Cereal Chemists Inc., St Paul, MN.

Pratt, R. C., Pauli, S. J. W., Miller, K., Nelsen, T. and Bietz, J. A. (1995). Association of zein classes with maize kernel hardness. *Cereal Chemistry* **72**, 162–167.

Raboy, V. and Gerbasi, P. (1996). Genetics of myo-inositol phosphate synthesis and accumulation. *In* "Subcellular Biochemistry: Myoinositol Phosphates, Phosphoinositides, and Signal Transduction" (B. B. Biswas and S. Biswas, eds), pp. 257–285. Plenum Press, New York.

Raeker, M. O., Gaines, C. S., Finney, P. L. and Donelson, T. (1998). Granule size distribution and chemical composition of starches from 12 soft wheat cultivars. *Cereal Chemistry* **75**, 721–728.

Raghavan, V. (1997). "Molecular Embryology of Flowering Plants". Cambridge University Press, Cambridge, 690pp.

Rahman, S., Jolly, C. J., Skerritt, J. H. and Wallosheck, A. (1994). Cloning of a wheat 15-kDa grain softness protein (GSP) GSP is a mixture of puroindoline-like polypeptides. *European Journal of Biochemistry* **223**, 917–925.

Rahman, S., Kosar-Hashemi, B., Samuel, M. S., Hill, A., Abbott, D. C., Skerritt, J. H., Preiss, J., Appels, R. and Morell, M. K. (1995). The major proteins of wheat endosperm starch granules. *Australian Journal of Plant Physiology* **22**, 793–803.

Rahman, S., Abrahams, S., Abbott, D., Mukai, Y., Samuel, M., Morell, M. and Appels, R. (1997). A complex arrangement of genes at a starch branching enzyme I locus in the D-genome donor of wheat. *Genome* **40**, 465–474.

Rahman, S., Li, Z., Abrahams, S., Abbott, D., Appels, R. and Morell, M. K. (1999). Characterisation of a gene encoding wheat endosperm starch branching enzyme-I. *Theoretical and Applied Genetics* **98**, 156–163.

Rahman, S., Li, Z., Batey, I. L., Cochrane, M. P., Appels, R. and Morell, M. (2000). Genetic alteration of starch functionality in wheat. *Journal of Cereal Science* **31**, 91–110.

Rasmussen, S. K. and Hatzack, F. (1998). Identification of two low-phytate barley (*Hordeum vulgare* L.) grain mutants by TLC and genetic analysis. *Hereditas* **129**, 107–112.

Reddy, I. and Seib, P. (1999). Paste properties of modified starches from partial waxy wheats. *Cereal Chemistry* **76**, 341–349.

Rédei, G. P. (1982). "Genetics". Macmillan Publishing Co., New York.

Repellin, A., Nair, R. B., Baga, M. and Chibbar, R. N. (1997). "Isolation of a starch branching enzyme I cDNA from a wheat endosperm library" (Accession No. Y12320). Plant Gene Register (http://www.tarweed.com/pgr/index.1977.html).

Riffkin, H. L., Duffus, C. M. and Bridges, I. C. (1995). Sucrose metabolism during endosperm development in wheat (*Triticum aestivum*). *Physiologia Plantarum* **93**, 123–131.

Riggs, T. J., Sanada, M., Morgan, A. F. and Smith, D. F. (1983). Use of acid gel electrophoresis in the characterisation of 'B' hordein protein in relation to malting quality and mildew resistance in barley. *Journal of the Science of Food and Agriculture* **34**, 576–586.

Rijven, A. H. G. C. (1984). Use of polyethylene glycol in isolation and assay of stable, enzymatically active starch granules from developing wheat endosperms. *Plant Physiology* **75**, 323–328.

Rijven, A. H. G. C. (1986). Heat inactivation of starch synthase in wheat endosperm tissue. *Plant Physiology* **81**, 448–453.

Rijven, A. H. G. C. and Gifford, R. M. (1983). Accumulation and conversion of sugars by developing wheat grains. 3. Non-diffusional uptake of sucrose, the substrate preferred by endosperm slices. *Plant Cell and Environment* **6**, 417–425.

Roesler, K. R. and Rao, A. G. (1999). Conformation and stability of barley chymotrypsin inhibitor-2 (CI-2) mutants containing multiple lysine substitutions. *Engineering* **12**, 967–973.

Rogers, G. S., Gras, P. W., Batey, I. L., Milham, P. J., Payne, L., and Conroy, J. P. (1998). The influence of atmospheric CO_2 concentration on the protein, starch and mixing properties of wheat flour. *Australian Journal of Plant Physiology* **25**, 387–393.

Rooke, L., Békés, F., Fido, R., Barro, F., Gras, P., Tatham, A. S., Barcelo, P., Lazzeri, P. and Shewry, P. R. (1999). Overexpression of a gluten protein in transgenic wheat results in highly elastic dough. *Journal of Cereal Science* **30**, 115–120.

Ross, A. S., Walker, C. E., Booth R.I., Orth, R. A. and Wrigley, C. W. (1987). The Rapid-ViscoAnalyzer: a new technique for the estimation of sprout damage. *Cereal Foods World* **32**, 827.

Rutenberg, M. W. and Solarek, D. (1984). "Starch derivatives: Production and Uses", pp. 312–388. Academic Press, Orlando, FL.

Saastamoinen, M., Plaami, S. and Kumpulainen, J. (1989). Pentosan and β-glucan content of Finnish winter rye varieties as compared with rye of six other countries. *Journal of Cereal Science* **10**, 199–207.

Safford, R., Jobling, S. A., Sidebottom, C. M., Westcott, R. J., Cooke, D., Tober, K. J., Strongitharm, B. H., Russell, A. L. and Gidley, M. J. (1998). Consequences of antisense RNA inhibition of starch branching enzyme activity on properties of potato starch. *Carbohydrate Polymers* **35**, 155–168.

Sandstedt, R. M. (1946). Photomicrographic studies of wheat starch. I. Development of the starch granules. *Cereal Chemistry* **23**, 337–359.

Sarker, D. K., Wilde, P. J. and Clark, D. C. (1998). Enhancement of protein foam stability by formation of wheat arabinoxylan-protein crosslinks. *Cereal Chemistry* **75**, 493–499.

Sasaki, T. and Matsuki, J. (1998). Effect of wheat starch structure on swelling power. *Cereal Chemistry* **75**, 525–529.

Saulnier, L., Peneau, N. and Thibault, J. F. (1995). Variability in grain extract viscosity and water soluble arabinoxylan content in wheat. *Journal of Cereal Science* **22**, 259–264.

Schondelmaier, J., Jacobi, A., Fischbeck, G. and Jahoor, A. (1992). Genetical studies on the mode of inheritance and localization of the *amo1* (High Amylose) gene in barley. *Plant Breeding* **109**, 274–280.

Schulman, A. H. and Ahokas, H. (1990). A novel shrunken endosperm mutant of barley. *Physiologia Plantarum* **78**, 583–589.

Schulman, A. H., Tomooka, S., Suzuki, A., Myllärinen, P. and Hizukuri, S. (1995). Structural analysis of starch from normal and *shx* (shrunken endosperm) barley (*Hordeum vulgare* L). *Carbohydrate Research* **275**, 361–369.

Shewry, P. R. (1999a). Avenins: the prolamins of oats. *In* "Seed Proteins" (P. R. Shewry and R. Casey, eds), pp. 79–92. Kluwer Academic Publishers, Dordrecht.

Shewry, P. R. (1999b). Enzyme inhibitors of seeds: types and properties. *In* "Seed Proteins" (P. R. Shewry and R. Casey, eds), pp. 587–615. Kluwer Academic Publishers, Dordrecht.

Shewry, P. R. and Lucas, J. A. (1997). Plant proteins that confer resistance to pests and pathogens. *Advances in Botanical Research* **26**, 135–192.

Shewry, P. R. and Casey, R. (1999). Seed proteins. *In* "Seed Proteins" (P. R. Shewry and R. Casey, eds), pp. 1–10. Kluwer Academic Publishers, Dordrecht.

Shewry, P. R. and Tatham, A. S. (1999). The characteristics, structures and evolutionary relationships of prolamins. *In* "Seed Proteins" (P. R. Shewry and R. Casey, eds), pp. 11–33. Kluwer Academic Publishers, Dordrecht.

Shewry, P. R., Pratt, H. M., Leggatt, M. M. and Miflin, B. J. (1979). Protein metabolism in developing endosperms of high-lysine and normal barley. *Cereal Chemistry* **56**, 110–117.

Shewry, P. R., Faulks, A. J., Parmar, S. and Miflin, B. J. (1980). Hordein polypeptide pattern in relation to malting quality and the varietal identification of malted grain. *Journal of the Institute of Brewing* **86**, 138–141.

Shewry, P. R., Wolfe, M. S., Slater, S. E., Parmar, S., Faulks, A. J. and Miflin, B. J. (1981). Barley storage proteins in relation to varietal identification, malting quality and mildew resistance. *In* "Barley Genetics IV" (R. N. H. Whitehouse, ed.), pp. 596–603. Edinburgh University Press, Edinburgh.

Shewry, P. R., Williamson, M. S. and Kreis, M. (1987). Effects of mutant genes on the synthesis of storage components in developing barley endosperms. *In* "Developmental Mutants in Higher Plants" (H. Thomas and D. Grierson, eds), pp. 95–118. Cambridge University Press, Cambridge.

Shewry, P. R., Tatham, A. S., Barro, F., Barcelo, P. and Lazzeri, P. (1995). Biotechnology of breadmaking: unravelling and manipulating the multi-protein gluten complex. *Bio/Technology* **13**, 1185–1190.

Shewry, P. R., Tatham, A. S. and Halford, N. G. (1999). The prolamins of the Triticeae. *In* "Seed Proteins" (P. R. Shewry and R. Casey, eds), pp. 35–78. Kluwer Academic Publishers, Dordrecht.

Shewry, P. R., Tatham, A. S. and Popineau, Y. (2000). The chemical basis of wheat grain quality. *In* " Wheat Science and Technology in the European Union" (J.-C. Autran and R. J. Hammer, eds). IRTAC, Paris (in press).

Shi, Y. C. and Seib, P. A. (1989). Properties of wheat starch compared to normal maize starch. *In* "Wheat is Unique" (Y. Pomeranz, ed.), pp. 161–176. American Association of Cereal Chemists, St Paul, MN.

Shi, Y. C., Seib, P. A. and Bernardin, J. E. (1994). Effects of temperature during grain-filling on starches from six wheat cultivars. *Cereal Chemistry* **71**, 369–383.

Simmonds, D. H. (1974). Chemical basis of hardness and vitreosity in the wheat kernel. *Bakers Digest* **48**, 16–18, 20, 22, 24, 26–29, 63.

Sissons, M. J. and MacGregor, A. W. (1994). Hydrolysis of barley starch granules by α-glucosidases from malt. *Journal of Cereal Science* **19**, 161–169.

Skerritt, J. H. and Janes, P. W. (1992). Disulphide-bonded "gel protein" aggregates in barley: Quality-related differences in composition and reductive dissociation. *Journal of Cereal Science* **16**, 219–235.

Smith, D. B. and Lister, P. R. (1983). Gel forming proteins in barley grain and their relationship with malting quality. *Journal of Cereal Science* **1**, 229–239.

Sørensen, S. B., Bech, L. M., Muldbjerg, M., Beenfeldt, T. and Breddam, K. (1993). Barley lipid transfer protein 1 is involved in beer foam formation. *MBAA Technical Quarterly* **30**, 136–145.

Stoddard, F. L. (1999a). Survey of starch particle-size distribution in wheat and related species. *Cereal Chemistry* **76**, 145–149.

Stoddard, F. L. (1999b). Variation in grain mass, grain nitrogen, and starch B-granule content within wheat heads. *Cereal Chemistry* **76**, 139–144.

Stone, B. A. (1996). Cereal grain carbohydrates. *In* "Cereal Grain Quality" (R. J. Henry and P. S. Kettlewell, eds), pp. 251–288. Chapman and Hall, London.

Sun, C. X., Sathish, P., Ek, B., Deiber, A. and Jansson, C. (1996). Demonstration of in vitro starch branching enzyme activity for a 51/50-kDa polypeptide isolated from developing barley (*Hordeum vulgare*) caryopses. *Physiologia Plantarum* **96**, 474–483.

Sun, C., Sathish, P. and Jansson, C. (1997). Identification of four starch-branding enzymes in barley endosperm: partial purification of forms I, IIa and IIb. *New Phytologist* **137**, 215–215.

Sun, C. X., Sathish, P., Ahlandsberg, S. and Jansson, C. (1998). The two genes encoding starch-branching enzymes IIA and IIB are differentially expressed in barley. *Plant Physiology* **118**, 37–49.

Sun, C. X., Sathish, P., Ahlandsberg, S. and Jansson, C. (1999). Analyses of isoamylase gene activity in wild-type barley indicate its involvement in starch synthesis. *Plant Molecular Biology* **40**, 431–443.

Svensson, B., Asano, K., Jonasses, I., Poulsen, F. M., Mundy, J. and Svendsen, I. (1986). A 10kD barley seed protein homologous with an α-amylase inhibitor from Indian finger millet. *Carlsberg Research Communications* **51**, 493–500.

Takaoka, M., Watanabe, S., Sassa, H., Yamamori, M., Nakamura, T., Sasakuma, T. and Hirano, H. (1997). Structural characterization of high molecular weight starch granule-bound proteins in wheat (*Triticum aestivum* L). *Journal of Agricultural and Food Chemistry* **45**, 2929–2934.

Takeda, Y., Takeda, C., Mizukami, H. and Hanashiro, I. (1999). Structures of large, medium and small starch granules of barley grain. *Carbohydrate Polymers* **38**, 109–114.

Tallberg, A. (1977). The amino-acid composition in endosperm and embryo of a barley variety and its high lysine mutant. *Hereditas* **87** 43.

Tatham, A. S. and Shewry, P. R. (2000). Elastomeric proteins: biological roles, structures and mechanisms. *Trends in Biochemical Science* (in press).

Tenhaken, R. and Thulke, O. (1996). Cloning of an enzyme that synthesises a key nucleotide sugar precursor of hemicellulose biosynthesis from soybean: UDP-glucose dehydrogenase. *Plant Physiology* **112**, 1127–1134.

Tester, R. F., Morrison, W. R., Ellis, R. H., Piggott, J. R., Batts, G. R., Wheeler, T. R., Morison, J. I. L., Hadley, P., and Ledward, D. A. (1995). Effects of elevated growth temperature and carbon dioxide levels on some physicochemical properties of wheat starch. *Journal of Cereal Science* **22**, 63–71.

Tetlow, I. J., Blissett, K. J. and Emes, M. J. (1993). A rapid method for the isolation of purified amyloplasts from wheat endosperm. *Planta* **189**, 597–600.

Tetlow, I. J., Blissett, K. J. and Emes, M. J. (1994). Starch synthesis and carbohydrate oxidation in amyloplasts from developing wheat endosperm. *Planta* **194**, 454–460.

Tetlow, I. J., Bowsher, C. G. and Emes, M. J. (1996). Reconstitution of the hexose phosphate translocator from the envelope membranes of wheat endosperm amyloplasts. *Biochemical Journal* **319**, 717–723.

Tetlow, I. J., Blissett, K. J. and Emes, M. J. (1998). Metabolite pools during starch synthesis and carbohydrate oxidation in amyloplasts isolated from wheat endosperm. *Planta* **204**, 100–108.

Thorbjornsen, T., Villand, P., Denyer, K., Olsen, O. A. and Smith, A. M. (1996). Distinct isoforms of ADPglucose pyrophosphorylase occur inside and outside the amyloplasts in barley endosperm. *Plant Journal* **10**, 243–250.

Topping, D. L. (1999). Physiological effects of dietary carbohydrates in the large bowel: is there a need to recognize dietary fibre equivalents? *Asia Pacific Journal of Clinical Nutrition* **8**, S22-S26.

Toufeili, I., Habbal, Y., Shadarevian, S. and Olabi, A. (1999). Substitution of wheat starch with non-wheat starches and cross-linked waxy barley starch affects sensory properties and staling of Arabic bread. *Journal of the Science of Food and Agriculture* **79**, 1855–1860.

Turnbull, K. M., Gaborit, T., Marion, D. and Rahman, S. (2000) The *Ha* locus of wheat: variation in puroindoline polypeptides in Australian cultivars. *Australian Journal of Plant Physiology* **27**, 153–158.

Turner, J. F. (1969). Starch synthesis and changes in UDPglucose pyrophosphorylase and ADPglucose pyrophosphorylase in the developing wheat grain. *Australian Journal of Biological Science* **22**, 1321–1327.

Tyson, R. H. and ap Rees, T. (1988). Starch synthesis by isolated amyloplasts from wheat endosperm. *Planta* **175**, 33–38.

Verity, J. C. K., Hac, L. and Skerritt, J. H. (1999). Development of a field enzyme-linked immunosorbent assay (ELISA) for detection of α-amylase in preharvest-sprouted wheat. *Cereal Chemistry* **76**, 673–681.

Villand, P., Olsen, O.-A., Kilian, A. and Kleczkowski, L. A. (1992). ADP-glucose pyrophosphorylase large subunit cDNA from barley endosperm. *Plant Physiology* **100**, 1617–1618.

Vinkx, C. J. A. and Delcour, J. A. (1996). Rye (*Secale cereale* L.) arabinoxylans: a critical review. *Journal of Cereal Science* **24**, 1–14.

Vrinten, P. L. and Nakamura, T. (2000). Wheat granule-bound starch synthase I and II are encoded by separate genes that are expressed in different tissues. *Plant Physiology* **122**, 255–263.

Vrinten, P., Nakamura, T. and Yamamori, M. (1999). Molecular characterization of waxy mutations in wheat. *Molecular and General Genetics* **261**, 463–471.

Walker, J. T. and Merritt, N. R. (1968). Genetic control of abnormal starch granules and high amylose content in a mutant of "Glacier" barley. *Nature* **221**, 482–484.

Wallace, J. C., Lopes, M. A., Paiva, E. and Larkins, B. A. (1990). New methods for extraction and quantitation of zeins reveal a high content of γ–zein in modified *opaque-2* maize. *Plant Physiology* **92**, 191–196.

Wallwork, M. A. B., Logue, S. J., Macleod, L. C. and Jenner, C. F. (1998). Effects of a period of high temperature during grain filling on the grain growth characteristics and malting quality of three Australian malting barleys. *Australian Journal of Agricultural Research* **49**, 1287–1296.

Wang, H. L., Patrick, J. W., Offler, C. E. and Wardlaw, I. F. (1993). A novel experimental system for studies of photosynthate transfer in the developing wheat grain. *Journal of Experimental Botany.* **44**, 1177–1184.

Wang, H. L., Offler, C. E. and Patrick, J. W. (1994a). Nucellar projection transfer cells in the developing wheat grain. *Protoplasma* **182**, 39–52.

Wang, H. L., Offler, C. E., Patrick, J. W. and Ugalde, T. D. (1994b). The cellular pathway of photosynthate transfer in the developing wheat grain. I. Delineation of a potential transfer pathway using fluorescent dyes. *Plant, Cell and Environment* **17**, 257–266.

Wang, H. L., Offler, C. E. and Patrick, J. W. (1995a). The cellular pathway of photosynthate transfer in the developing wheat grain. II. A structural analysis and histochemical studies of the pathway from the crease phloem to the endosperm cavity. *Plant, Cell and Environment* **18**, 373–388.

Wang, H. L., Patrick, J. W., Offler, C. E. and Wang, X.-D. (1995b). The cellular pathway of photosynthate transfer in the developing wheat grain. III. A structural analysis and physiological studies of the pathway from the endosperm cavity to the starchy endosperm. *Plant, Cell and Environment* **18**, 389–407.

Wang, T. L., Bogracheva, T. Y. and Hedley, C. L. (1998). Starch: as simple as A, B, C. *Journal of Experimental Botany* **49**, 481–502.

Wardlaw, I. F. (1994). The effect of high temperature on kernel development in wheat: variability related to pre-heading and post-anthesis conditions. *Australian Journal of Plant Physiology* **21**, 731–739.

Watson, S. A. (1987). Structure and composition. *In* "Corn: Chemistry and Technology" (S. A. Watson and P. E. Ramstad, eds), pp. 53–82. American Association of Cereal Chemists, Inc. St Paul, MN.

Weber, E. J. (1987). Lipids of the kernel. *In* "Corn: Chemistry and Technology" (S. A. Watson and P. E. Ramstad, eds), pp. 311–349. American Association of Cereal Chemists, St Paul, MN.

Welch, R. W. (1995). The chemical composition of oats. *In* "The Oat Crop: Production and Utilization" (R.W. Welch, ed.), pp. 279–320. Chapman and Hall, London.

Wilde, P. J., Clark, D. C. and Mrion, D. (1993). Influence of competitive adsorption of a lysopalmitoylphosphatidylcholine on the functional properties of puroindoline, a lipid binding protein isolated from wheat flour. *Journal of Agricultural and Food Chemistry* **41**, 1570–1576.

Williamson, M. S., J. Forde, B. Buxton and M. Kreis. (1987). Nucleotide sequence of barley chymotrypsin inhibitor-2 (CI-2) and its expression in normal and high-lysine barley. *European Journal of Biochemistry* **165**, 99–106.

Wood, P. J. (1992). Aspects of the chemistry and nutritional effects of non-starch polysaccharides of cereals. *In* "Developments in Carbohydrate Chemistry" (R. J. Alexander and H. F. Zobel, eds), pp. 293–314. American Association of Cereal Chemists, St Paul, MN.

Wood, P. J., Weisz, J. and Blackwell, B. A. (1994). Structural studies of $(1\rightarrow3, 1\rightarrow4)$-β-glucans by ^{13}C-nuclear magnetic resonance spectroscopy and by rapid analysis of cellulose-like regions using high performance anion-exchange chromatography of oligosaccharides released by lichenase. *Cereal Chemistry* **71**, 301–307.

Woodward, J. R., Fincher, G. B. and Stone, B. A. (1983). Water-soluble $(1\rightarrow3),(1\rightarrow4)$-β-D-glucans from barley (*Hordeum vulgare*) endosperm. II. Fine structure. *Carbohydrate Polymers* **3**, 207–225.

Wootton, M., Panozzo, J. F., Hong, S. U. and Hong, S. H. (1998). Differences in gelatinisation behaviour between starches from Australian wheat cultivars. *Starch* **50**, 154–158.

Yamamori, M. (1998). "Selection of a Wheat Lacking a Putative Enzyme for Starch Synthesis, SGP-1", pp. 300–302. University Extension Press, University of Saskatchewan, Saskatoon.

Yamamori, M. and Endo, T. R. (1996). Variation of starch granule proteins and chromosome mapping of their coding genes in common wheat. *Theoretical and Applied Genetics* **93**, 275–281.

Yasui, T., Sasaki, T. and Yamamori, M. (1997). Waxy endosperm mutants of bread wheat (Triticum aestivum L.) and their starch properties. *Breeding Science* **47**, 161–161.

Yoshigi, N., Okada, Y., Maeba, H., Sahara, H. and Tamaki, T. (1995). Construction of a plasmid used for the expression of a sevenfold-mutant barley β-amylase with increased thermostability in *Escherichia coli* and properties of the sevenfold-mutant β-amylase. *Journal of Biochemistry* **118**, 562–567.

You, S. G., Fiedorowicz, M. and Lim, S. T. (1999). Molecular characterization of wheat amylopectins by multiangle laser light scattering analysis. *Cereal Chemistry* **76**, 116–121.

Youngs, V. L. (1972). Protein distribution in the oat kernel. *Cereal Chemistry* **49**, 407–411.

Youngs, V. L. (1986). Oat lipids and lipid-related enzymes. *In* "Oats: Chemistry and Technology" (F. H. Webster ed.), pp. 205–226. American Association of Cereal Chemists, St Paul, MN.

Zeeman, S. C., Umemoto, T., Lue, W. L., Auyeung, P., Martin, C., Smith, A. M. and Chen, J. (1998). A mutant of *Arabidopsis* lacking a chloroplastic isoamylase accumulates both starch and phytoglycogen. *Plant Cell* **10**, 1699–1711.

Zeng, M., Morris, C. F. and Wrigley, C. W. (1997). Sources of variation for starch gelatinization, pasting, and gelation properties in wheat. *Cereal Chemistry* **74**, 63–71.

Zhao, X. C. and Sharp, P. J. (1996). An improved 1-D SDS–PAGE method for the identification of three bread wheat "waxy" proteins. *Journal of Cereal Science* **23**, 191–193.

Zhao, X. C. and Sharp, P. J. (1998). Production of all eight genotypes of null alleles at waxy loci in bread wheat, *Triticum aestivum* L. *Plant Breeding* **117**, 488–490.

Zhao, X. C., Batey, I. L., Sharp, P. J., Crosbie, G., Barclay, I., Wilson, R., Morell, M. K. and Appels, R. (1998). A single genetic locus associated with starch granule properties and noodle quality in wheat. *Journal of Cereal Science* **27**, 7–13.

Ziegler, E. and Greer, E.N. (1978). Principles of Milling. *In* "Wheat Chemistry and Technology" (Y. Pomeranz, ed.), pp. 115–200. American Association of Cereal Chemists, St Paul, MN.

Resistance to Abiotic Freezing Stress in Cereals

M. ALISON DUNN, GILLIAN O'BRIEN, ANTHONY P. C. BROWN,
SENAY VURAL and MONICA A. HUGHES

*School of Biochemistry and Genetics, University of Newcastle
upon Tyne, Newcastle upon Tyne NE2 4HH, UK*

I. INTRODUCTION

Overwintering cereals have evolved mechanisms that enable them to survive a winter freezing stress. A low-positive-temperature treatment, typically in the autumn, will acclimate these plants to withstand a subsequent freezing temperature and acclimation (sometimes called cold or frost hardening) can be viewed as an adaptive response that enables the plant to survive an environmental stress. The threshold temperature for the initiation of acclimation for freezing or frost tolerance in winter cereals is approximately 10°C and there is an inverse relationship between positive temperatures

Advances in Botanical Research Vol. 34
incorporating Advances in Plant Pathology
ISBN 0-12-005934-7

lower than 10°C and the rate of acclimation (Fowler *et al.*, 1999). The metabolic and cellular changes that occur during acclimation are acknowledged to be complex (Hughes and Dunn, 1996), and can be considered to be quantitative, such that the level of the response may depend both on the temperature range and the genotype of the plant involved (Hughes and Dunn, 1990; Fowler *et al.*, 1999).

Frost or freezing tolerance, which is acquired by genetically competent plants during a low-positive-temperature treatment, is lost when the temperature rises. This process of de-acclimation is usually more rapid than the process of acclimation (Pearce *et al.*, 1996). Maximum frost tolerance depends upon maintenance of the low-acclimating temperature and, since temperatures will fluctuate in the field, the level of frost tolerance of field-grown plants will depend upon their thermal history. A second phase of freezing acclimation has been recognized by some researchers (Livingston, 1996; Pearce *et al.*, 1996). Winter barley (cv Igri) plants acclimated by a low-positive-temperature treatment of +6°C day/+2°C night for 24 days, have an LT_{50} (temperature that kills 50% of plants in a regrowth test) of –11°C, compared with non-acclimated controls with an LT_{50} of –2°C. Transfer of the acclimated plants to a regime of +4°C day/–4°C night for 14 days resulted in further acclimation to an LT_{50} of –17°C (Pearce *et al.*, 1996).

The process of frost acclimation is complex, involving a number of biochemical and physiological changes. Cellular and metabolic changes that occur during cold acclimation include increased levels of sugars, soluble proteins, proline and organic acids as well as the appearance of new isoforms of proteins and altered lipid membrane composition (Hughes and Dunn, 1990). There is an increasing body of evidence that many of these biochemical and physiological changes are regulated through changes in gene expression (Hughes and Dunn, 1996).

The complexity of the plant low-temperature responses have contributed to the fact that cereal breeders have been unable to develop cultivars with a low-temperature tolerance superior to that found in the land races of winter cereals (Fowler *et al.*, 1999).

II.   VERNALIZATION AND INTERACTIONS

Those temperatures that acclimate winter cereals for frost tolerance, also vernalize winter cultivars (Fowler *et al.*, 1999) and there is a complex genetic and physiological relationship between vernalization and acclimation. It has been observed that, when winter cereals (rye, wheat) are maintained for long periods of time at temperatures optimum for the acclimation response (4°C), they gradually lose their ability to tolerate freezing temperatures (Fowler *et al.*, 1996). This decline in levels of acclimation has been related to the establishment of vernalization saturation. It is also associated with a decline in the levels of expression of the low-temperature-responsive (LTR) gene family, *wcs120*, and it has been suggested that vernalization genes may have a regulatory influence on LTR gene expression and frost acclimation. The suggestion is further supported by the observation that spring cultivars, which do not have a vernalization response, are unable to maintain high levels of LTR gene (*wcs120*) expression at 4°C and do not achieve the same levels of frost tolerance as winter cultivars (Fowler *et al.*, 1996).

There are also interactions between the effects of light and temperature that affect the plant acclimation or low-temperature response (Gray *et al.*, 1997). In addition, other abiotic stresses, such as drought, can induce a freezing acclimation response (Pearce *et al.*, 1996). However, despite these interactions temperature is the major environmental factor that determines expression of the genetic potential of a plant for freezing tolerance (Fowler *et al.*, 1999).

III.   GENETIC ANALYSIS OF FROST TOLERANCE IN CEREALS

Temperature will have a direct effect on many metabolic processes; this and the complexity of the acclimation process makes it difficult to separate genes responsible for frost acclimation from those associated with metabolic adjustments to the low-acclimating temperature *per se*.

There are considerable differences in the level of the maximum freezing tolerance, both between different cereal species and between different cultivars of the same species. Cereals, such as rice and maize, are chilling sensitive and are damaged by low-positive temperatures and do not acclimate for freezing tolerance. Winter rye acclimates more rapidly than other cereals and in a regrowth test hardy cultivars will acclimate to an LT_{50} of −33°C (Wilen *et al.*, 1998), whereas some winter cultivars of barley and wheat can acclimate to an LT_{50} of −20°C (Livingston, 1996; Pearce *et al.*, 1996). Variation in the level of frost acclimation, between different cultivars of the same crop, have been reported for oats (Livingston, 1996), barley (Doll *et al.*, 1989; Giorni *et al.*, 1999), wheat (Ouellet *et al.*, 1998) and rye (Zhang *et al.*, 1993).

The genetic analysis of freezing tolerance has been carried out primarily in cereals. Winter hardiness, that is the ability of plants to overwinter in the field and produce a crop in the following year, is composed of a number of interacting components: freezing or frost tolerance, vernalization requirement, and photoperiod sensitivity. It has been suggested that the change in growth habit caused by vernalization can contribute directly to winter hardiness of the crop. This is now not accepted: Doll *et al.* (1989) were able to separate vernalization requirement and good frost tolerance in a number of doubled haploid barley lines produced by crossing a winter, frost-hardy cultivar (Vogelsanger Gold) with four spring, frost-sensitive cultivars

(Tron, Tystofte Prentice, Clipper and Alf). In a major study to detect quantitative trait loci (QTL) in barley, Hayes *et al.* (1998) have analysed a cross between a winter (Dicktoo) and a spring (Morex) genotype taken from different North American germplasm pools. A 6.6 cM interval on the long arm of chromosome 7 was found to have a significant effect on the frost hardiness (LT_{50}), winter survival in the field and heading date (involving vernalization and photoperiod response) in the progeny of this cross.

The homologous wheat chromosome for barley chromosome 7 is chromosome 5. In hexaploid wheat five vernalization genes have been identified. *Vrn1*, *Vrn3* and *Vrn4* are located on chromosomes 5A, 5D and 5B, respectively (Xin *et al.*, 1988), *Vrn2* is located on chromosome 2B (homologous to barley chromosome 2) and *Vrn5* is located on chromosome 7B (homologous to barley chromosome 1) (Law, 1966). The wheat chromosomes most frequently identified using monosomic and substitution lines, as carrying genes involved in frost tolerance are 5A, 7A, 4B, 4D and the homologous group 5 chromosomes (Galiba *et al.*, 1998). Chromosomes 5A and 5D appear to carry major genes involved in the trait. The major gene for vernalization requirement (*Vrn1*) and a major gene for frost tolerance (*Fr1*) are linked on the distal portion of the long arm of chromosome 5A. Although there is close linkage between these genes, recombination between them has been recorded (Galiba *et al.*, 1998).

Conservation of gene synteny within the *Triticeae* is well documented (Moore *et al.*, 1995) and one of the barley genes that control the vernalization response (*Sh2*), is situated on barley chromosome 7. Restriction fragment length polymorphism (RFLP) mapping data (Laurie *et al.*, 1995) confirms that the barley *Sh2* intrachromosomal location is consistent with the position of *Vrn1* in wheat. Given the location of a QTL affecting both frost tolerance and vernalization in the same region (Hayes *et al.*, 1998), it has been proposed that an analogous gene complex to *Vrn1–Fr1* exists in barley (Galiba *et al.*, 1995). Further, a spring growth habit locus (*Sp1*), which is situated on the long arm of the homologous rye chromosome (5R), is closely linked to RFLP markers which are linked to *Vrn1* in wheat (Galiba *et al.*, 1998). This suggests that the same gene complex also exists in rye.

These genetic studies only reveal loci that are involved in frost tolerance when the parent cultivars used in the cross contain different alleles, which segregate in progeny of F_1 hybrids or alter the genotype of cytological substitution or addition lines. The power of these analyses to dissect the process of frost acclimation depends, therefore, on the level of variation available within the cereal species.

It is perhaps surprising that attempts to exploit the superior freezing tolerance of rye, through interspecific gene transfer have been unsuccessful (Fowler *et al.*, 1999). Thus, despite the similarities of the chromosomal organization and the fact that low-temperature-responsive genes of both parents are expressed in crosses (Limin *et al.*, 1995), the superior frost tolerance of rye is not independently expressed when combined in tetraploid (Limin *et al.*, 1985) and hexaploid wheat (Dvorak and Fowler, 1978).

IV.   GENES ISOLATED THROUGH DIFFERENTIAL SCREENING OF cDNA LIBRARIES

In recent years a number of molecular studies of freezing tolerance have been initiated. The expectation is that these will lead to a greater understanding of the complexity of the processes involved in the trait, and thus provide the information and tools for the development of biotechnological methods for manipulating frost tolerance.

Frost acclimation has been shown to be accompanied by altered gene expression (Hughes and Dunn, 1996) and differential screening of cDNA libraries has resulted in the isolation of a number of genes, which are up-regulated at the steady-state mRNA level by a low-acclimating temperature. In a recent review by Thomashow (1999), 38 of these low-temperature-responsive (LTR) genes, from both monocotyledon and dicotyledon plants, are listed. In the model plant, *Arabidopsis*, 21 different genes/gene families have been cloned, reflecting the complex reaction of plants to a low temperature. The cereal LTR genes that have been reported are shown in Table I.

A number of barley LTR genes were isolated from a shoot meristem cDNA library made from young (2-week, 2–3 leaves) plants of the winter cultivar, Igri, which had been given a low-temperature treatment (6°C day/ 2°C night, 10-hour day). This temperature regime will acclimate this cultivar for frost tolerance (Dunn *et al.*, 1990; Hughes *et al.*, 1993; Pearce *et al.*, 1996). In overwintering grasses, including cereals, the shoot meristem is at ground level and comprises the meristematic bases of leaves enclosing the stem apex. These parts of the plant are crucial for survival of winter frost (Pearce, 1980; Gusta *et al.*, 1982) and they are the parts of the plant, which perceive low temperature for growth acclimation (Thomas and Stoddart, 1984).

Twelve genes were isolated from a cDNA library using a differential screen based on cDNA sequences from low-temperature (6°C day/2°C night; LT) shoot meristem mRNA as the positive probe, and control temperature (20°C day/15°C night, 10-hour day; HT) mRNA as the negative probe (Dunn *et al.*, 1990). These genes, which have no homology to each other, represent genes that are up-regulated at the steady-state mRNA level by the acclimation treatment. Many of these genes are members of small multigene families and for some, the low-temperature-responsive (LTR) gene encodes a low-temperature isoform of the protein (Dunn *et al.*, 1993). Table II shows five of these gene families that have been the subject of further studies, together with four barley gene families isolated by other groups.

TABLE I
Cloned cereal genes up-regulated by low temperature

Plant	Gene (homologues)	Accession numbers	Class of gene product	Reference
Hordeum vulgare L. (barley)	blt801	X81974	RNA-binding protein	Dunn et al. (1996)
	blt4.1 (blt4.2, blt4.6, blt4.9)	X56547	Lipid-transfer protein	White et al. (1994)
	blt63	Z23130	Protein translation elongation factor EF1-α	Dunn et al. (1993)
	blt14.0 (blt14.1, blt14.2)	X57554 (X97916, X97917)	NK	Dunn et al. (1990)
	blt101.1 (blt101.2)	Z25537	Hydrophobic protein	Goddard et al. (1993)
	COR14 (wcs19)		Hydrophilic protein	Crossatti et al. (1995)
	CR1 (BLYCLDAA)	117309 (M60732)	Hydrophilic protein	Cattivelli and Bartels (1990)
	Dhn5	M95810	LEAII	Van Zee et al. (1995)
	HVA1	X78205	LEAIII	Hong et al. (1992)
Triticum aestivum L. (wheat)	wcor518	U73214	Proline-rich protein	Danyluk and Sarhan (DBS)
	wcor719	—	Actin-binding protein	Danyluk et al. (1996)
	wcs19 (CORI4)	481813	Hydrophilic protein	Chauvin et al. (1993)
	tacr7 (barley blt14)	L28093	NK	Gana et al. (1997)
	wcs120 (wcs66)	M93342 (L27516)	LEAII	Houde et al. (1992)
	wcs726 (wcor 80)	U73213 (U73212)	LEAII	Danyluk and Sarhan (DBS)
	wcor410 (cor410b, cor410 c)	1169018 (U73210, U73211)	Acidic dehydrin	Danyluk et al. (1994)
	wcor615	U73217	LEAIII	Danyluk and Sarhan (DBS)
Secale cereal L. (rye)	rlt1412 (rlt1421, blt14)	Z23257 (Z23258)	NK	Zhang et al. (1993)

NK, not known; DBS, database submission.

The relationship between levels of gene expression and the ability of different genotypes to acclimate for frost tolerance has been investigated. Danyluk *et al.* (1994, 1998) have shown that *wcor410* is only expressed in freezing tolerant *Gramineae*. Houde *et al.* (1992) have shown that the accumulation kinetics of *wcs120* mRNA during acclimation are positively correlated with the capacity of each wheat genotype to develop frost tolerance. A study of *blt14* gene family expression in a barley albino mutant (a_n) showed that cold-induced expression of *blt14* genes was strongly reduced in the albino mutant, which cannot cold acclimate (Grossi *et al.*, 1998). This reduction is not solely due to the absence of the photosynthetic apparatus since these genes are up-regulated by cold in etiolated plants. Zhang *et al.* (1993) have shown that the low-temperature-induced steady-state mRNA levels of two rye LTR genes (*rlt1412, rlt1421*), which are cognates of the *btl14* gene family, were higher in the relatively frost-hardy cultivar, Puma.

In order to investigate the relationship between barley LTR gene expression and frost acclimation, winter barley plants (cv Igri) were given a range of temperature treatments and analysed for both frost hardiness (measure as the LT_{50} survival in a regrowth test) and levels of LTR gene expression (measured as integrated densitometric values of Northern blot autoradiographs). A correlation between gene expression and frost acclimation was found for individual genes in the gene families, *blt4, blt14, blt101* and *blt801*, with estimated correlation coefficients (*r*) between 0.976[***] for the gene *blt4.9* and 0.848[**] for *blt801* (Pearce *et al.*, 1996). The effects of other stresses on the expression of these genes have also been investigated and here the relationship between acclimation for frost hardiness and levels of gene expression varies between gene families.

Studies of gene expression at the protein level have also shown a relationship between the level of gene product and the capacity to acclimate for frost tolerance. Giorni *et al.* (1999) have shown that all winter cultivars of barley accumulate high levels of COR14a and COR14b proteins, whereas spring cultivars accumulate varying levels, suggesting that selection for winter survival has fixed the capacity for high accumulation. In a study of pathogenesis-related proteins, which show antifreeze activity, Antikainen and Griffith (1997) show that these proteins only accumulate in the apoplast of freezing tolerant monocots (rye, wheat, barley) during acclimation. These comparative studies support the proposal that these LTR genes have a function in frost acclimation.

Given that differential screening of low-temperature cDNA libraries isolates the mRNA sequence for genes that are expressed at relatively high levels, the method is considered unlikely to identify genes encoding those low-abundant proteins that are part of the temperature signal transduction pathway, nor will it identify genes that encode proteins that are involved in the sensing of temperature changes, since these are likely to be constituitively expressed. Although these are real limitations, the technique can provide important information about critical processes in frost acclimation.

V.   FUNCTION OF LTR GENE PRODUCTS

The first area where studies of LTR genes are beginning to make a contribution to our understanding of plant responses to temperature changes is in investigations of the proteins encoded by these genes. Analysis of the biochemical and physiological function of genes, for which only deduced amino-acid sequences exist, is a major challenge for all genome projects and also for understanding LTR genes, which have been isolated solely on the basis of their expression pattern.

A.   LATE EMBRYOGENESIS ABUNDANT PROTEINS

Freezing of plant tissues produces a dehydrative stress for cells because ice formation is initiated in the intracellular spaces. The chemical potential of ice is less than liquid water at a given temperature, and this results in movement of unfrozen water out of cells (Steponkus et al., 1993). The commonest class of protein encoded by LTR genes is the group that have homology to the dehydrin class of late embryogenesis abundant (LEA) group of proteins (Table I). These proteins are commonly induced by a treatment such as drought or at a developmental stage (late embryogenesis) associated with cellular water deficit (Close et al., 1993). They have also been shown to be induced by the phytohormone, abscisic acid (ABA) (Close et al., 1993). Although their gross structure may vary, all of these proteins contain repeat sequences with an amino-acid composition that is conserved within the same class of LEA protein (Close et al., 1993). They also share the property of being strongly hydrophilic and can be "boiling stable". A number of other non-LEA LTR proteins also have these properties (Table I). It has been suggested that the novel hydrophilic LTR proteins and LEA proteins have a common role in stabilizing membranes during dehydrative stress (Thomashow, 1999). The accumulation of an acidic dehydrin (WCOR410) in the vicinity of the plasma membrane during cold acclimation supports this proposal (Danyluk et al., 1998).

B.   LIPID TRANSFER PROTEINS

Lipid transfer proteins (LTPs) are small, basic proteins, which are characterized by their ability to transfer phospholipids from a donor to an acceptor membrane *in vitro*. They are abundant proteins whose primary structure is highly conserved between species and have been isolated from the aerial vegetative tissues and seeds of a large number of monocotyledon and dicotyledon plant species. In general their location is restricted to peripheral cells, such as the epidermis of leaves. The biological function of LTPs is a puzzle since, although their involvement in the intracellular flux of lipids is suggested by their *in vitro* activity, analysis of proteins and cDNA clones in a number of species shows that they are synthesized with N-terminal signal sequences, which are cleaved from the mature protein, and this is consistent with their extracellular location (Kader, 1996a,b). Because of their extracellular location and gene expression in epidermal cells, it has been suggested that, *in vivo*, they may be involved in cutin and surface wax deposition rather than lipid transfer (Sterk *et al.*, 1991). LTPs extracted from barley and maize leaves have also been shown to be potent inhibitors of bacterial and fungal plant pathogens (Molina *et al.*, 1993), suggesting a third possible *in vivo* function for these proteins.

Increases in the levels of lipid unsaturation at low temperature have been documented in several plant species (Harwood *et al.*, 1994) and direct evidence that a change in phospholipids affects the cryobehaviour of plant protoplasts has been reported (Steponkus *et al.*, 1988). In view of the fact that changes in membrane lipid composition are known to be important in the acquisition of tolerance to low temperatures (Hughes and Dunn, 1996), the production of new LTPs during frost acclimation in barley suggests a possible role of these proteins in modifying plasma membranes.

A homology modelling technique was used to predict the tertiary structures of three members of the low-temperature-inducible barley vegetative shoot epidermal LTP family, BLT4, on the basis of the three-dimensional structure of the maize seedling LTP determined by X-ray crystallography (Keresztessy and Hughes, 1998). Differences between the maize LTP and the BLT4 family include amino-acid substitutions around the entrance and inside the predicted hydrophobic binding tunnels of these proteins. In addition, because of the deletion of the loop region corresponding to Val60–Gly62 of the maize LTP from all three BLT4 LTPs, their internal hydrophobic tunnels are longer. Molecular dynamics modelling shows that *cis,cis*-9,12-octadecandienoic acid (an unsaturated lipid) had a more favourable interaction with the BLT4.9 LTP than with the maize protein. In view of the increase in membrane lipid unsaturation at low temperature, these results are consistent with the proposal that BLT4 LTPs have a lipid transfer function associated with frost acclimation in barley (Keresztessy and Hughes, 1998).

C.   RNA-BINDING PROTEIN BLT801

The barley LTR gene *blt801*, has homology to plant glycine-rich RNA-binding proteins (GR-RNPs) (Dunn *et al.*, 1996). Since a number of the LTR winter barley genes, which have been isolated, have been shown to be post-transcriptionally regulated (Dunn *et al.*, 1994), the properties of this putative RNA-binding protein are particularly interesting since it may have a role in controlling the steady-state mRNA levels of other LTR genes.

The RNA binding motif of BLT801 is part of a universal nucleic acid binding domain, which is highly conserved from bacteria to man and which has been demonstrated to bind both single-stranded (ss) DNA and RNA. The single-stranded nucleic acid binding properties of a recombinant BLT801-GST (glutathione *S*-transferase) fusion protein (FP) were investigated using supported ssDNA and homo-ribopolymers in a modification of the pull-down assay described by Swanson and Dreyfuss (1988). The BLT801 FP demonstrated a greater affinity for ssDNA than for any of the homo-ribopolymers. Homo-ribopolymer poly(G) bound more protein than either poly(A) or poly(U) and no binding of BLT801 FP to poly(C) was found. Purified GST protein showed no affinity for nucleic acids under the conditions used (Dunn *et al.*, 1996).

The SELEX procedure (Tuerk and Gold, 1990) was used to select radiolabelled *in vitro* transcripts, containing a variable 20mer sequence within constant flanking sequences, from pools of RNA, which were increasingly enriched for species with affinity to BLT801-FP compared with the control GST (Dunn *et al.*, 1996). Selected RNA molecules were purified, amplified by reverse transcriptase polymerase chain reaction (RT-PCR), used as templates for *in vitro* transcription and the process was repeated for ten cycles. The final products were cloned and sequenced, and secondary structures of 57 of the selected RNA sequences determined using DNAsis (Pharmacia Biotech). The information produced by this experimental technique suggests that the preferred RNA ligand of BLT801 contains a high proportion of the nucleotide adenine within a stable loop (open) structure; however, further work is needed to identify the *in vivo* substrate of this protein.

VI. CONTROL OF GENE EXPRESSION BY LOW ACCLIMATING TEMPERATURES

A. ANALYSIS OF CONTROL MECHANISMS BY NUCLEAR RUN-ON TRANSCRIPTION

The regulation of nine barley LTR genes, all of which have been shown to have enhanced steady-state levels of mRNA in winter barley (cv. Igri), in response to a two-week low-temperature treatment of 6°C/2°C was investigated by Dunn *et al.* (1994). To determine whether the increased steady-state mRNA levels of these LTR genes represents transcriptional or post-transcriptional regulation, nuclei were isolated from low-temperature (6° C/2° C) and control (20°C/15°C) treated barley shoot meristems and used in an *in vitro* run-on transcription assay. Using the gene-specific 3' non-coding sequence of *blt4.9* (the most low-temperature-specific member of the *blt4* gene family; White *et al.*, 1994), the results show that both *blt4.9* and *bl101* have high levels of transcript produced by low-temperature nuclei but very low levels by the control-temperature nuclei, indicating that these genes are transcriptionally regulated in response to low temperature.

In contrast to the LT transcriptional response, the gene *blt14* has a very low level of detectable transcript in both temperature treatments, indicating low transcription rates, in both LT and control temperature. One gene (*blt801*), which also only accumulates high steady-state mRNA levels at LT, appears to be transcriptionally down-regulated at LT, although transcription rates at both temperatures are higher than those of *blt14*. The transcripts of both *blt14* and *blt801* accumulate at LT but not the control temperature, and therefore these results suggest that low temperature stabilizes the transcripts of these genes and that their regulation is post-transcriptional.

B.   ANALYSIS OF TRANSCRIPTIONAL CONTROL OF GENE EXPRESSION

In some species, the expression of the genes encoding LTPs is influenced by environmental factors (see reviews: Kader, 1996a,b). The diploid cereal barley genome contains at least eight LTP genes distributed on three chromosomes (Gausing, 1994; Hughes and Dunn, 1996; Kalla *et al.*, 1994). The LTP gene family (*blt4*), located on chromosome 3, was isolated because members are transcriptionally up-regulated in young winter barley plants by a low positive acclimating temperature (Dunn *et al.*, 1991, 1994; White *et al.*, 1994).

Sequence analysis of 1938 bases of the region upstream of the initiation codon of the member of this gene family designated, *blt4.9* has identified, in addition to CAAT and TATA boxes, a number of motifs implicated in environmental responses (White *et al.*, 1994). In addition to putative low-temperature-responsive elements (LTREs) identified in dicotyledon species, the promoter contains several ACGT G-box core sequences, which have been identified as ABA responsive elements in a number of studies and a G-box (CACGTG) (Marcotte *et al.*, 1989; Straub *et al.*,1994; Knight *et al.*, 1995). Five chimeric constructs were produced with varying lengths of promoter, and increasing numbers of sequence motifs, to investigate their contribution to transcriptional regulation. All constructs included the 5'-untranslated sequence of the *blt4.9* transcript and the initiation codon of *blt4.9* and were thus translational fusions (Fig. 1a). Each construct was assayed by transient expression of the *uidA* β-glucuronidase (GUS) reporter gene (Jefferson *et al.*, 1987) following Biolistic (Bio-Rad Laboratories Ltd) particle bombardment into shoot meristematic tissue dissected from low-temperature and control plants (Dunn *et al.*, 1998).

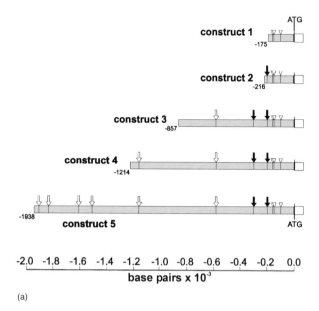

(a)

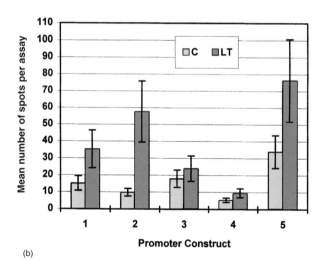

(b)

Fig. 1. Transient reporter gene expression analysis of *blt*4.9 promoter.
(a) Promoter constructs; ATG, *blt*4.9 initiation codon; ▽, position of CAAT
and TATA boxes; solid arrows, position of putative low-temperature response
elements (LTREs); open arrows, position of ACGT containing ABA response
elements and G-box. (b) Transient β-glucuronidase gene expression in barley shoot
meristems following particle bombardment with *blt*4.9 promoter constructs 1–5.
Expression is shown as the mean number of blue spots per assay; error bar, standard
error.

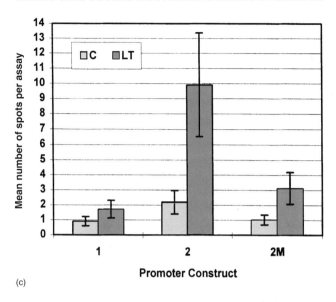

(c)

Fig. 1.   Contd. (c) Transient reporter gene expression analysis of mutant LTRE-1. Transient β-glucuronidase gene expression in barley leaf explants following particle bombardment with *blt4.9* promoter constructs 1, 2 and 2M. Expression is shown as mean number of blue spots per leaf; error bars, standard error. (b,c) C, control explants; LT, low-temperature-treated explants.

All of the *blt4.9* promoter constructs show a low-temperature response relative to a rice actin *uidA* control (Fig.1b); however, the enhanced LT response of construct 2 is statistically significant. These data also demonstrate the presence of negative regulatory elements in constructs 3 and 4, the responses of which are significantly different from those of constructs 2 and 5. Highest levels of expression are observed with construct 5, highlighting the potential importance of enhancer elements in a region of the promoter distal from the start of transcription (Dunn *et al.*, 1998).

The pentanucleotide CCGAC was identified as a LTRE in *Brassica napus* by Jiang *et al.* (1996) and the proximal promoter region of *blt4.9* contains two LTRE-like motifs, CCGAA (putative LTRE-1) at −195 and CCGAC (putative LTRE-2) at −295 (relative to ATG). The possibility that these LTRE-like motifs are target binding sites of nuclear proteins was investigated in electrophoretic mobility shift assays (EMSA) using synthetic double-stranded oligonucleotide probes, which include either these sequences or mutants of them (Dunn *et al.*, 1998).

The probe LT1 (42 bases)(Fig. 2a) was prepared from a synthetic oligonucleotide spanning the LTRE-1 motif and including all of the sequence (−218 to −177), which confers LT responsiveness to promoter reporter construct 2 in the transient expression assay (Fig. 1). The LT1 probe

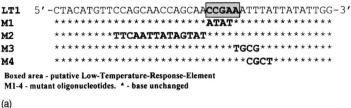

LT1   5′-CTACATGTTCCAGCAACCAGCAA⌈CCGAA⌉ATTTATTATATTGG-3′
M1   ************************ATAT***************
M2   ********TTCAATTATAGTAT*****************
M3   *****************************TGCG***********
M4   *****************************CGCT*********

Boxed area - putative Low-Temperature-Response-Element
M1-4 - mutant oligonucleotides. * - base unchanged

(a)

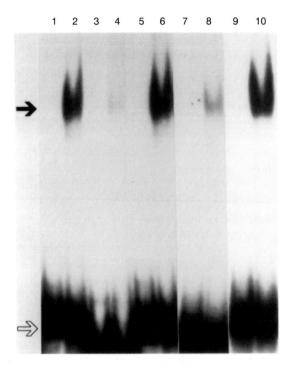

(b)

Fig. 2. Electrophoretic mobility shift analysis (EMSA) of synthetic oligo-nucleotide containing low-temperature response element 1 (LT1) and mutants (M1–4). (a) Synthetic oligonucleotides LT1, M1, M2, M3 and M4 were made double stranded, radiolabelled using Taq DNA polymerase and used as probes. (b) The effects of mutations of LT1 on binding to the low-temperature nuclear protein complex. Free probe (lanes 1, 3, 5, 7 and 9) and probe incubated with low-temperature nuclear proteins (lanes 2, 4, 6, 8 and 10), respectively, are shown in adjacent lanes; LT1, lanes 1 and 2; M1, lanes 3 and 4; M2, lanes 5 and 6; M3, lanes 7 and 8; M4, lanes 9 and 10. Solid arrow, retarded DNA–protein complexes; open arrow, free probe.

forms a complex, which is present in both LT and control nuclear extracts. The constitutive presence of a *trans*-acting factor in nuclear extracts, which is involved in the control of a stress-regulated gene, has also been demonstrated by Wilkinson *et al.* (1996). To determine the precise binding site of the protein complex, a series of mutant probes was made as shown in Fig. 2a. Testing each of these mutant probes in EMSA with LT nuclear extracts shows that the minimum recognition sequence for this protein complex lies within the hexamer CCGAAA (putative LTRE-1) (Fig. 2b). The sequence CCGAC (putative LTRE-2) is present 100 bases upstream of LTRE-1; however, EMSA analysis shows that the putative LTRE-2 is not a protein binding site and is not recognized by the protein complex associated with LTRE-1 (Dunn et al., 1998).

The results of promoter deletion analysis and EMSA suggest that a low-temperature response element lies within the hexanucleotide sequence LTRE-1 (CCGAAA) of reporter construct 2. The hypothesis that this hexanucleotide sequence is a LTRE was further tested by introducing the base mutation of M1 into promoter reporter construct 2 using site-directed mutagenesis. Transient expression of the mutant promoter construct, 2M, was assayed following Biolistic particle bombardment (Fig. 2c). As in the earlier experiment, there is a statistically significant LT induction by promoter construct 2, which is substantially reduced in the mutant promoter construct 2M (Dunn et al., 1998).

Several studies of dicotyledon species (Arabidopsis: Yamaguchi-Shinozaki and Shinozaki, 1994; Baker et al., 1994; Brassica napus: Jiang et al., 1996) have identified a conserved pentanucleotide (CCGAC) as a LTR element and a transcription factor (CBF1), which binds to this element has been identified in Arabidopsis (Stockinger et al., 1997). This dicotyledonary element is clearly related to the putative LTR element (CCGAAA) of the barley gene, blt4.9. However, recent work shows that the Arabidopsis CBF1 transcription factor (Stockinger et al., 1997), expressed in E.coli using the pET28a vector (Novagen Inc.), does not bind the barley CCGAAA putative LTR element (Senay, M. A. Dunn and M. A. Hughes, unpublished observations). The identification of LTR elements within the promoters of transcriptionally controlled LTR cereal genes will allow the isolation (via yeast one-hybrid cloning systems) of genes encoding transcription factors which bind these elements. This technique has already been successfully used to isolate the Arabidopsis transcription factor gene, CBF1 (Stockinger et al., 1997).

The Wcs120 gene from wheat is a member of a family of related genes and is specifically regulated by low temperature. The accumulation of Wcs120 mRNA and protein has been shown to correlate closely with the differential capacity of wheat cultivars to develop freezing tolerance (Limin et al., 1995). The Wcs120 promoter contains a number of potential control elements, including the CCGAC pentanucleotide at −175 and −337 bases (Vazquez-Tello et al., 1998). EMSA analysis of 860 bases of the Wcs120 promoter subdivided into six overlapping fragments between 100 and 160 bases long, shows a complex pattern of nuclear protein interaction (Vazquez-Tello et al., 1998), which compares with a similar study of the barley blt4.9 promoter

TABLE II

Summary of responses to other environmental factors by barley genes (blt) *which are induced by low temperature (data from Northern blot analysis)*

Gene	Low temperature	Mildew	Drought	ABA	Salt	Oxidative stress	Interaction with light
blt4.9	✓	✓	✓	✓	?	✓	✓
blt14.0	✓	✗	✗	✗	?	✓	✓
blt101.1	✓	✗	✗	✗	✗	✗	✗

(Dunn *et al.*, 1998). Interestingly, Vazquez-Tello *et al.* (1998) (unlike the barley gene study) showed that nuclear proteins from LT-acclimated plants were unable to bind to these promoter fragments unless the nuclear protein extract had been dephosphorylated with alkaline phosphatase. This result was paralleled by increased levels of both Ca^{2+}-dependent and Ca^{2+}-independent kinase activities in LT-acclimated nuclear extracts and the *in vivo* stimulation of the *Wcs120* protein family in plants treated with the phosphatase inhibitor, okadaic acid (Vazquez-Tello *et al.*, 1998). This result is unexpected since it implies that LT stimulation of *Wcs120* expression is due to inactivation of DNA-binding factors which interact with all six fragments of the promoter. No putative positive acting factor was seen in this study.

Since the promoters of a number of LT-regulated dicotyledon and monocotyledon genes contain elements that resemble the CCGAC LTRE element (Dunn *et al.*, 1998; Vazquez-Tello *et al.*, 1998; Thomashow, 1999), it has been suggested that this represents a common control element for all LTR genes. This suggestion is refuted by studies of the LT-specific transcriptionally controlled barley gene, *blt101.1* (Brown *et al.*, 2001). Steady-state mRNA levels of *blt101.1* are dramatically upregulated by low temperature and, unlike other barley LTR genes, a range of other factors (Table II) does not induce expression of this gene. The expression of the gene is also characterized by the fact that other factors do not interact with low-temperature induction and the observation that it is expressed in all organs of the plant, although high levels of expression (mRNA) have been documented in the vascular-transition zone of the crown (Pearce *et al.*, 1998). The promoter of *blt101.1* contains no CCGAC-like element; it does contain an element that resembles an ABA-response element (ABRE) but, consistent with the lack of ABA induction, this has been shown not to be functional (Brown *et al.*, 2001). Although the element responsible for the elevated low-temperature transcription of this gene has not yet been identified, an enhancer element (TCA), found in a number of stress-controlled genes, and chromatin remodelling have been shown to be important in the activation of the *blt101.1* promoter (Brown *et al.*, 2001).

The protein encoded by *blt101.1* is a small hydrophobic polypeptide (Goddard *et al.*, 1993), which is predicted to be a membrane protein. A homologue encoding an identical protein, *ESI3*, has been isolated from salt-tolerant tall wheatgrass (*Lophopyrum elongatum* (Host) Löve) (Gulick *et al.*, 1994), and this gene is strongly upregulated by high-salt conditions. However, homologues of *blt101* in *Arabidopsis* (Capel *et al.*, 1997) are responsive to low temperature, drought and ABA, whilst a homologue isolated from cultivated strawberry (*Fragaria* × *anannassa*) (Ndong *et al.*, 1997), which is preferentially up-regulated by low temperature, is also weakly up-regulated by ABA, suggesting that the membrane proteins that these genes encode play a conserved role in the plants' response to dehydration.

C. ANALYSIS OF POST-TRANSCRIPTIONAL CONTROL

Post-transcriptional mechanisms are important in plant development and in plant response to environmental change (see review: Gallie, 1993). It has been shown that the major control of *blt14.0* expression involves either the processing, transport or stability of mRNA (Dunn *et al.*, 1994) and therefore *blt14.0* mRNA stability during cold acclimation and de-acclimation was investigated by gel blot analysis using metabolic inhibitors (Phillips *et al.*, 1997). Since global effects of such inhibitors can give misleading results, the same Northern blots were also analysed for mRNA levels of the transcriptionally controlled barley LTR gene, *blt101.1* (Goddard *et al.*, 1993; Dunn *et al.*, 1994). Cordycepin, which inhibits plant RNA synthesis when used at high concentrations, has been successfully used in plant systems to analyse the stability of transcripts, including the small subunit of ribulose-1,5-bisphosphate carboxylase (Peters and Silverthorne, 1995) and the cold acclimation specific gene *cas18* (Wolfraim *et al.*, 1993). However, the precise mode of action of cordycepin is unclear; Millkarek (1974) suggests that it may inhibit poly(A) polymerase and thus polyadenylation. However, Walbot *et al.* (1974) investigated the effects of cordycepin on *in vivo* protein synthesis during the germination of cotton cotyledons and showed that cordycepin was effective at inhibiting both RNA and protein synthesis.

The relationship between *blt14.0* mRNA stability and translation was also investigated using cycloheximide. In eukaryotes, cycloheximide stabilizes most unstable transcripts, either through arresting translation of the transcript itself (a *cis* effect) or by preventing translation of an unstable *trans*-acting factor required for mRNA degradation (Ross, 1995). Cycloheximide treatment of the barley explants does lead to a slight stabilization of the *blt101.1* transcript at 2°C. The relationship between translation and mRNA stability, however, is not simple and is likely to be mRNA specific (Caponigro and Parker, 1996).

The mRNA stability of pre-existing *blt14.0* mRNA was studied in barley shoot explants, in the presence or absence of cordycepin or cycloheximide during de-acclimation and acclimation environments. There were marked differences between the effect of these inhibitors on *blt14.0* and *blt101.1*. Maintenance of high *blt14.0* steady-state mRNA levels at low temperature is impaired by both cordycepin and cycloheximide. In contrast, although cordycepin reduces *blt101.1* mRNA steady-state levels at this temperature, cycloheximide treatment causes a slight overall increase (Phillips *et al.*, 1997).

The markedly reduced levels of *blt14.0* mRNA in the presence of cordycepin and cycloheximide in a low-temperature environment, indicates that both transcription and translation are required to maintain *blt14.0* steady-state mRNA levels at 2°C. These data suggest that a stabilizing protein factor(s), which is critical for the increase in *blt14.0* steady-state mRNA levels, is transcribed and translated at low temperature. During deacclimation at 18°C, *blt14.0* steady-state mRNA levels are not significantly affected by the presence of cordycepin, whereas a dramatic reduction of *blt14.0* steady-state mRNA levels is observed in the presence of cycloheximide. This also suggests that the decay of *blt14.0* mRNA during de-acclimation is controlled by a labile protein factor(s), which is not synthesized in the presence of cycloheximide.

VII.   LOW-TEMPERATURE PERCEPTION AND SIGNAL TRANSDUCTION

Very little is known about low-temperature perception by plant tissues; however, it is presumed to involve constitutively expressed proteins or other cell constituents. Despite our lack of knowledge of temperature perception, a number of components of possible signal transduction pathways (or networks) have been investigated in a variety of plants, including cereals. The principal components under study are ABA, calcium and phosphorylation.

During frost acclimation of winter wheat seedlings, transient increases in the level of ABA can be measured; further, these increases in ABA content are positively correlated with the level of frost tolerance acquired by different genotypes (Dörffling *et al.*, 1998). Treatment with the ABA biosynthesis inhibitor, norflurazone, reduces both the transient ABA increase and levels of frost tolerance (Dörffling *et al.*, 1998). Many of the LTR genes isolated from cereals and other species are also responsive to exogenous applications of ABA (Hughes and Dunn, 1996) and this treatment confers a low level of frost tolerance on many plants, including cereals (Churchill *et al.*, 1998). However, the role of ABA in activation of LTR genes and in frost acclimation is not clear. Some LTR genes are not responsive to ABA and experiments with *Arabidopsis* ABA-insensitive mutants show that even the ABA-responsive LTR genes can be activated via an ABA-independent pathway (Hughes and Dunn, 1996). Reductions in the level of frost tolerance observed in many ABA mutants have to be interpreted with caution, since they could be due the pleiotropic effects of these mutations on plant growth and not a direct consequence of the loss of an ABA function in low-temperature signal transduction (Thomashow, 1999).

There is also evidence that calcium may play a role in cold acclimation, since low-temperature treatments lead to a rapid transient increase in cytoplasmic calcium levels, largely through influx from extracellular stores (Knight *et al.*, 1996). Cold acclimation has also been shown to increase Ca^{2+} transport activity in isolated plasma membranes from winter rye (Puhakainen *et al.*, 1999). However, although there is work with alfalfa, which links calcium influx during frost acclimation with inhibition of protein phosphatase 2A activity (Monroy *et al.*, 1998), the precise role of calcium as a second messenger and the nature of the signal transduction pathway(s) is not known.

It is clear from studies of barley LTR genes that there will be a complex network of interacting factors involved in low-temperature signal transduction. This is also illustrated by studies of the interaction of light and low temperature in wheat. *Wcs19* is a nuclear gene, encoding a chloroplast protein; it is specifically regulated by low temperature and requires light for maximal induction (Chauvin *et al.*, 1993). Studies of *wcs19* in winter rye given different combinations of light and temperature, indicate that the accumulation of *wcs19* mRNA and the change in growth habit associated with frost hardening, are associated with the relative reduction state of photosystem II rather than the temperature or irradiance *per se* (Gray *et al.*, 1997). In the same experiments, induction of the *wcs120* transcript appears to be dependent on both temperature and irradiance, but in an independent, additive manner. Recent studies of the *wcs19* homologue in barley (*cor14b*) also show that light has an effect on the accumulation of the WCS19 protein (Crosatti *et al.*, 1999).

VIII.   PROSPECTS FOR MANIPULATING FROST TOLERANCE IN CROP PLANTS

Atherly and Jenkins (1997), in a review of frost acclimation, conclude that the most productive approach to introducing frost tolerance into cold-sensitive species may be to manipulate the signal transduction pathways. It has been argued that this prospect has been made more realistic by the discovery that the *Arabidopsis* transcription factor, CBF1, when constitutively expressed in transgenic plants, activates a suite of LTR genes and leads to improved frost tolerance without acclimation (Jago-Ottosen *et al.*, 1998; Sarhan and Danyluk, 1998). There are some cautions that must be considered in developing this approach. Firstly, it is clear that the promoter element, through which CBF1 induces low-temperature gene expression, is not present in all LTR gene promoters, and CBF1 cannot therefore be the single "master switch" for frost acclimation responses. Secondly, the critical genes that encode proteins with a role in frost acclimation, which also have a CBF1 LTR element in the promoter, may not be present in frost-sensitive crop species. Finally, it is clear from attempts to introduce rye freezing-tolerance levels into *Triticale*, that the correct coordination of cellular signalling pathways may be important for full development of the trait.

Nevertheless, frost tolerance is quantitative and our increasing knowledge of the genetics, and molecular biology of the mechanisms involved, will lead to tools for the significant incremental improvement of frost tolerance both within cereal species as well as through interspecific gene manipulation.

REFERENCES

Antikainen, M. and Griffith, M. (1997). Antifreeze protein accumulation in freezing tolerant cereals. *Physiologia Plantarum* **99**, 423–432.

Atherly, A. G. and Jenkins, G. J. (1997). Mechanisms underlying plant acclimation to low temperatures. *AgBiotech* **9**, 77–80.

Baker, S. S., Wilhelm, K. S. and Thomashow, M. F. (1994). The 5′-region of *Arabidopsis thaliana cor15a* has *cis*-acting elements that confer cold-regulated, drought-regulated and ABA-regulated gene-expression. *Plant Molecular Biology* **24**, 701–713.

Brown *et al.* (2001). (Submitted)

Capel, J., Jarillo, J. A., Salinas, J. and Martinez-Zapater, J. M. (1997). Two homologous low-temperature-inducible genes from *Arabidopsis* encode highly hydrophobic proteins. *Plant Physiology* **115**, 569–576.

Caponigro, G. and Parker, R. (1996). Mechanisms and control of mRNA turnover in *Saccharomyces cerevisiae*. *Microbiological Reviews* **60**, 233–249.

Cattivelli, L. and Bartells, D. (1990). Molecular cloning and characterisation of cold-regulated genes in barley. *Plant Physiology* **93**, 1504–1510.

Chauvin, L -P., Houde, M. and Sarhan F. (1993). A leaf specific gene stimulated by light during wheat acclimation to low temperature. *Plant Molecular Biology* **23**, 255–265.

Churchill, G. C., Reaney, M. J. T., Abrams, S. R. and Gusta, L. V. (1998). Effects of ABA and ABA analogs in the induction of freezing tolerance of winter rye. *Plant Growth Regulation* **25**, 35–45.

Close, T. J., Fenton, R. D., Yang, A., Asghar, R., DeMason, D. A., Crone, D. E. Meyer, N. C. and Moonan, F. (1993). Dehydrin: the protein. *In* "Plant Responses to Cellular Dehydration during Environmental Stress" (T. J. Close and E. A. Bray eds), pp. 104–118. The American Society of Plant Physiologists, USA.

Crosatti, C., Soncini, C., Stanca, A. M. and Cattivelli, L. (1995). The accumulation of a cold-regulated chloroplastic protein is light-dependent. *Planta* **196**, 458–463.

Crosatti, C., Polverino de Laureto, P., Bassi, R. and Cattivelli, L. (1999). The interaction between cold and light controls the expression of the cold-regulated barley gene *cor14b* and the accumulation of the corresponding protein. *Plant Physiology* **119**, 671–680.

Danyluk, J., Houde, M., Rassart, E. and Sarhan, F. (1994). Differential expression of a gene encoding an acidic dehydrin in chilling sensitive and freezing tolerant gramineae species. *FEBS Letters* **344**, 20–24.

Danyluk, J., Carpentier, E. and Sarhan, F. (1996). Identification and characterisation of a low temperature regulated gene encoding an actin-binding protein from wheat. *FEBS Letters* **389**, 324–327.

Danyluk, J., Perron, A., Houde, M., Limin, A., Fowler B., Benhamov, N. and Sarhan, F. (1998). Accumulation of an acidic dehydrin in the vicinity of the plasma membrane during cold acclimation in wheat. *Plant Cell* **10**, 623–638.

Dörffling, K., Abromeit, M., Bradersen, U., Dörffling, H. and Melz, G. (1998). Involvement of abscissic acid and proline in cold acclimation of winter wheat. *In* "Plant Cold Hardiness" (P. H. Li and T. H. H. Chen, eds), pp. 283–292. Plenum Press, New York.

Doll, H., Haahr, V. and Søgaard, B. (1989). Relationships between vernalisation requirement and winter hardiness in doubled haploids of barley. *Euphytica* **42**, 209–213.

Dunn, M. A. Hughes, M. A., Pearce, R. S. and Jack, P. L. (1990). Molecular characterisation of a barley gene induced by cold treatment. *Journal of Experimental Botany* **41**, 1405–1413.

Dunn, M. A. Hughes, M. A., Zhang, L., Pearce, R. S., Quigley, A. S. and Jack, P. L. (1991). Nucleotide sequence and molecular analysis of the low-temperature induced cereal gene, *blt4*. *Molecular and General Genetics* **229**, 389–394.

Dunn, M. A., Morris, A., Jack, P. L. and Hughes, M. A. (1993). A low-temperature responsive translation elongation factor 1α from barley (*Hordeum vulgare* L.) *Plant Molecular Biology* **23**, 231–225.

Dunn, M. A., Goddard, N. J., Zhang, L., Pearce, R. S. and Hughes, M. A. (1994). Low-temperature-responsive barley genes have different control mechanisms. *Plant Molecular Biology* **24**, 879–888.

Dunn, M. A., Brown, K., Lightowlers, R. and Hughes, M. A. (1996). A low-temperature-responsive gene from barley encodes a protein with single stranded nucleic acid binding activity which is phosphorylated *in vitro*. *Plant Molecular Biology* **30**, 947–959.

Dunn, M. A., White, A. J., Vural, S. and Hughes, M. A. (1998). Identification of promoter elements in a low-temperature-responsive gene (*blt4.9*) from barley. *Plant Molecular Biology* **38**, 551–564.

Dvorak, J. and Fowler, D. B. (1978). Cold hardiness potential of tricale and tetraploid rye. *Crop Science* **17**, 477–478.

Fowler, D. B., Chauvin, L.-P. Limin, A. E. and Sarham, F. (1996). The regulatory role of vernalisation in the expression of low-temperature genes in wheat and rye. *Theoretical and Applied Genetics* **93**, 554–559.

Fowler, D. B., Limin, A. E. and Ritchie, J. T. (1999). Low-temperature tolerance in cereals: model and genetic interpretation. *Crop Science* **39**, 626–633.

Galiba, G., Quarrie, S. A., Sutka, J., Morgounov, A. and Snape, J. W. (1995). RFLP mapping of the vernalisation (*Vrn1*) and frost resistance (*Fr1*) genes in chromosome 5A of wheat. *Theoretical and Applied Genetics* **90**, 1174–1179.

Galiba, G., Kerepesi, I., Snape, J. W. and Sutka, J. (1998). Mapping of genes controlling cold hardiness in wheat 5A and its homologous chromosomes of cereals. *In* "Plant Cold Hardiness" (P. H. Li and T. H. H. Chen, eds), pp. 89–98. Plenum Press, New York.

Gallie, D. R. (1993). Post-transcriptional regulation of gene expression in plants. *Annual Review Plant Physiology and Plant Molecular Biology* **44**, 77–105.

Gana, J. A., Sutton, F. and Kenefick, D. G. (1997). cDNA structure and expression patterns of a low-temperature-specific wheat gene *tacr7*. *Plant Molecular Biology* **34**, 643–650.

Gausing, K. (1994). Lipid transfer protein genes specifically expressed in barley leaves and coleopiles. *Planta* **192**, 574–580.

Giorni, E., Crosatti, C., Baldi, P., Grossi, M., Maré C. Stanca, A. M. and Cattivelli, L. (1999). Cold-regulated gene expression during winter in frost tolerent and frost susceptible barley cultivars grown under field conditions. *Euphytica* **106**, 149–157.

Goddard, N. J., Dunn, M. A., Zhang, L., White, A. J., Jack, P. L. and Hughes, M. A. (1993). Molecular analysis and spatial expression pattern of a low temperature specific barley gene, *blt101*. *Plant Molecular Biology* **23**, 871–879.

Gray, G. R. Chauvin, L.-P., Sarhan, F. and Huner, N. P. A. (1997). Cold acclimation and freezing tolerance. *Plant Physiology* **114**, 467–474.

Grossi, M., Giorni, E., Rizza, F., Stanca, A. M. and Cattivelli, L. (1998). Wild and cultivated barley show differences in the expression pattern of a cold-regulated gene family under different light and temperature conditions. *Plant Molecular Biology* **38**, 1061–1069.

Gulick, P. J., Shen, W. and An, H. (1994). *ESI*3, a stress-induced gene from *Lophopyrum elongatum*. *Plant Physiology* **104**, 799–800.

Gusta, L. V., Fowler, D. B. and Tyler, N. J. (1982). Factors influencing hardening and survival in winter wheat. *In* "Plant Cold Hardiness and Freezing Stress", Vol. 2 (P. H. Li and A. Sakai, eds), pp. 23–40. Academic Press, New York.

Harwood, J. L., Jones, A. L., Perry, H. J., Rutter, A. J.,Smith, K. L. and Williams, M. (1994). Changes in plant lipids during temperature adaption. *In* "Temperature Adaptation of Biological Membranes" (A. R. Cossins, ed.), pp. 107–118. Portland Press Proceedings, London.

Hayes, P., Chen, F. Q., Corey, A. Pau, A., Chen, T., Baird, E., Powell, W., Thomas, W., Waugh, R., Bedo, Z., Karsai,I., Blake, T. and Oberthur, L. (1998). The Dicktoo and Morex. Population. *In* "Plant Cold Hardiness" (P. H. Li and T. H. H. Chen, eds), pp. 77–88. Plenum Press, New York.

Hong, B., Barg, R. and Ho, T. H. (1992). Developmental and organ-specific expression of an ABA- and stress-induced protein in barley. *Plant Molecular Biology* **18**, 663–674.

Houde, M., Danyluk, J., Laliberte, J. F., Rassart, E., Dhindsa, R. S. and Sarhan, F. (1992). Cloning, characterisation, and expression of a cDNA encoding a 50-kilodalton protein specifically induced by cold acclimation in wheat. *Plant Physiology* **99**, 1381–1387.

Hughes, M. A. and Dunn, M. A. (1990). The effect of temperature on plant growth and development. *Biotechnology and Genetic Engineering Reviews* **8**, 161–188.

Hughes, M. A. and Dunn, M. A. (1996). The molecular biology of plant acclimation to low temperature. *Journal of Experimental Botany* **47**, 291–305.

Hughes, M. A., Dunn, M. A., Zhang, L., Pearce, R. S., Goddard, N. J. and White, A. J. (1993). Long-term adaptation and survival. *In* "Plant Adaptation to Environmental Stress" (L. Fowden, T. Mansfield and J. Stoddart, eds), pp. 251–262. Chapman Hall, London.

Jefferson, R. A., Kavanagh, T. A. and Bevan, M. W. (1987). GUS fusions: β-glucuronidase is a sensitive and versatile gene fusion marker in higher plants. *EMBO Journal* **6**, 3901–3907.

Jaglo-Ottosen, K. R., Gilmour, S. J., Zarka, D. G., Schabenberger, O. and Thomashow, M. (1998). *Arabidopsis CBF1* overexpression includes COR genes and enhances freezing tolerance. *Science* **280**, 104–106.

Jiang, C., Iu, B. and Singh, J. (1996). Requirement of a CCGAC *cis*-acting element for cold induction of *BN*115 gene from *B. napus. Plant Molecular Biology* **30**, 679–684.

Kader, J. C. (1996a). Lipid-transfer proteins: a puzzling family of proteins. *Trends in Plant Science* **2**, 66–70.

Kader, J. C. (1996b). Lipid-transfer proteins in plants. *Annual Reviews of Plant Molecular Biology* **47**, 627–654.

Kalla, R., Shimamoto, K., Potter, R., Nielsen, P. S., Linnestad, C. and Olsen, O. A. (1994). The promoter of the barley aleurone-specific gene encoding a 7-kD lipid transfer protein confers aleurone cell-specific expression in transgenic rice. *Plant Journal* **6**, 849–860.

Knight, C. D., Sehgal, A., Atwal, K., Wallace, J. C., Cove, D. J., Coates, D., Quatrano, R. S., Bahadur, S., Stockley, P. G. and Cuming, A. C. (1995). Molecular responses to abscisic acid are conserved between moss and cereals. *The Plant Cell* **7**, 499–506.

Knight, M., Trewavas, A. J. and Knight, M. R. (1996). Cold calcium signalling in *Arabidopsis* involves two cellular pools and a change in calcium signature after acclimation. *Plant Cell* **8**, 489–503.

Keresztessy, Zs. and Hughes, M. A. (1998). Homology modelling and molecular dynamics aided analysis of ligand complexes demonstrates functional properties of lipid transfer proteins encoded by the barley low-temperature-inducible gene family, *blt4. Plant Journal* **14**, 523–533.

Laurie, D. A., Pratchett, N., Bezant, J. and Snape, J. W. (1995). RFLP mapping of five major genes and eight quantitative trait loci controlling flowering time in a winter and spring barley cross. *Genome* **38**, 575–585.

Law, C. N. (1966). The location of genetic factors affecting a quantitative character in wheat. *Genetics* **53**, 487–498.

Limin, A.E., Dvorak, J. and Fowler, D. B. (1985). Cold hardiness in hexaploid triticale. *Canadian Journal of Plant Science* **65**, 487–490.

Limin, A. E., Houde, M, Chauvin, L-P., Fowler, D. B. and Sarhan F. (1995). Expression of the cold-induced wheat gene *Wcs120* and its homologues in related species and interspecific combinations. *Genome* **38**, 1023–1031.

Livingston, D. P. (1996). The second phase of cold hardening. *Crop Science* **36**, 1568–1573.

Marcotte, W. R., Russell, S. H. and Quatrano, R. S. (1989). Abscisic acid-responsive sequences from the *Em* gene of wheat. *Plant Cell* **1**, 969–976.

Millkarek, C., Price, R. and Penman, S. (1974). The metabolism of poly(A) minus mRNA fraction in HeLa cells. *Cell* **3**, 1–10.

Molina, A., Segura, A. and Garcia-Olmedo, F. (1993). Lipid transfer proteins (LTPs) from barley and maize leaves are potent inhibitors of bacterial and fungal plant pathogens. *FEBS Letters* **316**, 119–122.

Monroy, A. F., Sangwan, V. and Dhindsa, R. S. (1998). Low temperature signal transduction during cold acclimation: protein phosphatase 2A as an early target for cold in activation. *Plant Journal* **13**, 653–660.

Moore,G., Devos, K. M., Wang, Z. and Gale, M.D. (1995). Grasses line up and form a circle. *Current Biology* **5**, 737–739.

Ndong, C., Ouellet, F., Houde, M. and Sarhan, F. (1997). Gene expression during cold acclimation in strawberry. *Plant Cell Physiology* **38**, 863–870.

Ouellet, F., Vasquez-Tello, A. and Sarhan, F. (1998). The wheat *wcs120* promoter is cold-inducible in both monocotyledonous and dicotyledonous species. *FEBS Letters* **423**, 324–328.

Pearce, R. S. (1980). Relative hardiness to freezing of laminae, roots and tillers of tall fescue. *New Phytologist* **84**, 449–463.

Pearce, R. S., Dunn, M. A., Rixon, J. A., Harrison, P. and Hughes, M. A. (1996). Expression of cold-inducible genes and frost hardiness in the crown meristem of young barley (*Hordeum vulgare* L. cv. Igri) plants grown in different environments. *Plant, Cell and Environment* **19**, 275–290.

Pearce, R. S., Houlston, C. E., Atherton, K. M., Rixon, J. E., Harrison, P., Hughes, M. A. and Dunn, M. A. (1998). Localisation of expression of three cold-induced genes *blt101*, *blt4.9*, and *blt14* in different tissues of the crown and developing leaves of cold acclimated cultivated barley. *Plant Physiology* **117**, 787–795.

Peters, J. L. and Silverthorne, J. (1995). Organ specific stability of two *Lemna rbcS* mRNAs is determined primarily in the nuclear compartment. *Plant Cell* **7**, 131–140.

Phillips, J. R., Dunn, M. A. and Hughes, M. A. (1997). mRNA stability and localisation of the low-temperature-responsive barley gene family *blt14*. *Plant Molecular Biology* **33**, 1013–1023.

Puhakainen, T., Pihakashi Maunsbach, K., Widell, S. and Sommarin M.N.A. (1999). Cold acclimation enhances the activity of plasma membrane Ca^{2+} ATPase in winter rye leaves. *Plant Physiology and Biochemistry* **37**, 231–239.

Ross, J. (1995). mRNA stability in mammalian cells. *Microbiological Reviews* **59**, 423–450.

Sarhan, F. and Danyluk, J. (1998). Engineering cold-tolerant plants – throwing the master switch. *Trends in Plant Science* **3**, 289–290.

Steponkus, P. L., Uemura, M., Balsamo, R. A., Arvinte, T. and Lynch, D. V. (1988). Transformation of cryobehaviour of rye protoplasts by modification of the plasma membrane lipid composition. *Proceedings of the National Academy of Sciences (USA)* **85**, 9026–9030.

Steponkus, P. L., Uemura, M. and Webb, M. S. (1993). Membrane destabilisation during freeze-induced dehydration. *Current Topics in Plant Physiology* **10**, 37–47.

Sterk, P. L., Booij, H., Scheleekens, G. A., van Kammen, A. and de Vries, S. C. (1991). Cell-specific expression of the carrot EP2 lipid transfer protein gene. *Plant Cell* **3**, 907–921.

Stockinger, E. J., Gilmour, S. J. and Thomashow, M. F. (1997). *Arabidopsis thaliana* CBF1 encodes an AP2 domain-containing transcriptional activator that binds to the C-repeat/DRE, a cis-acting DNA regulatory element that stimulates transcription in response to low temperature and water deficit. *Proceedings of the National Academy of Science (USA)* **94**,1035–1040

Straub, P. F., Shen, Q. and Ho, T.-H.D. (1994). Structure and promoter analysis of an ABA- and stress-regulated barley gene *HVA1*. *Plant Molecular Biology* **26**, 617–630.

Swanson, M. S. and Dreyfuss, G. (1988). Classification and purification of proteins of heterogeneous nuclear ribonucleoprotein particles by RNA-binding specificities. *Molecular Cell Biology* **8**, 2237–2241.

Thomas, H., and Stoddart, J. L. (1984). Kinetics of leaf growth in *Lolium temulentum* at optimal and chilling temperatures. *Annals of Botany* **53**, 341–347.

Thomashaw M.F. (1999). Plant cold acclimation: freezing tolerance genes and regulatory mechanisms. *Annual Review of Plant Physiology and Plant Molecular Biology* **50**, 571–599.

Tuerk, C. and Gold, L. (1990). Systematic evolution of ligands by exponential enrichment: RNA ligands to bacteriophage T4 DNA polymerase. *Science* **249**, 505–510.

Van Zee, K., Chen, F. Q., Hayes, P. M., Close, T. J. and Chen, T. H. H. (1995). Cold-specific induction of a dehydrin gene family member in barley. *Plant Physiology* **108,** 1233–1239.

Vazquez-Tello, A., Ouellet, F. and Sarham, F. (1998). Low temperature-stimulated phosphorylation regulates the binding of nuclear factors to the promoter of *Wcs120*, a cold-specific gene in wheat. *Molecular and General Genetics* **257,** 157–166.

Walbot, V., Capdevila, A. and Dure III, L. S. (1974). Action of 3′d adenosine (cordycepin) and 3′d cytidine on the translation of the stored mRNA of cotton cotyledons. *Biochemical and Biophysical Research Communications* **60,** 103–110.

White, A. J., Dunn, M. A., Brown, K. and Hughes, M. A. (1994). Comparative analysis of genomic sequence and expression of a lipid transfer protein gene family in winter barley. *Journal of Experimental Botany* **45,** 1885–1892.

Wilen, R. W., Fu, P., Robertson, A. J. and Gusta, L. V. (1998). A comparison of the cold hardiness potential of spring cereals and vernalised and non-vernalised winter cereals. *In* "Plant Cold Hardiness" (P. H. Li and T. H. H. Chen, eds), pp. 191–202 Plenum Press, New York.

Wilkinson, M. G., Samuels, M., Takeda, T., Toone, W. M., Shieh, J. C., Toda, T., Millar, J. B. A. and Jones, N. (1996). The Atf1 transcription factor is a target for the sty1 stress activated map kinase pathway in fission yeast. *Genes Development* **10,** 2289–2301.

Wolfraim, L. A., Langis, R., Tyson H. and Dhindsa, R. S. (1993). cDNA sequence, expression, and transcript stability of a cold acclimation specific gene, *casl8*, of alfalfa (*Medicago falcata*) cells. *Plant Physiology* **101,** 1275–1282.

Xin, Z. Y., Law, C. N. and Worland, A. J. (1988). Studies on the effects of the vernalisation-responsive genes on the chromosomes of homologous group 5 of wheat. *In* "Proceedings of the 7th International Wheat Genetics Symposium", Vol. 1, pp. 675–680. Institute of Plant Science Research, Cambridge.

Yamaguchi-Shinozaki, K. and Shinozaki, K. (1994). A novel *cis*-acting element in an *Arabidopsis* gene is involved in responsiveness to drought, low temperature, or high-salt stress. *The Plant Cell* **6,** 251–264

Zhang, L., Dunn, M. A., Pearce, R. S. and Hughes, M. A. (1993). Analysis of organ specificity of a low-temperature-responsive gene family in rye (*Secale cereals* L). *Journal of Experimental Botany* **44,** 1787–1793.

Genetics and Genomics of the Rice Blast Fungus *Magnaporthe grisea*: Developing an Experimental Model for Understanding Fungal Diseases of Cereals

NICHOLAS J. TALBOT and ANDREW J. FOSTER

School of Biological Sciences, University of Exeter, Washington Singer Laboratories, Perry Road, Exeter EX4 4QG, UK

I. INTRODUCTION: FUNGAL DISEASES OF CEREALS

Each year cereal production throughout the world is seriously affected by the action of fungal diseases. The effect is felt either by severe disease outbreaks, or by the cost of using fungicides and newly developed resistant cultivars. Fungi cause the most serious diseases of cereals and in the temperate cereal-growing regions of Europe and the USA their action is a significant constraint on production (Baker *et al.*, 1997; Hewitt, 1998). Powdery mildew of wheat and barley caused by *Erysiphe graminis* and *Mycosphaerella graminicola*, which causes Septoria blotch of wheat, for

Advances in Botanical Research Vol. 34
incorporating Advances in Plant Pathology
ISBN 0-12-005934-7

example, cause severe damage to cereal harvests and collectively represent fungicide markets worth in excess of $654 million per annum (Hewitt, 1998).

In recent years the mechanisms by which plants perceive and respond to pathogens have been studied intensively. Since the first cultivar-specific resistance genes were isolated in the early 1990s, there has been enormous excitement and rapid progress in identifying the signal transduction pathways that transmit perception of pathogen attack into the orchestrated deployment of plant defence compounds, and often programmed cell death, at the point of infection (Baker *et al.*, 1997). At the same time new methods to determine gene function and to isolate novel mutants have been applied to some of the most important fungal pathogens of plants and in the last few years there has been equally rapid progress in determining the underlying genetic control of fungal pathogenesis. Both fields of study have tremendous implications for developing durable control methods for plant disease, either by production of transgenic disease resistant cultivars or by development of novel broad-spectrum fungicides (Baker *et al.*, 1997).

This review aims to show how a concerted and multidisciplinary effort to understand the biology of a single disease, rice blast, has led to identification of conserved pathways for controlling pathogenic development and of gene products that are used by diverse pathogens, often for a directly orthologous, and yet pathogen-specific function.

II.   CLASSICAL AND MOLECULAR GENETICS OF THE RICE BLAST FUNGUS *MAGNAPORTHE GRISEA*

Research on the causal agent of rice blast *Magnaporthe grisea* (Hebert) Barr. (anamorph, *Pyricularia oryzae*) has been facilitated by the extensive use of classical genetic techniques and the development, during the 1980s and 1990s, of molecular genetic methods for manipulation of the fungus.

M. grisea is a heterothallic ascomycete and therefore genetic crossing of haploid strains allows direct analysis of the products of meiosis and collection of large numbers of progeny. *M. grisea* produces flask-shaped perithecia that develop and mature in approximately 3 weeks on oatmeal agar plates (Valent *et al.*, 1991). Perithecia can be obtained from plate cultures by gentle removal of aerial hyphae with a sterile needle and isolation of single perithecia from the agar. For classical genetic analysis, perithecia are then rubbed gently on distilled water agar to remove any adhering conidia and mycelial fragments, and left for 20–30 minutes. A perithecium will normally rupture within this period and large numbers of long sac-like asci can be separated using a glass needle. *M. grisea* asci contain eight ascospores, although the ascus is sufficiently wide that tetrads are not ordered as in *Neurospora crassa* (Griffiths *et al.*, 1996). Tetrad analysis is commonly carried out in *M. grisea* (see Valent *et al.*, 1991; Hamer *et al.*,

1989b) as well as random ascospore analysis for larger scale analysis of segregation patterns and genetic mapping. The use of classical genetic analysis in *M. grisea* was, however, originally constrained by the fact that the vast majority of rice pathogenic forms of the fungus are infertile (Ou, 1985). In contrast, isolates of *M. grisea* that are pathogenic on other grasses, such as weeping lovegrass (*Eragrostis curvula*) and finger millet (*Eleusine indica*), tend to be of much greater fertility and cross readily in the laboratory. Genetic analysis of rice pathogenic strains of *M.* grisea was therefore carried out by finding rare, fertile, pathogenic strains of *M. grisea*, a notable example being the widely used rice pathogenic strain Guy-11 isolated from French Guiana by Notteghem and Silué (1992). An alternative strategy was carried out by introgression of fertility from grass-pathogenic forms of *M. grisea* into rice pathogens by serial back-crossing (Valent *et al.*, 1991). This approach generated some of the most useful laboratory strains of *M. grisea* and paved the way for the significant molecular genetic work carried out later (for reviews, see Valent and Chumley, 1991, 1994). The first loci affecting pathogenicity were also identified in these experiments, showing the potential for genetic analysis of virulence. Genetic analysis has facilitated development of genetic restriction fragment length polymorphism (RFLP) maps for *M. grisea*. These have been utilized for map-based cloning of avirulence genes (Valent *et al.*, 1991; Laugé and De Wit, 1998) and have formed the physical framework for genomic characterization of the fungus.

Genetic transformation of *M. grisea* was originally carried out by complementation of an *argB* auxotrophic mutant (Parsons *et al.*, 1987). Development of transformation vectors for *M. grisea* was facilitated by promoter studies in *Aspergillus nidulans* (Hamer and Timberlake, 1987) and application of isolated *A. nidulans* and *N. crassa* promoters to control expression of antibiotic resistance genes, such as those bestowing resistance to hygromycin B (Leung *et al.*, 1989), phleomycin, sulfonylurea and bialophos (Carroll *et al.*, 1994). Transformation frequencies were originally quite low, in the order of 20–50 transformants per microgram of DNA used. Protocols for electroporation of protoplasts to improve transformation frequencies are now described, however (Wu *et al.*, 1997), and transformation has improved sufficiently such that it is now a widely used technique allowing gene disruption and gene expression studies to be readily carried out.

The development of transformation had two main impacts on *M. grisea* research. The first was the development of gene replacement techniques for functional characterization of genes (see Hamer and Talbot, 1998). The second was the application of restriction enzyme-mediated insertional mutagenesis (REMI) for identification of new pathogenicity genes (Shi *et al.*, 1995; Sweigard *et al.*, 1998; Balhadère *et al.*, 1999).

The first one-step gene replacement in *M. grisea* was the mutation of *CUT1*, encoding cutinase (Sweigard *et al.*, 1992). This showed that if long-flanking DNA sequences were added to a selectable marker gene cassettes, then detectable levels of homologous recombination could be observed, allowing selection of null mutants at a given locus in approximately 10–20% of transformants analysed. Gene replacement was soon carried out during analysis of further genes for investigation of fungal pathogenicity (Talbot *et al.*, 1993; Mitchell and Dean, 1995; Xu and Hamer, 1996; Choi and Dean, 1997; Liu and Dean, 1997; Xu *et al.*, 1997; Sweigard *et al.*, 1998; deZwaan *et al.*, 1999; Dixon *et al.*, 1999). Vectors used in these studies have used flanking sequence varying in size from 1 kb to approximately 5 kb and the relative advantages and frequencies of homologous recombination obtained are summarized in Table I.

III. LIFE HISTORY OF *M. GRISEA* AND EPIDEMIOLOGY OF RICE BLAST

The infection process of *M. grisea* starts when a conidium is carried to the surface of a rice leaf, normally in a dew drop. The spore immediately releases an adhesive from its apex (Hamer *et al.*, 1988), which tethers it to the waxy leaf cuticle. The spore germinates within 1 hour, forming a short germ tube. The germ tube appears to be sensitive to the leaf surface and is involved in perception of the topography and rigidity of the substrate, determining whether it is conducive for appressorium development (Bourett and Howard, 1990). A number of modifications of the germ tube occur that may be associated with leaf surface recognition, including the positioning of apical vesicles close to the substrate and flattening of the germ tube when in contact with the leaf surface (Bourett and Howard, 1990). After 4–5 hours, an appressorium differentiates from the end of the germ tube. This process starts with formation of a swollen hook that tightly adheres to the surface. Within the developing germ tube a mitotic cell division occurs and one nucleus migrates into the incipient appressorium. This swollen hyphal tip differentiates and after only 40–50 minutes, a septum forms separating the appressorium from the germ tube and conidium. These structures eventually collapse and the cytoplasm from which the subsequent infection process is initiated, resides solely in the appressorium itself (Howard, 1997).

Surface contact and free water are required for successful conidial germination and appressorium development (Howard, 1994; Xiao *et al.*, 1994a,b). Rice leaves are coated with a waxy cuticle (Uchiyama and Okuyama, 1990) that makes them extremely hydrophobic (Talbot, 1995). Surface hydrophobicity has been shown to induce appressorium formation (Hamer *et al.*, 1988; Lee and Dean, 1994; Jelitto *et al.*, 1994) on surfaces such as Teflon or wax-coated plastics (Hamer *et al.*, 1988; Lee and Dean, 1994). A cAMP-dependent signal transduction pathway appears to be involved in this process because exogenous application of cAMP is sufficient to induce appressorium formation on non-inductive (hydrophilic) surfaces such as glass (Lee and Dean, 1993). It is, however, becoming increasingly apparent that a number of other signals stimulate appressorium development in addition to surface hydrophobicity. Exposure to cutin monomers, for example, such as *cis*-9,10-epoxy-18-hydroxyoctadecanoic acid and *cis*-9-octadecen-1-ol can induce appressorium formation on normally non-inductive surfaces, even when added in nanomolar concentrations (Gilbert *et al.*, 1996). Plant lipid and other wax compounds have also been tested and 1,16-hexadecanediol has been shown to be a potent inducer of appressorium development (Uchiyama *et al.*, 1979; Gilbert *et al.*, 1996; deZwaan *et al.* 1999). A number of other factors also affect appressorium development, such as starvation stress (Talbot *et al.*, 1993; Howard, 1994; Jelitto *et al.*, 1994) light (Jellito *et al.*, 1994) and the presence of yeast α-factor pheromone (Beckerman *et al.*, 1997). Multiple signals therefore act to induce development of appressoria, although most experiments to date have been carried out using artificial inert surfaces. Perhaps the best evidence that multiple input signals influence appressorium development, comes from the consistent observation that higher numbers of appressoria are always produced by *M. grisea* strains when conidia are incubated on rice leaves rather than artificial substrates (Hamer *et al.*, 1988, 1989a,b; Lee and Dean, 1993; Jelitto *et al.*, 1994; Talbot *et al.*, 1996). This suggests that there are factors influencing development which are only partially simulated in *ex planta* assays. Furthermore, it accounts for the high level of variability in appressorium development assays frequently observed in many laboratories.

The structure of the *M. grisea* appressorium has been studied in detail by the pioneering ultrastructural analyses of Howard and co-workers. The outer cell wall has a high chitin content and during maturation of the appressorium its affinity to lectins changes, either due to the addition of new wall material or from the matrix becoming so dense that lectins no longer have access to corresponding haptens (Howard, 1994, 1997; Howard and Valent, 1996). A major constituent of the appressorium wall is melanin, which is synthesized via the polymerization of 1,8-dihydroxynaphthalene. Melanin biosynthesis has long been recognized as a target for antifungal agents (Sisler, 1986; Butler and Day, 1998) and tricyclazole (one of the most effective blast fungicides) inhibits polyhydroxy-naphthalene reductase, preventing infection by the fungus (Woloshuk and Sisler, 1982; Viviani *et* al., 1993). It therefore appears that melanin has an essential role in appressorium-mediated infection, although until recently its function was not fully characterized. Three colour mutants were identified that are deficient in melanin biosynthesis and have helped to define the role of melanin in the appressorium (Howard and Ferrari, 1989; Chumley and Valent, 1990). These mutants, named *alb1*, *rsy1* and *buf1* due to their albino, rosy and buff appearance were shown to produce non-functional appressoria, which failed to penetrate rice leaf cuticles and were consequently non-pathogenic

TABLE I

Selection of genes deleted or disrupted in Magnaporthe grisea using one-step gene replacement strategies, and frequencies of homologous recombination observed

Gene	Size of flanking regions (kb)	Linear or circular construct transformed	Transformant screened?	Number of deletants obtained	Reference
CUT1	4.4	Linear (argB)[a]	9	1	Sweigard et al. (1992b)
		Circular (argB)	42	2	
	11.0	Linear (hyg[R])	9	1	
		Circular (hyg[R])[b]	48	1	
MPG1	7 (3.5 + 3.5)	Linear (hyg[R])	60	2	Talbot et al. (1993)
CPKA	2.95 (1.25 3' + 1.7 5')	?	62	2	Mitchell and Dean. (1995)
PMK1	7.2 (2.7 3' + 4.2 5')	Linear (hyg[R])	104	11	Xu and Hamer (1996)
NUT1	3.3 (2.0 + 1.3)	Linear (hyg[R])	300	26	Froeliger and Carpenter (1996)
MAGA	About 2.5	Linear (hyg[R])	65	6	Liu and Dean (1997)
MAGB	Fragment internal to gene used	Circular (hyg[R])	92	4	Liu and Dean (1997)

MAGC	About 5	Linear (hyg^R)	54	6	Liu and Dean (1997)
XYL1	~9 (~6 + ~3)	Linear (hyg^R)	100 (PCR screen)	1	Wu et al. (1997)
XYL2	~2.5 (~0.5 + ~2.0)	Linear ($benomyl^R$)[c]	200 (PCR screen)	1	Wu et al. (1997)
MAC1	6.1 (3.7 + 2.4)	Not reported (hyg^R)	229	3	Choi and Dean (1997)
MAC1	About 5 kb	Linear (hyg^R)	160	26	Adachi and Hamer (1998)
MPS1	3.8 (1.4 3' + 2.4 5')	Not reported	Not reported	2	Xu et al. (1998)
PTH1	5.5 (1.5 3' + 4 5')	Not reported	Not reported	Not reported	Sweigard et al. (1998)
PTH2	~6 (~3 + ~3)	Not reported	Not reported	Not reported	Sweigard et al. (1998)
PTH3	~6 (~3 + ~3)	Not reported	Not reported	Not reported	Sweigard et al. (1998)
OSM1	2.8 (1.0 + 1.8)	Linear	68	6	Dixon et al. (1999)
PTH11	>10	Not reported	Not reported	Not reported	DeZwann et al. (1999)

[a] Selectable marker complemented arginine auxotrophy.
[b] Selectable marker bestowing resistance to hygromycin B.
[c] Selectable marker bestowing resistance to benomyl.

(Howard and Ferarri, 1989). *ALB1* shows homology to genes encoding polyketide synthases (Howard and Valent, 1996), while *RSY1* encodes scytalone dehydratase (Andersson *et al.*, 1996; Motoyama *et al.*, 1998) and *BUF1* encodes NADPH-dependent polyhydroxy-naphthalene reductase (Vidal-Cros *et al.*, 1994; Motoyama *et al.*, 1998).

The first clue to the likely role of melanin in appressorium function came from experiments, which showed that appressoria generate enormous turgor during the infection process. Appressoria were incubated in hyperosmotic solutions of polyethylene glycol to a point where cytorrhysis (cell collapse) occurred. An estimate of intracellular hydrostatic pressure could therefore be made and suggested that appressoria were capable of generating up to 8 MPa of turgor (Howard *et al.*, 1991). Artificially lowering the turgor, by suspending appressoria in polyethylene glycol solutions of varying molarities, prevented penetration, indicating that appressorium turgor is required for infection (Howard *et al.*, 1991).

The mechanism by which turgor is generated in *M. grisea* appressoria has since been investigated by biochemical analysis of the contents of conidia during germination and subsequent appressorium differentiation (de Jong *et al.*, 1997). Glycerol accumulates rapidly during conidial germination and decreases in abundance during germ tube elongation, before accumulating once more to extremely high concentrations during appressorium maturation. The concentration of glycerol in mature appressoria has been estimated, based on an enzymatic assay, and found to be as high as 3.2 M. This is theoretically sufficient to generate turgor of 8.7 MPa, assuming that glycerol acts as an ideal solute. However, as glycerol is unlikely to act in this way at such high concentrations, turgor was experimentally determined using vapour pressure psychrometry and predicted to be 5.8 MPa. Estimates of glycerol concentration have been validated using the incipient cytorrhysis assay of Howard and colleagues (1991), but using glycerol as the external solute to collapse appressoria. The frequency of cell collapse was found to be dependent on external glycerol concentrations. A concentration of 1.75 M glycerol (−3.7 MPa) caused collapse of 52% of appressoria, supporting the link between accumulation of molar concentrations of glycerol and turgor generation. These experiments also indicated that the appressorium wall must be relatively impermeable to glycerol. Appressorium walls show very reduced porosity compared to those in which the melanin wall layer is absent (Howard *et al.*, 1991) and this prompted a series of experiments to determine the role of melanin pigmentation in glycerol-dependent turgor generation. Cell collapse assays were carried out using the wild type and a melanin-deficient mutant, *alb1*⁻. Appressoria of both strains collapsed in hyperosmotic concentrations of glycerol, but the non-melanized mutant appressoria recovered from this treatment immediately (within a minute) whereas the wild-type strain showed only limited recovery after 48 hours incubation. The melanin-pigmented appressorium wall is thus essential for maintenance of glycerol during the infection process and this is probably the principal reason that melanin-deficient mutants are non-pathogenic. The result also indicates why the antipenetrant fungicide tricyclazole – a known inhibitor of melanin biosynthesis – is such an effective fungicide for protection against rice blast; appressoria are unable to generate sufficient turgor for plant infection in its presence due to lack of glycerol retention (Woloshuk *et al.*, 1982; Viviani *et al.*, 1993; de Jong *et al.*, 1997).

Appressorium maturation and turgor generation lead to production of a penetration peg. This is a thin tip-growing protruberance produced at the base of the appressorium. Penetration pegs exert mechanical force as a result of turgor generation and the action of cytoskeletal components. The force generated by penetration pegs has been measured in a related fungus *Colletotrichum graminicola*, the causal agent of anthracnose disease of maize. *C. graminicola* also produces melanized appressoria and is susceptible to loss of melanization and tricyclazole treatment. Using a novel optical waveguide technique, Bechinger and co-workers (1999) showed that penetration pegs can produce a force of 16.8 μN, a very considerable force that is certainly capable of breaching a plant cuticle (Talbot, 1999b). The rice leaf cuticle is approximately 0.77 μm in thickness and is composed primarily of long-chain aliphatic waxes, which can be seen as thin platelets across the leaf surface (see Fig. 1), and insoluble cutin (Post-Beittenmiller, 1996). Below this, the plant cell wall is composed of cellulose microfibrils embedded in a protein and polysaccharide matrix (Carpita and Gibeaut, 1993). Mechanical force appears to be the primary means by which the penetration peg breaches the cuticle and wall layers. Cutinase is dispensable for pathogenicity (Sweigard *et al.*, 1992) and appressoria will puncture inert plastic surfaces (Howard *et al.*, 1991). It is possible, however, that as the penetration peg is forced through the surface leaf layers, its passage is aided by secretion of enzymes that soften the wall layers enhancing the rate of infection.

As the penetration peg forms, its cytoplasm is initially devoid of organelles and appears to be filled at the tip by a concentration of actin (Bourett and Howard, 1992). During penetration peg elongation, microtubules can be observed within the hypha aligned parallel to its long axis (Howard, 1997). The penetration peg widens into a primary infection hypha within the first epidermal cell encountered and the fungus then takes on a beaded, bulbous appearance as it grows throughout the first and adjacent epidermal cells. Subsequently, long runner hyphae spread out from the initial infection site and a large amount of the leaf becomes colonized. Large spreading lesions are visible some 3–4 days after inoculation and coalesce to form senescent patches on the leaf surface. In heavy blast infections the leaves become desiccated and senescent and in young seedlings can result in death of the plant. In older rice plants, *M. grisea* can spread to the neck and panicle causing blast symptoms that can rot away the grain-bearing structures resulting in complete harvest loss (Ou, 1985).

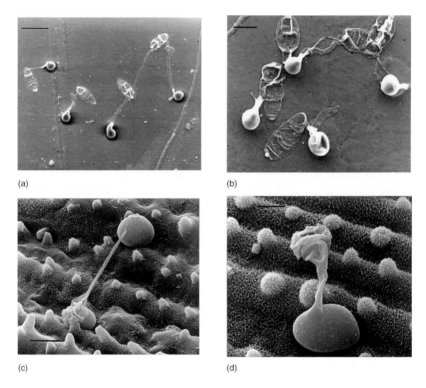

(a)   (b)

(c)   (d)

Fig. 1. Appressorium formation by the rice blast fungus *Magnaporthe grisea*. The fungus elaborates a specialized infection cell with which it penetrates the rice leaf cuticle using predominantly mechanical force. (a) Appressoria forming on a thin cellophane membrane. Apressoria are dome-shaped cells differentiated from the ends of fungal germ tubes. *M. grisea* appressoria can cause indentation of this thin surface. Scale bar = 25 μm. (b) Appressorium formation on PTFE–Teflon. Appressoria develop turgor pressure while the conidium and germ tube collapse following appressorium maturation. Scale bar = 10 μm. (c, d) Appressorium formation on rice leaves. Appressoria form within the waxy cuticle, decorated with small wax plates. Appressoria often form close to the prominent epidermal cell wall protruberances that occur on rice leaves. Scale bars = (c) 10 μm, (d) 6 μm. Understanding the molecular basis of appressorium formation and function are primary aims of contemporary rice blast research.

IV. IDENTIFICATION OF GENES INVOLVED IN APPRESSORIUM FORMATION

A number of approaches have been used for identification of genes involved in the pathogenicity of *M. grisea,* including identifying pathogenicity mutants using conventional or insertional mutagenesis (Hamer *et al.*, 1989b; Sweigard *et al.*, 1998; Balhadère *et al.*, 1999), differential cDNA screens (Talbot *et al.*, 1993; McCafferty and Talbot, 1998), or identification of conserved genes based on homologous sequences from other organisms (Mitchell and Dean, 1995; Choi and Dean, 1997; Liu and Dean, 1997; Xu *et al.*, 1997; Adachi and Hamer, 1998; Dixon *et al.*, 1999). Many of the genes identified to date (Table II) have been found to have functions in appressorium development and this has led to rapid progress in defining important protein components that regulate or play a structural role in their formation.

The hydrophobin-encoding gene *MPG1* was identified in a differential cDNA screen aimed at identifying genes expressed during rice plant infection. An abundantly expressed gene, *MPG1*, was deleted using a one-step gene replacement (as described earlier) and mutants were found to be affected in their ability to elaborate appressoria (Talbot *et al.*, 1993). *MPG1* was found to encode a fungal hydrophobin, a class of protein implicated in diverse fungal developmental processes (Kershaw and Talbot, 1998; Wessels, 1997). Hydrophobins are widespread in fungi and are expressed during conidiation and fruitbody formation. They are small secreted hydrophobic proteins and are involved in the ability of fungal hyphae to grow into the air (Wösten *et al.*, 1999; Talbot, 1999a) or adhere to hydrophobic surfaces (Wösten *et al.*, 1994). Hydrophobins have been found to undergo spontaneous polymerization when exposed to interfaces between air and water, or hydrophobic surfaces (Wessels, 1997). As a consequence of this biochemical characteristic, abundant secretion of the MPG1 hydrophobin during appressorium formation appears to act as a primer for appressorium formation, the process being extremely inefficient in the absence of the protein (Talbot *et al.*, 1996). There appears to be little specifically that is important about the MPG1 hydrophobin except its ability to aggregate in response to secretion onto a hydrophobic surface because quite diverse class I hydrophobin genes can substitute for *MPG1* if expressed under the *MPG1* promoter (Kershaw *et al.*, 1998).

Identification of *MAC1* encoding adenylate cyclase confirmed the role of cyclic AMP in appressorium development because a targeted deletion results in Δ*mac1* mutants that are non-pathogenic and do not form appressoria (Choi and Dean, 1997). The mutation is, however, unstable in certain strain backgrounds and can be overcome by an extragenic suppressor mutation in the *SUM1* gene, which encodes the regulatory subunit of protein kinase A (Adachi and Hamer, 1998). This is consistent with protein kinase A activity being required for appressorium elaboration. However, a mutation in the CPKA gene, which encodes the catalytic subunit of PKA, does not affect appressorium formation, although the process is delayed in mutants (Mitchell and Dean, 1995; Xu *et al.*, 1997). Instead Δ*cpkA* mutants produce non-functional appressoria and are completely non-pathogenic (see below). Taken together these results suggest that divergent cAMP signalling pathways exist for appressorium formation and function (Adachi and Hamer, 1998; Hamer and Talbot, 1998). The upstream signalling components of this pathway are likely to include include heterotrimeric G-proteins, which must

TABLE II
Genes known to be involved in pathogenicity of M. grisea

Gene	Product	Pathogenicity	Phenotypes	Reference
MPG1	Hydrophobin	Reduced	Reduced conidiation. Easily wetted	Talbot *et al.* (1993, 1996)
NPR1	Unknown	—	Nitrogen metabolism regulator	Lau and Hamer (1996)
NPR2	Unknown	—	As above	As above
CPKA	Protein kinase A	—	Penetration defective	Mitchell and Dean (1995)
MAGB	G-protein α-subunit	—	No appressorium development	Liu and Dean (1997)
MAC1	Adenylate cyclase	—	No appressorium development	Choi and Dean (1997)
PMK1	MAPK	—	No appressorium development	Xu and Hamer (1996)
MPS1	MAPK	—	Penetration defective	Xu *et al.* (1998)
ABC1	ATP-efflux pump	Reduced		Urban *et al.* (1999)
APF1	Unknown	—		Silué *et al.* (1998)
APP1	Unknown	—		Zhu *et al.* (1996)
CON1-7	Unknown	—	Conidial morphology mutants	Shi and Leung (1995)
SMO1	Unknown	Reduced	Conidial morphology	Hamer *et al.* (1989)
PDE1	Unknown	—	Penetration defective	Balhadère *et al.* (1999)
PDE2	Unknown	—	as above	As above
IGD1	Unknown	Reduced	Invasive defect	As above
MET1	Unknown	Reduced	Methionine auxotroph	As above
GDE1	Unknown	Reduced on rice		As above
PTH1	*GRR1* homologue	Reduced		Sweigard *et al.* (1998)
PTH2	Carnitine acetyl transferase	—		As above
PTH3	Imidazole glycerol-P dehydratase	Reduced	Histidine auxotroph	As above
PTH8	Unknown yeast homologue			As above
PTH9	Trehalase	Reduced		as above
PTH10	Unknown			As above
ACR	*medA* homologue	Reduced	Conidial morphology	Lau and Hamer (1998)

act in concert with a membrane-bound surface receptor. Three G-protein α-subunit encoding genes have been isolated from *M. grisea* and one of these, *MAGB*, has been shown to be required for appressorium formation: Δ*magB* mutants are non-pathogenic and blocked in appressorium elaboration (Liu and Dean, 1997).

A gene encoding a putative membrane-bound receptor for appressorium development has recently been described (deZwaan *et al.*, 1999). The *PTH11* gene encodes a novel transmembrane protein that has been localized to vacuoles and plasma membranes using a green fluorescent protein tag. Targeted mutation of *PTH11* results in strains that are reduced in their ability to differentiate appressoria. However, the role of *PTH11* appears to be different in diverse strain backgrounds of *M. grisea*. Although *PTH11* wild-type alleles complemented the original *pth11* mutation in strain 4091-5-8, the role of these active alleles in their original hosts was quite different. In certain strain backgrounds, such as 4091-5-8, *PTH11* was required for appressorium differentiation and pathogenicity, while in other strains, such as Guy-11, *PTH11* was a pathogenicity factor but could also act as a repressor of appressorium formation on non-inductive (hydrophillic) surfaces (deZwaan *et al.*, 1999). *PTH11* appears to be involved in surface recognition, although if the gene encodes a membrane-bound receptor – as seems likely from the predicted amino-acid sequence – then it is able to respond to both cutin monomers and surface hydrophobicity, because both stimuli fail to induce appressorium formation in *pth11* mutants. In addition, *PTH11* appears likely to act upstream of diverse signalling pathways because diacylglycerol isable to remediate appressorium formation in *pth11* mutants, but not pathogenicity, while cAMP can remediate both appressorium differentiation and pathogenicity (deZwaan *et al.*, 1999).

Appressorium formation clearly involves MAP kinase signalling, which may occur as part of the general cAMP response pathway described previously. The *PMK1* gene (Xu and Hamer, 1996) is a functional homologue of the *S. cerevisiae FUS3/KSS1* genes involved in the pheromone response pathway (for review, see Herskowitz, 1995). A targeted replacement of *PMK1* causes loss of the ability to make appressoria. Significantly, the ability of Δ*pmk1* mutants subsequently to colonize plant tissue – even when conidia are injected directly into rice leaves – is also affected by the mutation. Consistent with this involvement of elements of the pheromone response pathway in *M. grisea* appressorium formation, yeast α-factor has been shown to inhibit *M. grisea* appressorium development in a mating-type specific manner (Beckerman *et al.*, 1997).

Another group of genes that have been found to be required for completion of the pre-penetration phase of development of *M. grisea* are those that mediate development of conidia. Mutations in these genes tend to have pleiotropic effects on appressorium development and thereby prevent disease. The involvement of these genes – which include *CON1–CON7*,

SMO1 and *ACR1* – in both morphogenetic processes highlights the underlying similarity between conidiogenesis and appressorium formation (Hamer *et al.*, 1989b; Shi and Leung, 1995). The *ACR1* gene plays a role in conidiogenesis and the normal sympodial patterning of conidial production. In *acr1* mutants conidia are formed in rows arranged tip to base, or acropetally (Lau and Hamer, 1998). Recent evidence suggests that *ACR1* may be related to the *MedA* gene from *Aspergillus nidulans*, which encodes a transcription factor that plays a role in spatial regulation of conidiogenesis (J. Hamer, personal communication; Adams *et al.*, 1998).

Collectively, the genes so far identified in *M. grisea* that play roles in appressorium differentiation have revealed a number of important characteristics of the developmental process. The most apparent is the linkage between appressorium differentiation and conidiogenesis with the involvement in both processes of a number of structural components, such as hydrophobins and melanin, and regulators such *ACR1*. Another obvious conclusion is that divergent signalling pathways regulate appressorium function in response to more than one signal. These signals include perception of cutin monomers, starvation stress and surface hydrophobicity, and their recognition and stimulation of appressorium differentiation involves elements of a cyclic AMP pathway, a MAP kinase signalling pathway, and a protein kinase C–diacylglycerol pathway (Thines *et al.*, 1997a,b; Hamer and Talbot, 1998). It is also possible from pharmacological data that calcium and calmodulin signalling take place during this process (Lee and Lee, 1998). Finally, the distinct behaviour of different *M. grisea* strains suggests considerable plasticity in the action of these signalling pathways and the likely complexity of appressorium differentiation.

V. IDENTIFICATION OF GENES INVOLVED IN APPRESSORIUM-MEDIATED PLANT INFECTION

Genes involved in appressorium function have been found in mutant screens and by analysis of genes encoding conserved signalling components. A second MAP kinase-encoding gene *MPS1* has been described, which is homologous to the *SLT2/MPK1* MAP kinase from *S. cerevisiae* (Gustin *et al.*, 1998). *MPS1* is required for appressorium function (Xu *et al.*, 1998) and Δ*mps1* mutants are unable to penetrate plant epidermis, although their ability to generate turgor pressure appears to be unaffected (Xu *et al.*, 1998). Interestingly plant defence responses appeared to be unaffected after inoculation of plants by Δ*mps1* mutants, suggesting that perception of *M. grisea* occurs during growth of the fungus on the rice leaf surface. The third MAP kinase, *OSM1* is a functional homologue of the *S. cerevisiae HOG1* MAPK gene, which is required for osmoregulation by *M. grisea* but does not play a role in appressorium turgor generation or subsequent pathogenicity

(Dixon *et al.,* 1999). Thus appressorium-mediated penetration in *M. grisea,* although requiring a large increase in intracellular turgor, appears to be independent of the conserved MAP kinase pathway known to regulate this process in most eukaryotic cells. Remarkably, the synthesis of large amounts of glycerol is therefore regulated in a novel manner in the appressorium.

VI. IDENTIFICATION OF GENES REQUIRED FOR EFFECTIVE COLONIZATION OF PLANT TISSUES

The adaptation of *M. grisea* to growth within plant tissue has not been extensively studied, but a number of mutant screens have identified processes that may be required for effective colonization of plant tissue. Certain basic processes in intermediary metabolism, for example, appear to be required for effective disease progression. A histidine auxotrophic mutant, *pth3,* is reduced in pathogenicity. The *PTH3* gene encodes imidazole glycerol phosphate dehydratase, indicating a requirement for synthesis of this histidine independent of plant-derived nutrition (Sweigard *et al.*, 1998). Similarly, a methionine auxotroph, *met1,* was discovered in another REMI mutant hunt and showed significant reductions in disease development (Balhadère *et al.*, 1999).

Mobilization of the disaccharide trehalose appears to be significant for disease progression and plant tissue colonization. A neutral trehalase-encoding gene *PTH9/NTH1* has been shown to be required for full pathogenicity of *M. grisea*. A Δ*nth1* mutant produces significantly fewer lesions in seedling infections. Trehalose accumulates under conditions of environmental stress including osmotic stress (Dixon *et al.*, 1999) and during conidiogenesis (A. J. Foster and N. J. Talbot, unpublished). Neutral trehalase activity occurs during conidial germination, but the mutant phenotype suggests that trehalose mobilization subsequent to initial infection may be of greater importance for disease progression. The fact that carnitine acetyl transferase, involved in fatty acid oxidation, is also required for pathogenicity of *M. grisea* (Sweigard *et al.*, 1998) indicates that metabolism *in planta* may be quite distinct from when *M. grisea* is grown in axenic culture.

Another way in which *M. grisea* has to adapt to growth within plant tissues is by protecting itself from plant defence mechanisms. Consistent with this an ATP-efflux pump has recently been discovered in *M. grisea* that is important for pathogenicity. The *ABC1* gene was discovered by screening an insertional mutant collection (Urban *et al.*, 1999). The *abc1* mutant was found to have a plasmid insertion in the promoter of a ATP-efflux pump-encoding gene. A targeted gene replacement of *ABC1* confirmed its requirement for pathogenicity and it was found to be transcriptionally regulated in response to a variety of toxic compounds. This result indicates that one way that *M. grisea* reacts to plant defence mechanisms is to activate pumps to rid the growing mycelium of plant antimicrobial compounds, such as phytoalexins that may be produced in response to infection (Urban et al., 1999).

VII.   GENE-FOR-GENE RESISTANCE TO M. GRISEA

Genetic studies of M. grisea have shown that a gene-for-gene interaction exists between the fungus and rice (for review, see Laugé and De Wit, 1998). Cultivated rice cultivars have major dominant genes for blast resistance, which have been introgressed from land races of rice and geographically distinct germplasm collections. This varietal resistance has formed the main method by which rice blast is controlled, although the field life of rice cultivars has often been curtailed by disease outbreaks caused by newly virulent forms of the fungus (Ou, 1985). In gene-for-gene interactions, the products of plant resistance genes recognize fungal proteins (directly or indirectly) that are encoded by avirulence genes. Fungal avirulence genes are genetically dominant and the evolution of newly virulent forms, therefore, occurs by mutation, or loss of avirulence gene products. The retention of avirulence genes in fungal species suggests that they have additional roles in pathogenesis or fitness, but also encode proteins that plants have the ability to perceive when they carry an appropriate resistance gene (De Wit, 1992). Fungal avirulence genes have been isolated from only a small number of fungi so far including M. grisea, the tomato pathogens, Cladosporium fulvum and Phytophthora infestans, and the barley scald fungus, Rhynchosporium secalis (Van der Ackerveken et al., 1992; Joosten et al., 1994; Rohe et al., 1995; Kamoun et al., 1998). In M. grisea, a large number of avirulence genes have been predicted based on genetic studies and are in the process of being cloned and characterized (Silué et al., 1992; Dioh et al., 1996; Farman and Leong, 1998).

The first avirulence gene characterized from M. grisea was the Avr2-YAMO gene, which prevents infection of the rice cultivar Yashiro-Mochi. Avr2-YAMO was isolated by positional cloning after being found to map to the end of chromosome 1 of M. grisea. The gene is located in the subtelomeric region and required construction of a telomere-containing gene library, to be cloned. Perhaps due to its position, Avr2-YAMO is subject to frequent mutation, including deletion and insertions at the locus. Avr2-YAMO encodes a protein showing homology to a neutral Zn^{2+} protease (Valent, 1997) and the active site of this protease appears to be required for it to function as an avirulence gene. The putative enzymatic action of the Avr2-YAMO product is unusual compared with the avirulence genes identified in other pathogens. For example, avr9 and avr4 from C. fulvum are both cysteine-rich secreted proteins found in apoplastic fluids during infection of tomato. They may therefore be perceived extracellularly and the structure of the corresponding Cf4 and Cf9 resistance gene products is consistent with this (Jones and Jones, 1996; Laugé and De Wit, 1998).

Avirulence genes have also been isolated from *M. grisea* which control species specificity. Although *M. grisea* can infect more than 50 species of grass, individual isolates of the fungus are normally restricted to a single or small number of species. The *PWL* gene family controls the ability of *M. grisea* to infect weeping lovegrass (*Eragrostis curvula*). *PWL1* prevents infection of weeping lovegrass and originates from isolates of *M. grisea* that are virulent on finger millet (*Eleusine coracana*). *PWL2* meanwhile was found in rice pathogenic strains of the fungus. When *PWL1* or *PWL2* are transformed into strains of *M. grisea* that normally infect weeping lovegrass, the resulting transformants are avirulent (Kang *et al.*, 1995; Sweigard *et al.*, 1995). This indicates that species-specificity works in a similar fashion to cultivar-specificity in *M. grisea* – by recognition of the products of dominant avirulence genes. *PWL2* putatively encodes a 145-amino-acid hydrophilic protein with no significant homologies to known proteins. *PWL1* was isolated by homology to *PWL2* and shares 75% amino acid identity. A number of other members of the PWL gene family have since been isolated based on homology. *PWL3* is from a fingermillet pathogen and shows 51% identity with *PWL2*. *PWL4* originates from a weeping lovegrass pathogen and shows 72% identity with *PWL3* and 57% identity with *PWL2*. Interestingly, *PWL4* appears to be non-functional as an avirulence gene (it was found in a weeping lovegrass pathogen) due to a promoter mutation. Expression of the *PWL4* open reading frame under control of the *PWL1* or *PWL2* promoter, however, leads to transformants that are avirulent on weeping lovegrass (Kang *et al.*, 1995).

Recognition of *M. grisea* by resistant hosts therefore results from perception of fungal proteins encoded by avirulence genes, although whether these fungal proteins are pathogenicity factors, predominantly secreted proteins, or contribute to infection-related development is not yet clear (Dioh *et al.*, 1996; Farman and Leong, 1998). The site and timing of fungal perception by host plants is also an unresolved question and will not only require identification of rice blast resistance genes (Ronald, 1998), but also integration of the study of avirulence and resistance as initiated in the *C. fulvum*–tomato interaction (Laugé and De Wit, 1998).

VIII.   GENOMIC APPROACHES TO THE STUDY OF *M. GRISEA*

The completion of the first eukaryotic genome sequencing project in 1995 heralded a new era for genetic investigations. The complete budding yeast genome has provided an invaluable resource for characterizing potentially orthologous genes in fungi such as *M. grisea*, but has also highlighted the urgent need to obtain the full genomic sequence of an actual pathogen. Genome level analysis in *M. grisea* has therefore advanced in the last 4 years to a point where large-scale genomic sequencing is beginning to take place.

The *M. grisea* genome is approximately 40 Mb in size (~2.5 × the yeast genome) and the seven chromosomes can be resolved using pulse-field gel electrophoresis (Hamer *et al.*, 1989a; Talbot *et al.*, 1993). Several laboratories have collaborated to produce a linkage map that contains in excess of 200 molecular markers. In addition to single-site RFLPs, the map contains the positions of repetitive DNA families including long interspersed nuclear elements (LINEs), short interspersed nuclear elements (SINEs) and retrotransposon elements (Romao and Hamer, 1992; Nitta *et al.*, 1997). Current research is aimed at using the physical map of *M. grisea* to anchor a minimum set of overlapping genomic clones comprising the whole *M. grisea* genome. This will provide the framework for sequence analysis. Bacterial artificial chromosome (BAC) libraries have been constructed from rice pathogenic strains of *M. grisea* in a number of laboratories (Zhu *et al.*, 1997). One of the libraries, constructed by Dean and colleagues has been characterized in detail and shown to contain 9216 clones with large inserts of approximately 130 kb, giving >25-fold coverage of the genome. Recently this library has been used to produce a physical map of overlapping clones that span chromosome 7 of *M. grisea*. Five anchored contigs have been shown to give greater than 95% coverage of the chromosome and the length of chromosome 7 can be covered by a minimal tiling path of 42 BAC clones (Zhu *et al.*, 1999).

Large-scale sequence analysis of *M. grisea* has also been initiated in both private and public sector laboratories. Approximately 2700 non-redundant expressed sequence tags (ESTs) are deposited in public databases at the Clemson Genome Center (www.genome.clemson.edu) with 12, 674 BAC-end sequences also deposited. BAC-end sequences are used to link BACs using informatic clustering software. Many more EST sequences are present in private sector laboratories and are likely to be made publicly available as more public-sector genomic information become available.

Utilizing the colossal genetic resources already available and the entire gene set that is likely to be produced within the next 5 years represents a new and exciting challenge to plant pathologists. The possibility of investigating the orchestrated action of entire gene families during appressorium morphogenesis and plant infection will provide new insights into the developmental biology and pathology of the organism. Coupled with this will be the opportunity to compare genomes of distinct pathogenic and non-pathogenic fungi and evaluate the importance of genes involved in both core biological functions and specific pathogenic functions. There will, however, be a number of very significant challenges in this post-genomic era. Perhaps the most important is devising a more rapid means of producing null mutants such that mutating every predicted gene in the *M. grisea* genome becomes an achievable and realistic goal. Additionally, the development of means of transiently silencing genes will be vital in order to determine precise functions of genes involved in the plant–pathogen interaction, particularly those causing lethality when mutated. Finally, developing informatic tools to utilize genome-wide expression data in association with production of large sets of baseline expression control experiments will be critical in interpreting this form of analysis (see Tao *et al.*, 1999).

IX.   OPPORTUNITIES FOR DURABLE CONTROL MECHANISMS FOR CEREAL DISEASES

Unravelling the mechanisms of pathogenicity in *M. grisea* will afford significant benefits for controlling cereal pathogens. Defining conserved signalling pathways for pathogenesis, such as the PMK1-type MAPK signalling pathway, has already provided a novel target for disease intervention but perhaps more importantly has provided a means of identifying key components of the pathogenicity process that are regulated by this pathway or act upstream of it. Systematic functional analysis of the *M. grisea* genome and effective comparative analysis with other cereal pathogens will doubtless prove the most effective means of identifying new targets for development of fungicides. Similarly, defining the recognition events between pathogen avirulence gene products and plant resistance gene products will prove useful in developing durably resistant cereal hosts. This area is, however, not without commercial risk, as recent public concerns regarding genetically modified food in Europe have highlighted. Agricultural biotechnology companies are currently in a process of merger and redefinition of goals. It is, however, very apparent that new and effective disease control measures will be needed long into the 21st century and the commercial incentives for development of broad-spectrum fungicides and resistant cereal cultivars will be as strong as ever.

ACKNOWLEDGEMENTS

The authors are funded by the Biotechnology and Biological Sciences Research Council (BBSRC, UK), the Royal Society, The Nuffield Foundation and the European Union. We are indebted to members of the rice blast community for communicating results to us prior to publication and for stimulating discussions.

REFERENCES

Adachi, K. and Hamer, J. E. (1998). Divergent cAMP signaling pathways regulate growth and pathogenesis in the rice blast fungus *Magnaporthe grisea*. *Plant Cell* **10**, 1361–1373.

Adams, T. H., Wieser, J. K. and Yu, J. H. (1998). Asexual sporulation in *Aspergillus nidulans*. *Microbiology and Molecular Biology Reviews* **62**, 35–54.

Andersson, A., Jordan, D., Schneider, G., Valent, B. and Linqvist, Y. (1996). Crystallization and preliminary X-ray diffraction study of 1,3,8-trihydroxynaphthalene reductase from *Magnaporthe grisea*. *Proteins – Structure Function and Genetics* **24**, 525–527.

Baker, B., Zambryski, P., Staskawicz, B, and Dinesh-Kumar, S. P. (1997). Signalling in plant–microbe interactions. *Science* **276**, 726–733.

Balhadère, P. V., Foster, A. J. and Talbot, N. J. (1999). Identification of pathogenicity mutants of the rice blast fungus *Magnaporthe grisea* by insertional mutagenesis. *Molecular Plant Microbe Interactions* **12**, 129–142.

Bechinger, C., Giebel, K.-F., Schnell, M., Leiderer, P., Deisinger, H. B. and Bastmeyer, M. (1999). Optical measurement of invasive forces exerted by appressoria of a plant pathogenic fungus *Science* **285**, 1896–1899.

Beckerman, J. L., Naider, F. and Ebbole, D. J. (1997). Inhibition of pathogenicity of the rice blast fungus by *Saccharomyces cerevisiae* α-factor. *Science* **276**, 1116–1119.

Bourett, T. M. and Howard, R. J. (1990). *In vitro* development of penetration structures in the rice blast fungus *Magnaporthe grisea*. *Canadian Journal of Botany* **68**, 329–342.

Bourett T. M. and Howard R. J. (1992). Actin in penetration pegs of the fungal rice blast pathogen, *Magnaporthe grisea*. *Protoplasma* **168**, 20–26.

Butler, M. J. and Day, A. W. (1998). Fungal melanins: a review. *Canadian Journal of Microbiology* **44**, 1115–1136.

Carpita, N. C. and Gibeaut D. M. (1993). Structural models of primary cell walls in flowering plants: consistency of molecular structure with the physical properties of the walls during growth. *Plant Journal* **3**, 1–30

Carroll, A. M., Sweigard, J. A. and Valent, B. (1994). Improved vectors for selecting resistance to hygromycin. *Fungal Genetics Newsletter* **41**, 22.

Choi, W. and Dean, R. A. (1997). The adenylate cyclase gene *MAC1* of *Magnaporthe grisea* controls appressorium formation and other aspects of growth and development. *Plant Cell* **9**, 1973–1983.

Chumley, F. G. and Valent, B. (1990). Genetic analysis of melanin deficient non-pathogenic mutants of *Magnaporthe grisea*. *Molecular Plant Microbe Interactions* **3**, 135–143.

de Jong, J. C., McCormack, B. J., Smirnoff, N. and Talbot, N. J. (1997). Glycerol generates turgor in rice blast. *Nature* **389**, 244–245.

De Wit, P. J. G. M. (1992). Molecular characterization of gene-for-gene systems in plant–fungus interactions and the application of avirulence genes in control of plant pathogens *Annual Review of Phytopathology* **30**, 391–418.

Dixon, K. P., Xu, J. R., Smirnoff, N. and Talbot, N. J. (1999). Independent signaling pathways regulate cellular turgor during hyperosmotic stress and appressorium-mediated plant infection by *Magnaporthe grisea*. *Plant Cell* **11**, 2045–2058.

DeZwaan, T. M., Carroll, A. M., Valent, B. and Sweigard, J. A. (1999). *Magnaporthe grisea* Pth11p is a novel plasma membrane protein that mediates appressorium differentiation in response to inductive substrate cues. *Plant Cell* **11**, 2013–2030.

Dioh, W., Tharreau, D., Gomez, R., Roumen, E., Orbach, M., Notteghem, J. L. and Lebrun, M. H. (1996). Mapping avirulence genes in the rice blast fungus *Magnaporthe grisea*. *In* "Rice Genetics III" (G. S. Khush, ed.), pp. 916–920. IRRI, Philippines.

Farman, M. L. and Leong, S. A. (1998). Chromosome walking to the *AVR1-CO39* avirulence gene of *Magnaporthe grisea*: discrepancy between the physical and genetic maps. *Genetics* **150**, 1049–1058.

Froeliger, E. H. and Carpenter, B. E. (1996). *NUT1*, a major nitrogen regulatory gene in *Magnaporthe grisea*, is dispensable for pathogenicity. *Molecular and General Genetics* **251**, 647–656.

Gilbert, R. D., Johnson, A. M. and Dean, R. A. (1996). Chemical signals responsible for appressorium formation in the rice blast fungus. *Physiological and Molecular Plant Pathology* **48**, 335–346.

Griffiths, A. J., Miller, J. H., Suzuki, D., T., Lewontin, R. C. and Gelbart, W. M. (1996). "Introduction to Genetics Analysis". W.H. Freeman, New York.

Gustin, M. C., Albertyn, J., Alexander, M. and Davenport, K. (1998). MAP kinase pathways in the yeast *Saccharomyces cerevisiae*. *Microbiology and Molecular Biology Reviews* **62**, 1264–1300.

Hamer, J. E. and Talbot, N. J. (1998). Infection-related development in the rice blast fungus *Magnaporthe grisea*. *Current Opinions in Microbiology* **1**, 693–697.

Hamer, J. E. and Timberlake, W. E. (1987). Functional organization of the *Aspergillus nidulans trpC* promoter. *Molecular Cell Biology* **7**, 2352–2359.

Hamer, J. E., Howard, R. J., Chumley, F. G. and Valent, B. (1988). A mechanism for surface attachment of spores of a plant pathogenic fungus. *Science* **239**, 288–290.

Hamer, J. E., Farrall, L., Orbach, M. J., Valent, B. and Chumley, F. G. (1989a). Host species-specific conservation of a family of repeated DNA-sequences in the genome of a fungal plant pathogen. *Proceedings of the National Academy of Sciences (USA)* **86**, 9981–9985.

Hamer, J. E., Valent, B. and Chumley, F. G. (1989b). Mutations at the *SMO* locus affect the shapes of divers cell types in the rice blast fungus. *Genetics* **122**, 351–361.

Herskowitz, I. (1995). MAP kinases in yeast: for mating and more. *Cell* **80**, 187–197.

Hewitt, H. G. (1998). "Fungicides in Crop Protection". CAB International, Wallingford.

Howard, R. J. (1994). Cell biology of pathogenesis. *In* "The Rice Blast Disease" (R. S. Zeigler, S. A. Leong and P. S. Teng, eds), pp. 3–22. CAB International, Wallingford.

Howard, R. J. (1997). Breaching the outer barriers – cuticle and cell wall penetration. *In* "The Mycota V. Part B, Plant Relationships" (G. C. Carroll and P. Tudzynski eds), pp. 43–60. Springer-Verlag, Berlin.

Howard, R. J. and Ferrari, M. A. (1989). Role of melanin in appressorium formation. *Experimental Mycol*ogy **13**, 403–418.

Howard R.J. and Valent, B. (1996). Breaking and entering – host penetration by the fungal rice blast pathogen *Magnaporthe grisea*. *Annual Review of Microbiology* **50**, 491–512.

Howard, R. J., Ferrari, M. A., Roach, D. H. and Money, N. P. (1991). Penetration of hard substrates by a fungus employing enormous turgor pressures. *Proceedings of the National Academy of Sciences (USA)* **88**, 11281–11284.

Jelitto, T. C., Page, H. A. and Read, N. D. (1994). Role of external signals in regulating the pre-penetration phase of infection by the rice blast fungus, *Magnaporthe grisea*. *Planta* **194**, 471–477.

Jones, D. A. and Jones, J. D. G. (1996). The role of leucine-rich repeat proteins in plant defences. *Advances in Botanical Research* **24**, 91–167.

Joosten, M. H. A. J., Cozijnsen, T. J. and De Wit, P. J. G. M. (1994). Host resistance to a fungal tomato pathogen lost by a single base-pair change in an avirulence gene. *Nature* **367**, 384–386.

Kamoun, S., Van West, P., Vleeshouwers, V. G. A. A., De Groot, K. E. and Govers, F. (1998). Resistance of *Nicotiana benthamiana* to *Phytophthora infestans* is mediated by the recognition of the elicitor protein *INF1*. *Plant Cell* **10**, 1413–1426.

Kang, S., Sweigard, J. A. and Valent, B. (1995). The *PWL* host specificity gene family in the blast fungus *Magnaporthe grisea*. *Molecular Plant–Microbe Interactions* **8**, 939–948.

Kershaw, M. J. and Talbot, N. J. (1998). Hydrophobins and repellents: proteins with fundamental roles in fungal morphogenesis. *Fungal Genetics and Biology* **23**, 18–33.

Kershaw, M. J., Wakley, G. E. and Talbot, N. J. (1998). Complementation of the *Mpg1* mutant phenotype in *Magnaporthe grisea* reveals functional relationships between fungal hydrophobins. *EMBO Journal* **17**, 3838–3849.

Lau, G. W. and Hamer, J. E. (1996). Regulatory genes controlling *MPG1* expression and pathogenicity in the rice blast fungus *Magnaporthe grisea*. *Plant Cell* **8**, 771–781.

Lau, G. W. and Hamer, J. E. (1998). *Acropetal*: a genetic locus required for conidiophore architecture and pathogenicity in the rice blast fungus. *Fungal Genetics and Biology* **24**, 228–239.

Laugé, R. and De Wit, P. J. G. M. (1998). Fungal avirulence genes: structure and possible functions. *Fungal Genetics and Biology* **24**, 285–297.

Lee, S. C. and Lee, Y. H. (1998). Calcium/calmodulin-dependent signalling for appressorium formation in the plant pathogenic fungus *Magnaporthe grisea*. *Molecules and Cells* **8**, 698–704.

Lee, Y.-H. and Dean, R. A. (1993). cAMP regulates infection structure formation in the plant pathogenic fungus *Magnaporthe grisea*. *Plant Cell* **5**, 693–700.

Lee, Y.-H. and Dean, R. A. (1994). Hydrophobicity of contact surface induces appressorium formation in *Magnaporthe grisea*. *FEMS Microbiology Letters* **115**, 71–75.

Leung, H., Lehtinen, U., Karjalainen, R., Skinner, D., Tooley, P., Leong, S. and Ellingboe, A. (1989). Transformation of the rice blast fungus *Magnaporthe grisea* to hygromycin B resistance. *Current Genetics* **17**, 409–411.

Liu, S. and Dean R. A. (1997). G protein α subunit genes control growth, development, and pathogenicity of *Magnaporthe grisea*. *Molecular Plant–Microbe Interactions* **10**, 1075–1086.

McCafferty, H. R. K. and Talbot, N. J. (1998). Identification of three ubiquitin genes of the rice blast fungus *Magnaporthe grisea*, one of which is highly expressed during initial stages of plant colonisation. *Current Genetics* **33**, 352–361.

Mitchell, T. K. and Dean, R. A. (1995). The cAMP-dependent protein kinase catalytic sub-unit is required for appressorium formation and pathogenesis by the rice blast fungus *Magnaporthe grisea*. *Plant Cell* **7**, 1869–1878.

Motoyama, T., Imanishi, K. and Yamaguchi, I. (1998). cDNA cloning, expression, and mutagenesis of scytalone dehydratase needed for pathogenicity of the rice blast fungus, *Pyricularia oryzae*. *Bioscience Biotechnology and Biochemistry* **62**, 564–566.

Notteghem, J. L. and Silué, D. (1992). Distribution of the mating type alleles in *Magnaporthe grisea* populations pathogenic on rice. *Phytopathology* **82**, 421–424.

Nitta, N., Farman, M. L. and Leong, S. A. (1997). Genomic organization of *Magnaporthe grisea*: integration of genetic maps, clustering of transposable elements and identification of genome duplications. *Theoretical and Applied Genetics* **95**, 20–32.

Ou, S. H. (1985). "Rice Diseases", pp. 109–201. Commonwealth Mycological Institute, CABI, Kew.

Parsons, K. A., Chumley, F. G. and Valent, B. (1987). Genetic transformation of the fungal pathogen responsible for rice blast disease. *Proceedings of the National Academy of Sciences (USA)* **84**, 4161–4165.

Post-Beittenmiller, D. (1996). Biochemistry and molecular biology of wax production in plant. *Annual Review of Plant Physiology and Plant Molecular Biology* **47**, 405–430.

Rohe, M., Gierlach, A., Hermann, H., Hahn, M., Schmidt, B., Rosahl, S. and Knogge, W. (1995). The race-specific elicitor, *NIP1*, from the barley pathogen *Rhynchosporium secalis* determines avirulence on host plants of the Rrs1 resistance genotype. *EMBO Journal* **14**, 4168–4177.

Romao, J. and Hamer, J. E. (1992). Genetic organization of a repeated DNA sequence family in the rice blast fungus. *Proceedings of the National Academy of Sciences (USA)* **89**, 5316–5320.

Ronald, P. C. (1998). Resistance gene evolution. *Current Opinions in Plant Biology* **1**, 294–298.

Shi, Z. and Leung, H. (1995). Genetic analysis of sporulation in *Magnaporthe grisea* by chemical and insertional mutagenesis. *Molecular Plant–Microbe Interactions* **8**, 949–959.

Shi, Z., Christian, D. and Leung, H. (1995). Enhanced transformation in *Magnaporthe grisea* by chemical and insertional mutagenesis. *Phytopathology* **85**, 329–333.

Silué, D., Notteghem, J. L. and Tharreau, D. (1992). Evidence for a gene-for-gene relationship in the *Oryza sativa–Magnaporthe grisea* pathosystem. *Phytopathology* **82**, 577–580.

Silué, D., Tharreau, D., Talbot, N. J., Clergeot, P.-H., Notteghem, J.-L. and Lebrun, M.-H. (1998). Identification and characterisation of *apf1⁻* a non-pathogenic mutant of the rice blast fungus *Magnaporthe grisea* which is unable to differentiate appressoria. *Physiological and Molecular Plant Pathology* **53**, 239–251.

Sisler, H. D. (1986). Control of fungal diseases by compounds acting as anti-penetrants. *Crop Protection* **5**, 306–313.

Sweigard, J. A., Chumley, F. G. and Valent, B. (1992). Disruption of a *Magnaporthe grisea* cutinase gene. *Molecular and General Genetics* **232**, 183–190.

Sweigard, J. A., Carroll, A. M., Kang S., Farrall, L., Chumley F.G. and Valent, B. (1995). Identification, cloning, and characterization of *PWL2*, a gene for host species specificity in the rice blast fungus. *Plant Cell* **7**, 1221–1233.

Sweigard, J. A., Carroll, A. M., Farrall, L., Chumley, F. G. and Valent B. (1998). *Magnaporthe grisea* pathogenicity genes obtained through insertional mutagenesis. *Molecular Plant–Microbe Interactions* **11**, 404–412.

Talbot, N. J. (1995). Having a blast: exploring the pathogenicity of *Magnaporthe grisea*. *Trends in Microbiology* **3**, 9–16.

Talbot, N. J. (1999a). Fungal biology – coming up for air and sporulation. *Nature* **398**, 295–296.

Talbot, N. J. (1999b). Forcible entry. *Science* **285**, 1860–1861.

Talbot, N. J., Ebbole, D. J. and Hamer, J. E. (1993). Identification and characterization of *MPG1*, a gene involved in pathogenicity from the rice blast fungus *Magnaporthe grisea*. *Plant Cell* **5**, 1575–1590.

Talbot, N. J., Kershaw, M. J., Wakley, G. E., de Vries, O. M.H., Wessels, J. G. H. and Hamer, J. E. (1996). *MPG1* encodes a fungal hydrophobin involved in surface interactions during infection-related development of *Magnaporthe grisea*. *Plant Cell* **8**, 985–999.

Tao, H., Bausch, C., Richmond, C., Blattner, F. R. and Conway, T. (1999). Functional genomics: expression analysis of *Escherichia coli* growing on minimal and rich media. *Journal of Bacteriology* **181**, 6425–6440.

Thines, E., Eilbert, F., Sterner, O. and Anke, H. (1997a). Glisoprenin A, an inhibitor of the signal transduction pathway leading to appressorium formation in

Impact of Biotechnology on the Production of Improved Cereal Varieties

ROBERT G. SOLOMON and RUDI APPELS

CSIRO Plant Industry, PO Box 1600, Canberra, ACT 2601, Australia

I. INTRODUCTION

The application of DNA-based technologies to plant molecular biology has introduced a range of powerful techniques for tracing genetic changes in plants. Biotechnology is also revolutionizing the development of novel and improved cereal varieties through the introduction of genes from a much wider range of organisms than previously possible, as well as chimeric or chemically synthesized genes that do not exist in nature. The new phase of high-technology plant breeding at the DNA level is being matched by the introduction of technologies to measure the traits of specific interest to end users. The combination of these developments is improving the efficiency of the procedures for breeding cultivars, and is also providing the basis for the identification and certification of cereals by grain handling and marketing facilities and purchasers, and for defining the criteria for quality assurance.

The impact of biotechnology on the breeding of new cereal varieties, through advances in transgenic methods for modifying quality and agronomic traits, has been covered at some length by other authors in this volume and will not be discussed here. We will consider transformation technologies as a special example of methods used for inducing mutations,

Advances in Botanical Research Vol. 34
incorporating Advances in Plant Pathology
ISBN 0-12-005934-7

and will thus address some issues pertinent to them. In particular, we will consider the role of biotechnology in helping to decide advantageous targets for transformation, through pre-transformation assays of modified quality traits. In addition, the potential for biotechnological solutions to the challenges of cultivar identification and determination of seed purity/quality assurance will be reviewed.

II.   "HIGH-TECH" BREEDING

A major impact of molecular studies on cereals has been the ability to define agronomically significant regions of chromosomes through the presence of closely linked DNA sequences. Polymorphic classes of DNA sequences, such as microsatellites (Roder *et al.*, 1998), restriction fragment length polymorphisms (RFLPs; Paull *et al.*, 1998) and amplification fragment length polymorphisms (AFLPs; Becker *et al.*, 1995), are now available for distinguishing between cultivars of wheat and barley. Many of these DNA sequences are also utilized to trace the respective, linked, genes in segregating populations derived from crosses used in breeding programmes. These developments have provided the basis for the molecular/genetic mapping of wheat to identify major genes controlling important characteristics of the grain, such as flour processing and end-product qualities. In addition, they provide the opportunity to develop varieties better adapted to particular environmental conditions by identifying DNA sequences linked to genes involved in these adaptations.

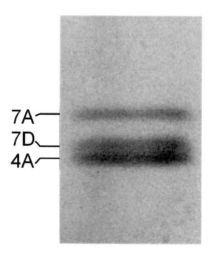

Fig. 1.   The granule bound starch synthase proteins in starch granules (cf. Zhao and Sharp, 1996). (Photograph kindly supplied by Dr P. Sharp, Plant Breeding Institute, Sydney University, Cobbitty, NSW, Australia.)

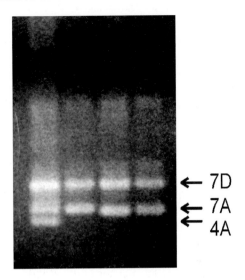

← 7D
← 7A
← 4A

Fig. 2.   PCR-based assays of the GBSS genes on chromosomes 7D, 7A and 4A, the *Wx-D1*, *Wx-A1* and *Wx-B1* loci, respectively.

The development of PCR-based technologies has allowed some genotypes to be readily assessed. As examples, the determination of allelic variation at the granule-bound starch synthase (GBSS) loci and the *Rht* (dwarfing) loci will be discussed in detail.

In some crosses, variation (presence or absence) at a single GBSS locus accounts for much of the variation in starch properties that are associated with quality traits for white salted noodle production (Zhao *et al.*, 1998). The allelic variation at the GBSS loci has been assayed by denaturing polyacrylamide gel electrophoresis of starch granule proteins (Zhao and Sharp, 1996), allowing for the identification of mutant lines that are null for GBSS at one or more of the three loci (Fig. 1).

The sequence of the genes coding for GBSSs have been determined in a number of plants, including maize (Shure *et al.*, 1983; Klösgen *et al.*, 1986), potato (Hovenkamp-Hermelink *et al.*, 1987; Visser *et al.*, 1989; van der Leij *et al.*, 1991), barley (Rohde *et al.*, 1988), rice (Wang *et al.*, 1990; Hirano and Sano, 1991; Okagaki, 1992), pea (Dry *et al.*, 1992) and common wheat (Clark *et al.*, 1991; Mason-Gamer *et al.*, 1998; Murai *et al.*, 1999; Yan *et al.*, 2000). Based on these sequence data, it is possible to design primers that span the most variable region of the gene in genomic DNA, namely exon 4 through to exon 6. Sensitive PCR-based assays of the GBSS alleles in wheat have, as a result, been formulated (Briney *et al.*, 1998), and provide the basis for the assay of the *Wx-B1* null genotype, where the GBSS gene on chromosome 4A is missing (Fig. 2). Extensive comparisons between laboratories in Australia (Southern Cross University, Lismore, NSW; University of Sydney, Cobbitty,

NSW; CSIRO Plant Industry, Canberra, ACT) have established the reliability of the PCR-based assay for assessing the genotypes of breeders' lines of wheat.

Using this ability to determine their *Wx-B1* status quickly, 170 doubled haploid lines of a cross between cultivars Cranbrook (*Wx-B1* plus) and Halberd (null *Wx-B1*) were assayed for starch swelling power (SWP) and peak starch viscosity (PV). The results showed a strong correlation between high SWP and high PV for lines null at the *Wx-B1* locus (Fig. 3).

Detailed statistical analyses of the data using MAPMANAGER (Manly, 1994), indicate that when treated as quantitative traits, much of the variation in the starch attributes assayed could be accounted for by the presence or absence of the *Wx-B1* locus on chromosome 4A. Assaying the status of the *Wx-B1* alleles is now being integrated into breeding programmes targeting the white salted noodle market and is being combined with other DNA probes to improve the efficiency of early generation screening for traits of interest.

The utility of microsatellite markers may be illustrated by the example of attempts to separate the dwarfing phenotype associated with various *Rht* loci from the long coleoptile phenotype, which is of particular importance in the Australian environment. The dwarfing phenotype is important in many crops, since it allows the plant to use more of its energy for seed set rather than vegetative growth. The interest in the long coleoptile length/ establishment phenotype of wheats has focused on the concept that the dwarfing genes that are now extremely widespread world-wide, *Rht1* and *Rht2*, not only reduce plant height but also coleoptile length. It has been

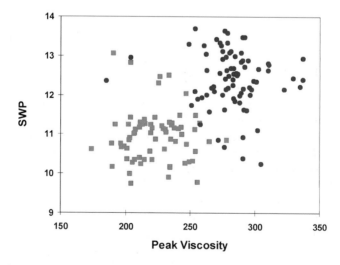

Fig. 3. Plot of starch swelling power (SWP) versus starch peak viscosity for doubled haploid lines of a Cranbrook/Halberd cross. *Wx-B1* plus lines are shown as squares, while null *Wx-B1* lines are indicated by circles.

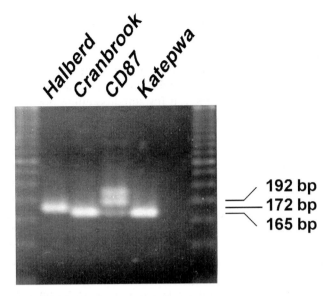

Fig. 4.   The gwm261 satellite screening for *Rht* alleles. The polymorphism in the gwm261 microsatellite DNA has been used to map the DNA sequence and the very closely linked (0.6 cM) *Rht8* gene to chromosome 2D (Korzun *et al.*, 1998). Different alleles of the microsatellite can be assigned to different alleles of *Rht8* that vary in their adaptive significance (Worland *et al.*, 1998). The photograph shows three alleles of gwm261 in Australian wheat lines. The 192bp allele of the gwm261 satellite is linked to the allele of *Rht8* that allows the long coleoptile phenotype to be expressed.

proposed that the short coleoptile lengths of *Rht1* and *Rht2* plants may be detrimental to plant establishment in dry Australian environments and this has led to the investigation of alternate, "gibberellic acid sensitive" *Rht* genes (Rebetzke and Richards, 2000) for reducing plant height. It is evident that the *Rht8* and *Rht9* genes can produce plants of reduced height without reducing coleoptile length because they do not suppress cell size and, hence, do not inhibit coleoptile cell elongation in the crucial establishment phase. Sources of *Rht8* and *Rht9* that have been investigated include cultivars Mara (*Rht8/Rht9*) and Chuan-Mai (*Rht8*).

Recently, Worland *et al.* (1998) described a microsatellite marker, gwm261, which is tightly linked (within 0.6 cM) to the dwarfing locus, *Rht8*. It has been found that this locus does not suppress coleoptile length, and thus the linked marker has allowed a search to be made for the genes controlling that phenotype. Variations in the size of the gwm261 microsatellite have been further used to distinguish various allelotypes of *Rht8*, some of which are particularly good at allowing the expression of the long coleoptile phenotype (Fig. 4).

III.   QUALITY ASSURANCE

An additional contribution to improved cereal quality coming from the broad area of biotechnology is the ability of DNA sequence tags to provide criteria for essential derivation, distinctiveness, uniformity and genetic distance that are independent of environmental effects. Under the broad banner of quality assurance, these tags are being used for cultivar identification, post-harvest segregation, and to assess seed purity. They are thus useful tools for farmers, breeders and end users.

The establishment of unique DNA sequence tags for characterizing grain deliveries of barley is providing a breakthrough in tracing the origins of impurities due to cross-contamination of different cultivars. The feed/malting quality segregation of barley grain is a high-value exercise and the application of these DNA tags (based on AFLP and microsatellite sequences) is providing a level of resolution that is not possible with protein-based techniques. Currently, AFLPs are proving to be the most powerful method for discrimination of genotype in this regard. The presence of a specific AFLP marker may be used to detect the presence of a particular cultivar in a seed sample (Fig. 5, right panel) or to detect the presence of contaminants in a sample of seed from a particular cultivar (Fig. 5, left panel).

IV.   PRE-TRANSFORMATION PROOF OF CONCEPT

The recent advances in genetic engineering of cereals have opened up a vast array of possible manipulations of quality traits that would be difficult to achieve by conventional breeding techniques. Since cereal transformation is still a lengthy and labour-intensive process, this increased scope for modification of cereals has engendered the necessity for the development of ways of assaying the effects of the newly expressed protein before the transformation process. This pre-transformation proof of concept has been established for anticipated changes in the content and properties of the endogenous grain components, or where there are natural mutations available capable of providing some indication as to the expected effect of the genetic modification. The development of sensitive small-scale mixing, extension and baking tests (Rath *et al.*, 1990; Bekes and Gras, 1991; Rath *et al.*, 1994; Gras *et al.*, 1995, 1997) has provided the basis for these studies. The amount of the protein of interest required for the assays is now within the range commonly produced by expression in heterologous hosts such as bacterial, yeast or fungal cells.

Studies on the dough-strengthening effects of increasing the high-molecular-weight glutenin subunit (HMW-GS) content of the grain and on the possible protective role of an antifreeze protein in frozen bread doughs

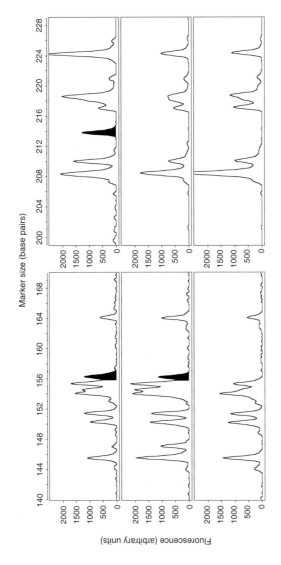

Fig. 5. Electrophoretograms of fluorescently labelled AFLP markers with the potential to discriminate between cultivars of barley.

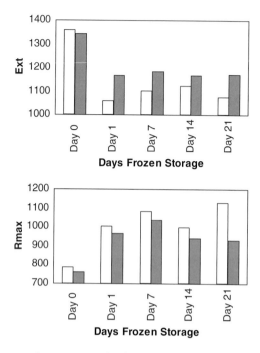

Fig. 6.   Extensograph parameters for frozen doughs stored for various lengths of time. Effect of rAFP at 0.1% by mass (flour basis). Unshaded bars, control; shaded bars, anti-freeze protein.

are two good examples of this methodology. In the case of the HMW-GS proteins, the strengthening effects of incorporation of bacterially produced proteins were directly assayed *in vitro* (Bekes *et al.*, 1993). Subsequent transformation studies (Barro *et al.*, 1997; Blechl *et al.*, 1997; He *et al.*, 1998) demonstrated that the *in vitro* studies provided a valuable predictor of the functional properties of the same proteins when they were expressed *in vivo* in the endosperm during grain fill.

The use of an antifreeze protein to enhance product quality in frozen-dough applications has recently been studied in our laboratories. It was observed that the industry's solutions to dough weakening in frozen doughs, addition of yeast, gluten, oxidants and improvers, were aimed at treating the physical symptoms rather than the primary cause of the problem, viz. ice crystal formation and growth. A novel and highly active recombinant antifreeze protein (rAFP) was designed, based on those found in winter flounder. The gene was synthesized by the polymerase chain reaction (PCR) and expressed in bacteria, and the rAFP was purified (Solomon and Appels, 1999). In small-scale testing, addition of the rAFP was found to have small effects on dough-mixing properties (Solomon *et al.*, 1997), but only when present at high levels relative to those needed for its antifreeze activity to be manifest. At low levels, addition of the rAFP protein partly ameliorated the adverse effect of freezing on dough extension properties (Fig. 6).

The utilization of this antifreeze protein in frozen-dough applications may now be examined further through its expression in wheat under control of an endosperm-specific promoter. Alternatively, for other uses (e.g. in frozen confectionery), it may be economically feasible to utilize a fermentation route for its production, perhaps using one of the GRAS (generally regarded as safe) organisms commonly used in production of food additives (e.g. *Aspergillus* species).

V. CONCLUSION

The experiments described in this paper provide a proof of concept of the application of DNA-based technologies in delivery of better cereal cultivars to growers, processors and consumers. Applications such as the use of DNA markers for improving the efficiency of breeding and for quality assurance of grain supplies offer immediate benefits. However, in this area it is anticipated that the generation of technologies, viz. assaying single nucleotide polymorphisms, will provide major challenges and further opportunities in cereals such as wheat. Applications such as engineering new proteins to modify flour functionality and processing properties offer benefits in the longer term. The exploitation of these approaches will be greatly accelerated by the application of knowledge from model systems, such as rice and *Arabidopsis*, as well as the development of small-scale *in vitro* technologies for pre-testing the properties of engineered proteins.

ACKNOWLEDGEMENTS

The authors are grateful to the following colleagues for their generosity: Peter Sharp (Sydney University) for allowing the reproduction of Fig. 1; Greg Rebetzke, Matthew Morell, Wes Keys and Lynette Preston for valuable discussions associated with Figs 2, 4 and 5; Ian Batey for providing Fig. 3; and Frank Bekes and Sandra Partridge for their help in producing the data for Fig. 6.

REFERENCES

Barro, F., Rooke, L., Bekes, F., Gras, P., Tatham, A. S., Fido, R., Lazzeri, P. A., Shewry, P. R. and Barcelo, P. (1997). Transformation of wheat with high molecular weight subunit genes results in improved functional properties. *Nature Biotechnology* **15**, 1295–1299.

Becker, J., Vos, P., Kuiper, M., Salamini, F. and Heun, M. (1995). Combined mapping of AFLP and RFLP markers in barley. *Molecular and General Genetics* **249**, 65–73.

Bekes, F. and Gras, P. W. (1991). Demonstration of the 2-gram mixograph as a research tool. *Cereal Chemistry* **69**, 229–230.

Bekes, F., Anderson, O., Gras, P. W., Gupta, R. B., Tam, A., Wrigley, C. W. and Appels, R. (1993). Effects of individual *Glu-1D* HMW glutenin subunits on mixing properties. *In* "Proceedings of the 43rd Australian Cereal Chemistry Conference" (C. W. Wrigley, ed.), pp. 102–104. Sydney, NSW, 1993.

Blechl, A. E., Le, H. G., Bekes, F., Gras, P. W., Shimoni, Y., Galili, G. and Anderson, O. D. (1997). Applications of molecular biology in understanding and improving wheat quality. *In* "Proceedings of the International Wheat Quality Conference" (J. L. Steele and O. K. Chung, eds), pp. 205–212. Manhattan, USA, 1997.

Briney, A., Wilson, R., Potter, R. H., Barclay, I., Crosbie, G., Appels, R. and Jones, M. G. K. (1998). A PCR-based marker for selection of starch and potential noodle quality in wheat. *Molecular Breeding* **4**, 427–433.

Clark, J. R., Robertson, M. and Ainsworth, C. C. (1991). Nucleotide sequence of a wheat (*Triticum aestivum* L.) cDNA clone encoding the *waxy* protein. *Plant Molecular Biology* **16**, 1099–1101.

Dry, I., Smith, A., Edwards, A., Bhattacharyya, M., Dunn, P. and Martin, C. (1992). Characterization of cDNAs encoding two isoforms of granule-bound starch synthase which show differential expression in developing storage organs of pea and potato. *Plant Journal* **2**, 193–202.

Gras, P. W., Rath, C. R., Bekes, F. and Zhen, Z. (1995). A method for determining dough extensibility using 2.5 grams of flour. *In* "Cereals '95. Proceedings of the 45th Australian Cereal Chemistry Conference" (Y. A. Williams and C. W. Wrigley, eds), pp. 407–410. Adelaide, SA, 1995.

Gras, P. W., Ellison, F. W. and Bekes, F. (1997). Quality evaluation on a micro-scale. *In* "Proceedings of the International Wheat Quality Conference" (J. L. Steele and O. K. Chung, eds), pp. 161–172. Manhattan, USA, 1997.

He, G. Y., Rooke, L., Cannell, M., Rasco-Gaunt, S., Sparks, C., Lamacchia, C., Bekes, F., Tatham, A. S., Barcelo, P., Shewry, P. R. and Lazzeri, P.A. (1998). Current status of transformation in bread and durum wheats and modifications of gluten quality. *Acta Agronomica Hungarica* **46**, 449–462.

Hirano, H. and Sano, Y. (1991). Molecular characterization of the *waxy* locus of rice (*Oryza sativa*). *Plant Cell Physiology* **32**, 989–997.

Hovenkamp-Hermelink, J. H. M., Bijmolt, E. W., de Vries, J. N., Witholt, B. and Feenstra, W. J. (1987). Isolation of an amylose-free starch mutant of the potato (*Solanum tuberosum* L.). *Theoretical and Applied Genetics* **75**, 217–221.

Klösgen, R. B., Gierl, A., Schwarz-Sommer, Z. S. and Saedler, H. (1986). Molecular analysis of the *Waxy* locus of *Zea mays*. *Molecular and General Genetics* **203**, 237–244.

Korzun, V., Roder, M. S., Ganal, M. W., Worland, A. J. and Law, C. N. (1998). Genetic analysis of the dwarfing gene (*Rht8*) in wheat. Part I. Molecular mapping of *Rht8* on the short arm of chromosome 2D of bread wheat (*Triticum aestivum* L). *Theoretical and Applied Genetics* **96**, 1104–1109.

Manly, K. F. (1994). Map Manager: genetic mapping software. *Rice Genome* **3**, 9.

Mason-Gamer, R. J., Weil, C. F. and Kellogg, E. A. (1998). Granule-bound starch synthase: structure, function, and phylogenetic utility. *Molecular Biology of Evolution* **15**, 1658–1673.

Murai, J., Taira, T. and Ohta, D. (1999). Isolation and characterisation of the three waxy genes encoding the granule-bound starch synthase in hexaploid wheat. *Gene* **234**, 71–79.

Okagaki, R. J. (1992). Nucleotide sequence of a long cDNA from the *waxy* gene. *Plant Molecular Biology* **19**, 513–516.

Paull, J. G., Chalmers, K. J., Karakousis, A., Kretschmer, J. M., Manning, S. and Langridge, P. (1998). Genetic diversity in Australian wheat varieties and breeding material based on RFLP data. *Theoretical and Applied Genetics* **96**, 435–446.

Rath, C. R., Gras, P. W., Wrigley, C. W. and Walker, C. E. (1990). Evaluation of dough properties from two grams of flour using the mixograph principle. *Cereal Foods World* **35**, 572–574.

Rath, C. R., Gras, P. W., Zhen, Z., Appels, R., Bekes, F. and Wrigley, C. W. (1994). A prototype extension tester for two-gram dough samples. *In* "Proceedings of the 44th Australian Cereal Chemistry Conference" (J. F. Panozzo and P. G. Downie, eds), pp. 122–126. Ballarat, VIC, 1994.

Rebetzke, G. J. and Richards, R. A. (2000). Gibberellic acid-sensitive dwarfing genes reduce plant height to increase kernel number and grain yield of wheat. *Australian Journal of Agricultural Research* **51**, 235–245.

Roder, M. S., Korzun, V., Wendehake, K., Plaschke, J., Tixier, M. H., Leroy, P. and Ganal, M. W. (1998). A microsatellite map of wheat. *Genetics* **149**, 2007–2023.

Rohde, W., Becker, D. and Salamini, F. (1988). Structural analysis of the *waxy* locus from *Hordeum vulgare*. *Nucleic Acids Research* **16**, 7185–7186.

Shure, M., Wessler, S. R. and Federoff, N. (1983). Molecular identification and isolation of the *waxy* locus in maize. *Cell* **35**, 225–233.

Solomon, R. G. and Appels, R. (1999). Stable, high-level expression of a type I antifreeze protein in *Escherichia coli*. *Protein Expression and Purification* **16**, 53–62.

Solomon, R. G., Partridge, S., Bekes, F., Gras, P. and Appels, R. (1997). Alleviating the adverse effects of ice-crystal growth in frozen bread doughs. *In* "Cereals '97. Proceedings of the 47th Australian Cereal Chemistry Conference" (A. W. Tarr, A. S. Ross and C. W. Wrigley, eds), pp. 235–238. Perth, WA, 1997.

van der Leij, F. R., Visser, R. G. F., Ponstein, A. S., Jacobsen, E. and Feenstra, W. J. (1991). Sequence of the structural gene for granule-bound starch synthase of potato (*Solanum tuberosum* L.) and evidence for a single point deletion in the *amf* allele. *Molecular and General Genetics* **228**, 240–248.

Visser, R. G. F., Hergersberg, M., Van der Leij, F. R., Jacobsen, E., Witholt, B. and Feenstra, W. J. (1989). Molecular cloning and partial characterization of the gene for granule-bound starch synthase from a wild type and an amylose-free potato (*Solanum tuberosum* L.). *Plant Science* **64**, 185–192.

Wang, Z., Wu, Z., Xing, Y., Zheng, F., Guo, X., Zhang, W. and Hong, M. (1990). Nucleotide sequence of rice *waxy* gene. *Nucleic Acids Research* **18**, 5898.

Worland, A. J., Korzun, V., Roder, M. S., Ganal, M. W. and Law, C. N. (1998). Genetic analysis of the dwarfing gene *Rht8* in wheat. Part II. The distribution and adaptive significance of allelic variants at the *Rht8* locus of wheat as revealed by microsatellite screening. *Theoretical and Applied Genetics* **96**, 1110–1120.

Yan, L., Bhave, M., Fairclough, R., Konik, C., Rahman, S. and Appels, R. (2000). The genes encoding granule-bound starch synthases at the *waxy* loci of the A, B and D progenitors of common wheat. *Genome* **43**, 264–272.

Zhao, X. C. and Sharp, P. J. (1996). An improved 1-D SDS-PAGE method for the identification of three bread wheat 'waxy' proteins. *Journal of Cereal Science* **23**, 191–193.

Zhao, X. C., Batey, I. L., Sharp, P. J., Crosbie, G., Barclay, I., Wilson, R., Morell, M. K. and Appels, R. (1998). A single genetic locus associated with starch granule properties and noodle quality in wheat. *Journal of Cereal Science* **27**, 7–13.

Overview and Prospects for Cereal Biotechnology

PETER R. SHEWRY,[1] PAUL A. LAZZERI[2] and
KEITH J. EDWARDS[1]

[1] *IARC–Long Ashton Research Station, Department of Agricultural
Sciences, University of Bristol, Long Ashton, Bristol BS41 9AF, UK*
[2] *Dupont Wheat Transformation Laboratory, c/o Rothamsted
Experimental Station, Harpenden, Hertfordshire AL5 2JQ, UK*

I. INTRODUCTION

The previous chapters clearly show that cereal biotechnology is at a crucial stage of development. Basic technologies for mapping, genomic analysis and transformation have been established and further refinement and improvement of these can be anticipated over the next few years. In addition, success has already been achieved in manipulating and improving some simple target traits, although others still need to be defined at the molecular and biochemical levels.

II. MAPPING AND GENOMICS

Cereal biotechnology has advanced rapidly in the last 5 years. Today the cereal biotechnologist can call upon a wide range of molecular tools to map and characterize the various species. All cereals have well-developed genetic maps consisting of several thousand restriction fragment length polymorphism (RFLP), amplified fragment length polymorphism (AFLP) and microsatellite markers. In most cases a significant number of these markers are in the public

Advances in Botanical Research Vol. 34
incorporating Advances in Plant Pathology
ISBN 0-12-005934-7

domain. Current large-scale efforts and the development of numerous large insert libraries for most of the cereals, means that these genetic maps could soon be merged with the corresponding physical maps. Given these resources, the mapped-based cloning of genes, currently reserved for model species such as *Arabidopsis* and rice, could soon become commonplace for all the large genome cereals. However, resource-intensive map-based cloning is itself likely to be superseded by the generation and public availability of large numbers of mapped expressed sequence tags (ESTs).

It is probable that future developments will focus on two areas. Firstly, the detailed characterization of the available genetic resources within the existing seed banks. The availability of both EST information and details of the growth characteristics of all seed bank accessions, will be a necessary prerequisite for the efficient management, breeding and industrial exploitation of the existing biodiversity in both natural and man-made populations. The challenge is that such characterization will not only require cross-discipline collaborations, but it will also require faithful and long-term collaborations between First and Third World farmers and the newly emerging agricultural biotechnology and seed companies. Secondly, the enhanced exploitation of genetic resources. It is relatively easy to describe the variation that exists between accessions within the seeds banks using molecular markers. However, it will require considerably more imagination (and experience) to translate this information into strategies for the further enhancement of the cereal gene pool.

Clearly, the available technology is teaching us much about the complexities of the cereal genome. Used wisely and in conjunction with conventional genetic manipulation, the current pool of knowledge should mean that the next 50 years will be both an extremely interesting time for cereal biotechnology and also a productive time for world agriculture.

III.   TRANSFORMATION TECHNOLOGY

Research in cereal transformation is at present in a stage of transition. After a decade of intense focus on the development of procedures for reproducible and efficient production of transgenic plants, this basic technology is now available for all the major cereal crops. Consequently, the focus of research is moving more towards understanding and controlling the "quality" of transgenic plants, in areas such as the complexity of transgene insertions, the stability of transgene expression and the presence of unwanted sequences. In addition, in company laboratories, there is an emphasis on achieving routine and effective integration of genetic modification (GM) technology into breeding systems. This involves streamlining the process from the creation of new transgenic events through laboratory-based analyses to field evaluation and the incorporation of selected events in crossing programmes for varieties and hybrids.

This is not to say that there is no need for continued development of improved transformation procedures: there are still a number of technology bottlenecks that reduce efficiency and limit the breadth of application. Also, there are number of important new technologies, which have been demonstrated in model plants and need to be further developed to function in crop species.

The routine use of *Agrobacterium* transformation for rice and maize has been a highly significant development during the last few years. The first reports were confined to amenable genotypes but the techniques are now being extended to elite varieties. *Agrobacterium* transformation of model genotypes of wheat and barley is possible today and we can expect to see application to an expanded range of germplasm as these systems become better understood.

Current cereal transformation techniques depend on the use of tissue cultures as targets for gene delivery. These cultures are typically derived by laborious manual isolation of explants from donor plants maintained under strict growth conditions. In rice, a simpler approach of targeting cultures derived from mature seed can now be used and there may be potential to extending this approach to other species. However, in the longer term, the ideal approach would be to target cells in whole cereal plants by adapting the *in planta Agrobacterium* infiltration techniques first used in *Arabidopsis* and now being applied in other amenable dicot species.

Public concern about potential harmful effects of GM crops has focused attention on the development of "clean" transformation technology to produce plants containing the minimum transgene DNA needed to modify the target trait and on using plant genes in preference to prokaryotic or viral sequences wherever possible. Advances have been made in this area, such as the use in direct gene transfer techniques of transformation vector fragments containing only the trait or marker gene and the essential control sequences, thus avoiding the integration of bacterial selection markers and plasmid backbone sequences. Alternative selection markers are now available, such as those used in positive selection systems, which are free of the health or environmental impact concerns associated with antibiotic or herbicide resistance genes. In addition, marker-free transgenic cereal plants are being produced, either by recombinase-mediated transgene excision mechanisms or by the use of *Agrobacteria* containing two Ti plasmids, which integrate at unlinked loci and are separated by meiotic segregation in progeny. These methods are still relatively new and undeveloped, and we can expect to see steady improvements in efficiency as we gain experience in their use and later generations of technology emerge.

While great progress has been made in developing technology to produce transgenic cereals, we still face major challenges in the areas of understanding and controlling transgene integration, and in achieving predictable and stable expression of integrated sequences. Advances in the development of efficient gene targeting technology for higher plants has been slow and transgene insertion is still essentially a random process. This leads to variation in levels and stability of transgene expression because of the influence of the chromosomal environment, although approaches such as the use of transgenes flanked by matrix attachment region elements to modify chromatin structure at the integration site show some promise for stabilizing expression.

In DNA delivery by direct gene transfer, in addition to the effects of random insertion sites, it is common for multiple and rearranged transgene copies to be integrated. Such complex insertions, containing repeated homologous sequences and features, such as inverted repeats, which may form secondary structures, such as hairpin loops. These structures are increasingly recognized as being triggers for gene-silencing mechanisms, which may prevent expression not only of the transgene but also of homologous native genes.

There is currently considerable research interest in understanding the mechanisms of undesired transgene silencing and in the related areas of targeted (native) gene silencing by antisense or co-suppression approaches. These are subjects of primary importance for applied cereal biotechnology as better understanding of the underlying mechanisms will clearly underpin the development of technology for predictable transgene expression and down-regulation of endogenous gene expression.

IV.   TARGETS FOR MANIPULATION

Plant biotechnology can be regarded as a "tool kit" rather than a scientific discipline, which has the ability to increase our understanding of plant processes and their regulation and to provide germplasm with novel or improved properties to the plant breeder.

Consequently, the major targets for most plant biotechnologists, at least in the short term, will reflect those identified by plant breeders over many years: yield; resistance to biotic and abiotic stress; and end-use quality. Leaving aside herbicide resistance, the most readily identified targets for manipulation include resistance to specific insect pests (e.g. Bt corn and cotton) and aspects of end-use quality, such as gluten elasticity and starch composition. In contrast, other resistances (e.g. to fungal pathogens, nematodes, and abiotic stresses) and quality parameters are still not sufficiently well understood at the molecular level and the initial impact of plant biotechnology will be to elucidate mechanisms and identify potential strategies for future improvement. Similarly, yield is a complex multigenic characteristic and biotechnology will allow the individual components to be dissected.

In the longer term, the major impact of plant biotechnology may well be on novel crops and products rather than traditional crops. Cereal seeds provide an attractive system as hosts for novel compounds, being adapted to all climatic zones with high yields. In addition, the grain are easily harvested, stable during long-term storage, and fractionated using current milling and separation technology (e.g. for starch or starch/gluten separation). This should allow the production of high-value/low-volume compounds, such as pharmaceuticals, and low-value/high-volume compounds, such as plastics and packaging materials.

V. PUBLIC ACCEPTANCE – A EUROPEAN PERSPECTIVE

The advent of GM technology initially seemed to promise unlimited opportunities to the plant breeder, farmer and grain utilizer, with visions of high yielding crops with improved and stable end-use quality growing with minimal inputs of fertilizers and agrochemicals. This vision has become threatened over the past 2 years by increasing public concern, with scientists in the UK being at the forefront of the debate. It is easy to lay the blame for this situation at the door of multinational companies, who have introduced GM crops without sufficient consultation and education, and the media and professional pressure groups that have generated an atmosphere of panic. However, it is clear that the situation would not have arisen without genuine public interest in food safety and the environmental impact of intensive farming methods.

Food safety is a major concern in Europe, and particularly in the UK, owing to a number of food scares extending back for a decade. These include various types of food poisoning (*Listeria* in milk products, *Salmonella* in eggs and chickens, *E. coli* in cooked meat products, *Campylobacter*, etc.). However, the most dramatic impact has been from bovine spongiform encephalitis (BSE), which changed the face of the British livestock farming and meat processing industries. The BSE crisis also acted to affect public confidence in scientific research severely. Against this background, it is not surprising that the introduction of an entirely new technology into the food chain has raised questions of safety.

Similarly, the countryside in Europe is not regarded as a "food factory" but is the place where many city workers live and spend their leisure time. Consequently, diversity in the landscape and in the floristic composition of both farmed and adjacent semi-natural areas (hedgerows, headlands, coppices, etc.) is treasured, and new technologies, which appear to threaten this, are approached with caution.

Impacts on health and the environment are the two major concerns of the general public. Others are more ethical than practical in nature, relating to freedom of choice, provision of information (labelling), and the role of multinational companies in dictating what we grow and eat. Allied to the latter is concern about the role of such companies in controlling agricultural production in developing countries leading to reliance on high-cost varieties and agrochemicals.

This concern cannot be countered by purely scientific arguments, in the way that scientists debate problems among themselves. What is needed is a "hearts and minds" approach, in which scientists form closer links with the general public, and communicate in terms which can be understood by those with little or no scientific background. The ultimate aim must be to enable the consumer to make his or her own decision, based on an understanding of the risks and benefits. This is an ambitious goal but one which must be achieved if the full benefits of GM technology are to be realized.

AUTHOR INDEX

Numbers in *italics* refer to pages on which full references are listed

SUBJECT INDEX